Methods in Cell Biology

VOLUME 40
A Practical Guide to the Study of Calcium in
Living Cells

Series Editors

Leslie Wilson
Department of Biological Sciences
University of California, Santa Barbara
Santa Barbara, California

Paul Matsudaira
Whitehead Institute for Biomedical Research and
Department of Biology
Massachusetts Institute of Technology
Cambridge, Massachusetts

Methods in Cell Biology

VOLUME 40
A Practical Guide to the Study of Calcium in
Living Cells

Edited by

Richard Nuccitelli

Section of Molecular and Cellular Biology
Division of Biological Sciences
University of California
Davis, California

ACADEMIC PRESS

A Harcourt Science and Technology Company

San Diego   San Francisco   New York   Boston   London   Sydney   Tokyo

Cover photograph: From Chapter 10 by Diliberto *et al.* For details see Color Plate 6.

Academic Press

A Harcourt Science and Technology Company
525 B Street, Suite 1900, San Diego, California 92101-4495, USA
http://www.academicpress.com

Academic Press

Harcourt Place, 32 Jamestown Road, London NW1 7BY, UK
http://www.academicpress.com

International Standard Serial Number: 0091-679X

International Standard Book Number: 0-12-564141-9 (Hardcover)

International Standard Book Number: 0-12-522810-4 (Paperback)

Printed and bound by CPI Group (UK) Ltd, Croydon, CR0 4YY
Transferred to Digital Print 2011

CONTENTS

PART II Microelectrode Techniques for Measuring $[Ca^{2+}]_i$ and Ca^{2+} Fluxes

4. How to Make and to Use Calcium-Specific Mini- and Microelectrodes

Stéphane Baudet, Leis Hove-Madsen, and Donald M. Bers

5. The Vibrating Ca^{2+} Electrode: A New Technique for Detecting Plasma Membrane Regions of Ca^{2+} Influx and Efflux

Peter J. S. Smith, Richard H. Sanger, and Lionel F. Jaffe

6. Application of Patch Clamp Methods to the Study of Calcium Currents and Calcium Channels

Colin A. Leech and George G. Holz, IV

PART III   Fluorescence Techniques for Imaging [Ca^{2+}]$_i$

PART IV   Use of Aequorin for $[Ca^{2+}]_i$ Imaging

CONTRIBUTORS

Numbers in parentheses indicate the pages on which the authors' contributions begin.

Stéphane Baudet (93), Laboratoire de Cardiologie Expérimentale, URA CNRS 1340, Hôpital Laënnec BP 1005, 44035 Nantes, France

Donald M. Bers (3, 93), Department of Physiology, Stritch School of Medicine, Loyola University at Chicago, Maywood, Illinois 60153

Marisa Brini (339), Department of Biomedical Sciences and CNR Center for the Study of Mitochondrial Physiology, University of Padova, 35121 Padova, Italy

Bibie M. Chronwall (183), Division of Cell Biology and Biophysics, School of Biological Sciences, University of Missouri at Kansas City, Kansas City, Missouri 64110

David E. Clapham (263), Department of Pharmacology, Mayo Foundation, Rochester, Minnesota 55905

R. John Cork (221, 287), Department of Biological Sciences, Purdue University, West Lafayette, Indiana 47907

Pamela A. Diliberto (243), Department of Cell Biology and Anatomy, University of North Carolina School of Medicine, Chapel Hill, North Carolina 27599

Steven Girard (263), Department of Pharmacology, Mayo Foundation, Rochester, Minnesota 55905

Michael R. Hanley (65), Department of Biological Chemistry, School of Medicine, University of California at Davis, Davis, California 95616

Brian Herman (243), Department of Cell Biology and Anatomy, University of North Carolina School of Medicine, Chapel Hill, North Carolina 27599

George G. Holz, IV (135), Laboratory of Molecular Endocrinology, Massachusetts General Hospital, Harvard Medical School, Boston, Massachusetts 02114

Leis Hove-Madsen (93), Cardiologie Cellulaire et Moléculaire, Faculté de Pharmacie, INSERM 92-11, F-92296 Chatenay-Malabry, France

Lionel F. Jaffe (115, 305), Calcium Imaging Laboratory, Marine Biological Laboratory, Woods Hole, Massachusetts 02543

Joseph P. Y. Kao (155), Medical Biotechnology Center and Department of Physiology, University of Maryland School of Medicine, Baltimore, Maryland 21201

Eric Karplus (305), Calcium Imaging Laboratory, Marine Biological Laboratory, Woods Hole, Massachusetts 02543

Thomas J. Keating (221, 287), Department of Biological Sciences, Purdue University, West Lafayette, Indiana 47907

Colin A. Leech (135), Laboratory of Molecular Endocrinology, Massachusetts General Hospital, Harvard Medical School, Boston, Massachusetts 02114

Andrew L. Miller (305), Calcium Imaging Laboratory, Marine Biological Laboratory, Woods Hole, Massachusetts 02543

Stephen J. Morris (183), Division of Molecular Biology and Biochemistry, School of Biological Sciences, University of Missouri at Kansas City, Kansas City, Missouri 64110

Richard Nuccitelli (3), Section of Molecular and Cellular Biology, University of California at Davis, Davis, California 95616

Chris W. Patton (3), Hopkins Marine Station, Stanford University, Pacific Grove, California 93950

Tullio Pozzan (339), Department of Biomedical Sciences and CNR Center for the Study of Mitochondrial Physiology, University of Padova, 35121 Padova, Italy

Rosario Rizzuto (339), Department of Biomedical Sciences and CNR Center for the Study of Mitochondrial Physiology, University of Padova, 35121 Padova, Italy

Kenneth R. Robinson (287), Department of Biological Sciences, Purdue University, West Lafayette, Indiana 47907

Richard H. Sanger (115), NIH National Vibrating Probe Facility, Marine Biological Laboratory, Woods Hole, Massachusetts 02543

Peter J. S. Smith (115), NIH National Vibrating Probe Facility, Marine Biological Laboratory, Woods Hole, Massachusetts 02543

David Thomas (65), Department of Biological Chemistry, School of Medicine, University of California at Davis, Davis, California 95616

Xue Feng Wang (243), Department of Cell Biology and Anatomy, University of North Carolina School of Medicine, Chapel Hill, North Carolina 27599

Larry W. Welling (183), Research Service, Veterans Affairs Hospital, Kansas City, Missouri 64128

Thomas B. Wiegmann (183), Renal Section, Veterans Affairs Hospital, Kansas City, Missouri 64128

Robert Zucker (31), Molecular and Cell Biology Department, University of California at Berkeley, Berkeley, California 94720

PREFACE

During the past two decades changes in intracellular Ca^{2+} ($[Ca^{2+}]_i$) have been found to be associated with a wide variety of cellular processes. Indeed, it is more difficult to find cellular functions that do not involve a change in $[Ca^{2+}]_i$ than those that do. We now know that cellular events such as secretion, fertilization, cleavage, nuclear envelope breakdown and many others are all accompanied by changes in $[Ca^{2+}]_i$. This important involvement of $[Ca^{2+}]_i$ has prompted the development of techniques to measure $[Ca^{2+}]_i$ in living cells and there are now several quite sensitive techniques for detecting submicromolar levels of $[Ca^{2+}]_i$. These techniques fall into three main categories, Ca^{2+}-sensitive microelectrodes, fluorescent dyes, and luminescent proteins. Each of these is presented and discussed at length in this volume in a manner that includes the problems and pitfalls that the researcher will encounter when applying the technique to study his or her favorite cell. In addition, we have included a practical guide to the preparation of Ca^{2+} buffers and techniques for perturbing intracellular Ca^{2+}.

Since $[Ca^{2+}]_i$ is typically at submicromolar levels in living cells, it is quite important to utilize similarly low levels in any buffer solution meant to mimic the cytoplasmic milieu. Thus, biochemists studying reactions in cell-free extracts should buffer Ca^{2+} to about $0.1\ \mu M$ and cell biologists injecting molecules into cells should buffer their injection solutions to similarly low Ca^{2+} levels. The preparation of solutions buffered to well defined Ca^{2+} levels is not trivial because most Ca^{2+} buffers are also quite sensitive to pH, ionic strength and other divalent cations. The first chapter of this volume presents a rather straightforward explanation of the important variables to consider and describes an easy-to-use computer program that makes such calculations a snap. This program allows you to determine exactly what Ca^{2+} buffer concentration to use for the desired free Ca^{2+} concentration for any pH and ionic strength.

The study of the role of $[Ca^{2+}]_i$ in cellular processes can be greatly aided by perturbing $[Ca^{2+}]_i$ at will. The next two chapters present techniques for imposing such changes. The first by Robert Zucker describes photorelease techniques for raising or lowering $[Ca^{2+}]_i$. This is the best description of the photorelease methodology available to date. Another way to perturb $[Ca^{2+}]_i$ is to inhibit the pumps used to sequester it in cells. A very detailed discussion of those pumps and their inhibitors is presented by Thomas and Hanley.

The only Ca^{2+} detection technique that allows an accurate measurement of absolute $[Ca^{2+}]_i$ in cells or in buffer solutions is the Ca^{2+}-specific microelectrode. A complete and detailed description of how to make and use such electrodes is included here by Baudet *et al.* and Smith *et al.* The latter group has adapted this electrode to measure Ca^{2+} gradients just outside the cell to detect regions of Ca^{2+} influx and efflux. Another technique for measuring Ca^{2+} currents is the patch clamp. Leech and Holz describe how the patch clamp can be used to study Ca^{2+} currents and channels.

The other two popular $[Ca^{2+}]_i$ detection techniques involve either the fluorescence imaging of Ca^{2+}-sensitive probes or the imaging of luminescence from Ca^{2+}-sensitive photoproteins. Both of these techniques are described here in detail. Five chapters address the practical aspects of measuring $[Ca^{2+}]_i$ with fluorescent indicators, including confocal imaging. Both computational confocal and laser scanning confocal techniques are described at length. The use of aequorin for $[Ca^{2+}]_i$ detection has been growing in popularity; two reasons are presented here. The first is the recent achievement of targeting aequorin to specific organelles using molecular techniques described in Chapter 14 and the second is the rather inexpensive technology required to detect aequorin signals as described by Robinson *et al.* in chapter 12.

Throughout this volume the authors have emphasized the detailed methodology that is critical for the implementation of these techniques. Therefore, this material will be very useful to the investigator who wants to apply any of these methods to his or her research. I thank all of the contributors for their care and attention to detail and hope that this volume proves helpful to investigators of cell biology.

Richard Nuccitelli

PART I

Ca^{2+} Buffers and $[Ca^{2+}]_i$ Perturbation Techniques

CHAPTER 1

A Practical Guide to the Preparation of Ca^{2+} Buffers

Donald M. Bers,★ Chris W. Patton,† Richard Nuccitelli‡

★ Department of Physiology
Stritch School of Medicine
Loyola University at Chicago
Maywood, Illinois 60153
† Hopkins Marine Station
Stanford University
Pacific Grove, California 93950
‡ Section of Molecular and Cellular Biology
University of California at Davis
Davis, California 95616

I. Introduction

Cell biologists quickly learn in the laboratory the importance of controlling the ionic composition of the solutions used when studying cellular biochemistry and physiology. Buffering the pH of solutions we use has become standard procedure and we have paid much attention to the selection of appropriate pH buffers. By far the most popular buffers currently are the zwitterionic amino acids introduced by Good *et al.* (1966). In contrast, we have paid much less attention to buffering Ca^{2+} because extracellular Ca^{2+} levels are typically in the millimolar range, and such concentrations are measured out and prepared easily. However, intracellular $[Ca^{2+}]$ ($[Ca^{2+}]_i$) is quite another matter because these levels are more typically in the 100 nM range, which is not prepared as easily. For example, a laboratory source of distilled water could easily have trace Ca^{2+} contamination in the range of 500 nM to 3 μM, as can commonly used chemicals and biochemicals. Indeed, often a considerable amount of endogenous Ca^{2+} exists in biological samples that is not removed easily. Therefore, when we are interested in studying intracellular reactions, Ca^{2+} buffering is extremely important.

In this chapter, we present a practical guide to the preparation of Ca^{2+} buffer solutions. Our goal is to emphasize the methods and the important variables to consider, while making the procedure as simple as possible. We also introduce two computer programs that may be of practical use to many workers in this field. One program is a spreadsheet that is useful in making and validating simple Ca^{2+} calibration solutions. The other is a more powerful and extensive program for the calculation of $[Ca^{2+}]$ (and that of other metals and chelators) in complex solutions with multiple equilibria. These programs have been developed and described with maximum ease of use in mind.

II. Which Ca^{2+} Buffer Should You Use?

When selecting the appropriate Ca^{2+} buffer for your application, the main consideration is a dissociation constant (K_d) close to the desired free $[Ca^{2+}]$. The ability of a buffer to absorb or release ions and, thus, to maintain the solution at a given concentration of that ion is greatest at the K_d. Just as choosing PIPES (pK_a = 6.8) to buffer a solution at pH 7.8 is an error, choosing a Ca^{2+} buffer with a K_d far from the desired [Ca^{2+}] set point is a mistake. As a rule of thumb, the K_d of the buffer should not lie more than a factor of 10 from the desired [Ca^{2+}]. In addition, the buffer should exhibit a much greater affinity for Ca^{2+} than for Mg^{2+} since intracellular [Mg^{2+}] is typically 10,000-fold higher than [Ca^{2+}]$_i$. Fortunately, about a dozen suitable buffers are available, spanning the range from 10 nM to 100 μM (Table I). Ethylene glycol bis(β-aminoethylether)-N,N,N',N'-tetraacetic acid (EGTA) can be a reliable buffer in the range of 10 nM to 1 μM at the typical intracellular pH of 7.2. However, if the goal is making buffers in the 1–10 μM range, dibromo-1,2-bis-(o-aminophenoxy)ethane-N,N,N',N'-tetraacetic acid (BAPTA) would be a much better choice.

A. EGTA: The Workhorse of Biological Ca^{2+} Chelators

By far the most popular Ca^{2+} buffer has been EGTA. This molecule has been used extensively because its apparent dissociation constant (K_d) at pH 7

Table I
Mixed Stability Constants for Useful Ca^{2+} Buffers at 0.15 M Ionic Strength in Order of Ca^{2+} Affinity

Ca^{2+} Buffer[a]	log K'_{Ca} at pH 7.4	K_d	$\frac{K'_{Ca}(\text{pH } 7.4)}{K'_{Ca}(\text{pH } 7.0)}$	K'_{Ca}/K'_{Mg} at pH 7.4	Reference
CDTA	7.90	13 nM	2.7	120	Martell and Smith (1974,1977); Bers and MacLeod (1988)
EGTA	7.18	67 nM	6.2	72,202	Martell and Smith (1974,1977); Bers and MacLeod (1988)
Quin 2	6.84	144 nM	1.15	25,114	Tsien (1980)
BAPTA	6.71	192 nM	1.14	158,244	Tsien (1980)
Fura-2	6.61	242 nM	1.14	72,373	Grynkiewicz et al. (1985)
Dibromo-BAPTA	5.74	1.83 μM	1.02	63,000	Tsien (1980)
4,4'-Difluoro-BAPTA	5.77	1.7 μM[b]	—	—	Pethig et al. (1989)
Nitr-5 photolysis product	5.2	6.3 μM[b]	—	—	Tsien and Zucker (1986)
5-Methyl-5'-nitro-BAPTA	4.66	22 μM[b]	—	—	Pethig et al. (1989)
6-Mononitro-BAPTA	4.4	40 μM[b]	—	—	Pethig et al. (1989)
NTA	3.87	134 μM	2.5	8	Martell and Smith (1974,1977); Bers and MacLeod (1988)
ADA	3.71	191 μM	1.24	32	Nakon (1979)
Citrate	3.32	471 μM	1.03	1.3	Martell and Smith (1974,1977); Bers and MacLeod (1988)
5,5'-Dinitro-BAPTA	2.15	7 mM[b]	—	—	Pethig et al. (1989)

[a] Abbreviations: CDTA, cyclohexylenedinitrilo-N,N,N',N'-tetraacetic acid; EGTA, ethylene glycol bis(β-aminoethylether),-N,N,N',N'-tetraacetic acid; BAPTA, 1,2-bis(o-aminophenoxy)ethane-N,N,N',N'-tetraacetic acid; NTA, nitrilotriacetic acid; ADA, acetamidominodiacetic acid.

[b] Measured at pH 7 and 0.1 M ionic strength.

(0.4 μM) is close to intracellular Ca^{2+} levels and it has a much higher affinity for Ca^{2+} than for Mg^{2+} (~100,000 times higher around neutral pH). However, the preparation of Ca^{2+} buffers using EGTA is complicated by the strong pH dependence of its Ca^{2+} affinity (see Fig. 1). Thus, although the free $[Ca^{2+}]$ would be ~0.4 μM when EGTA is half saturated with Ca^{2+} at pH 7, the free $[Ca^{2+}]$ in this same solution would decrease nearly 10-fold to 0.06 μM simply by raising the pH to 7.4! Therefore, the pH of Ca^{2+} buffers made with EGTA must be controlled very carefully, and the calculation of the appropriate amounts of EGTA and Ca^{2+} to use must be made at the desired pH. The purity of the EGTA is also a variable that can cause substantial errors, as large as 0.2 pCa units in the free $[Ca^{2+}]$ (Bers, 1982; Miller and Smith, 1984).

Many papers are available in the literature that explain how to calculate the proper amounts of EGTA and Ca^{2+} to combine to obtain a given free $[Ca^{2+}]$ (some are referred to in this chapter). Because of the steep pH dependence and slight Mg^{2+} sensitivity, both pH and Mg^{2+} must be considered in the calculation, which is best accomplished by computer. We provide a program for such calculations and describe it in a subsequent section. Systematic errors in EGTA purity and pH can be real practical problems (Bers, 1982), even with the best calculations for solution preparation. Thus, we also recommend measuring the free $[Ca^{2+}]$ whenever possible (see subsequent discussion and Chapter 4).

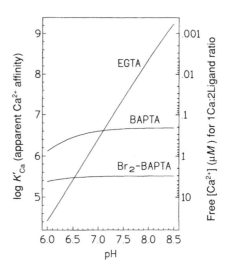

Fig. 1 The pH dependence of apparent affinities (K'_{Ca}) for EGTA, BAPTA, and Br$_2$-BAPTA at 20°C and 150 mM ionic strength.

B. BAPTA Family of Ca^{2+} Buffers

Tsien developed an analog of EGTA in which the methylene links between oxygen and nitrogen atoms were replaced with benzene rings to yield a compound called BAPTA (Tsien, 1980; Fig. 2). This compound exhibits a much lower pH sensitivity and much higher rates of Ca^{2+} association and dissociation (Hellam and Podolsky, 1969; Smith *et al.*, 1977). These characteristics may be due to the almost complete deprotonation of BAPTA at neutral pH. Moreover, modifications of BAPTA have been made to provide Ca^{2+} buffers with a range of K_d values covering the biologically significant range of 0.1 μM to 10 mM (see Table I; Pethig *et al.*, 1989). However, one disadvantage of BAPTA compared with EGTA is that this family of buffers exhibits a greater ionic strength dependence (see Figs. 3–5). In particular, increasing ionic strength from 100 to 300 mM decreases the apparent affinity constant, K'_{Ca}, for BAPTA or Br$_2$-BAPTA almost 3-fold, whereas the EGTA affinity constant is reduced only by about 30%. In contrast, raising temperature from 1 to 36°C approximately doubles the apparent affinity of all three of the Ca^{2+} buffers shown in Figs. 3–5 (i.e., EGTA, BAPTA, and Br$_2$-BAPTA).

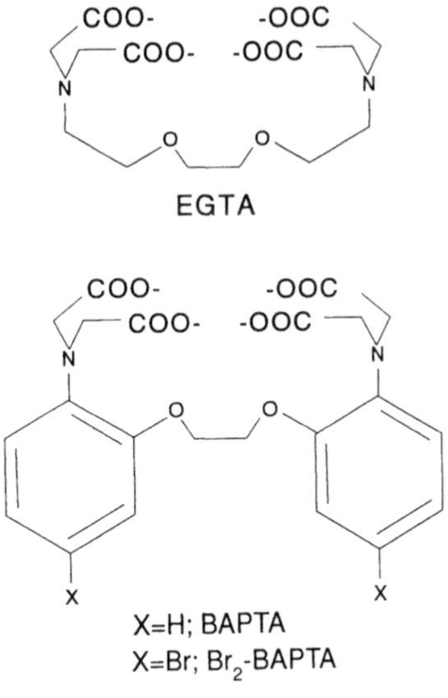

Fig. 2 Structural formulas for the Ca^{2+} chelators EGTA (*top*) and BAPTA and Br$_2$-BAPTA (*bottom*).

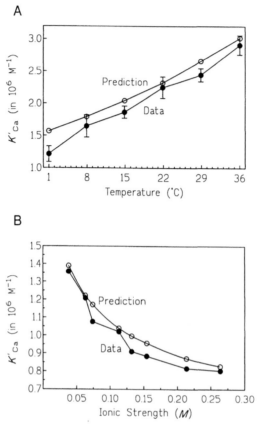

Fig. 3 The effect of temperature (A) and ionic strength (B) on the apparent Ca^{2+} affinity (K'_{Ca}) of EGTA. The experimental data are (A) from Harrison and Bers (1987), at pH 7.00 and 0.19 M ionic strength and (B) from Harafuji and Ogawa (1980), at pH 6.8 and 22°C. Predicted values are based on the temperature and ionic strength corrections described in the text.

III. Basic Mathematical Relationships

The preceding discussion and the data shown in Figs. 1 and 3–5 make it clear that one must know quantitatively how the buffers being used are altered by the typical range of experimental conditions (e.g., pH, temperature, and ionic strength). Although we do not want to belabor the equations, some readers may find it useful if we lay out some of the basic mathematical relationships. Readers who are not interested in the equations are referred to Section V and can use the programs as a black box.

In the preceding sections, we used K_d to address Ca^{2+} affinity. That K_d was the apparent overall dissociation constant, to which we will return in Eq. 5.

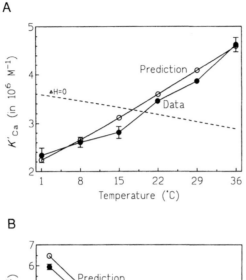

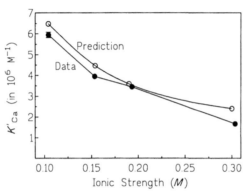

Fig. 4 The effect of temperature (A) and ionic strength (B) on the apparent Ca^{2+} affinity (K'_{Ca}) of BAPTA. The experimental data are from Harrison and Bers (1987), (A) at pH 7.00 and 0.19 M ionic strength and (B) at pH 7.00 and 22°C. Predicted values are based on the temperature and ionic strength corrections described in the text.

More traditionally, the mathematical expressions begin with the simple definition of the Ca^{2+} association constant, K_{Ca},

$$K_{Ca} = \frac{[CaR]}{[Ca] \cdot [R]} \tag{1}$$

where R is the Ca^{2+} buffer. This expression is not too useful directly, because we do not know any of the variables on the right side. Generally, having [Ca^{2+}] or bound Ca^{2+} ([CaR]) in terms of known quantities, such as total Ca^{2+} ([Ca$_t^{2+}$]) or total ligand ([R$_t$]), is more useful. One of the complicating factors is that Ca^{2+} buffers such as EGTA and BAPTA exist in multiple unbound forms

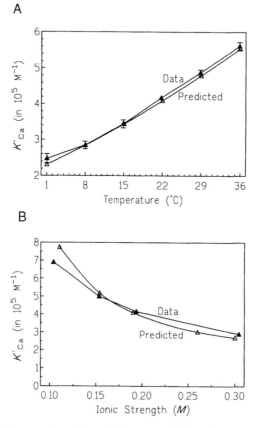

Fig. 5 The effect of temperature (A) and ionic strength (B) on the apparent Ca^{2+} affinity (K'_{Ca}) of Br_2-BAPTA. The experimental data are from Harrison and Bers (1987), (A) at pH 7.00 and 0.19 M ionic strength and (B) at pH 7.00 and 22°C. Predicted values are based on the temperature and ionic strength corrections described in the text.

in different states of protonation. Then, for a tetravalent Ca^{2+} buffer such as EGTA, the total of the non-Ca^{2+}-bound forms of the buffer is

$$[R_t] - [CaR] = [R] + [HR] + [H_2R] + [H_3R] + [H_4R] \qquad (2)$$

We have omitted valency for simplicity. R (or R^{4-}) is the form that binds Ca^{2+} most avidly. Transforming Eq. 1 to one with an apparent affinity constant for Ca^{2+}(K'_{Ca}) for a given pH is convenient

$$K'_{Ca} = \frac{[CaR]}{[Ca] \cdot [R]} \cdot \frac{[R]}{[R_t] - [CaR]} \qquad (3)$$

or, using Eqs. 1 and 2,

$$K'_{Ca} = K_{Ca} \frac{[R]}{[R] + [HR] + [H_2R] + [H_3R] + [H_4R)} \tag{4}$$

Then, simply

$$K'_{Ca} = \frac{K_{Ca}}{1 + [H]K_{H1} + [H]^2K_{H1}K_{H2} + [H]^3K_{H1}K_{H2}K_{H3} + [H]^4K_{H1}K_{H2}K_{H3}K_{H4}} \tag{5}$$

where K_{H1}–K_{H4} are the four acid association constants for the buffer. Now, if we know K_{Ca}, the pH, and K_{H1}–K_{H4}, we can calculate K'_{Ca}. This K'_{Ca} is the apparent affinity for a given $[H^+]$, where pH = $-\log_{10}([H^+]/\gamma_H)$, and γ_H is the activity coefficient for protons under the experimental conditions (see subsequent discussion). This K'_{Ca} is the reciprocal of the dissociation constant, K_d, discussed in the previous section.

Equation 3 also can be manipulated to yield

$$[CaR]/[Ca] = K'_{Ca}[R_t] - K'_{Ca}[CaR] \tag{6}$$

which is the linearization for Scatchard plots of bound/free ($[CaR]/[Ca^{2+}]$) vs bound ($[CaR]$), where the slope is $-K'_{Ca}$ and the x intercept is $[R_t]$. One also can solve for $[CaR]$, obtaining the familiar Michaelis–Menten form:

$$[CaR] = \frac{[R_t]}{1 + 1/(K'_{Ca} \cdot [Ca])} \tag{7}$$

Solving for free $[Ca^{2+}]$ is more complicated because we do not know $[CaR]$ a priori, but substituting $[CaR] = [Ca_t^{2+}] - [Ca^{2+}]$, we obtain a quadratic solution

$$[Ca]^2 + ([R_t] - [Ca_t] + 1/K'_{Ca})[Ca] - [Ca_t]/K'_{Ca} = 0 \tag{8}$$

Similar equations can be developed for Ca^{2+} binding to the protonated form (e.g., H-EGTA), which also binds Ca^{2+} with a lower affinity (e.g., see Harrison and Bers, 1987). For example, when we include Ca^{2+} binding to the protonated form of EGTA (or HR), the following term must be added to the apparent affinity expression on the right hand side of Eq. 5:

$$\frac{K_{Ca2}}{1/([H]K_{H1}) + 1 + [H]K_{H2} + [H]^2K_{H2}K_{H3} + [H]^3K_{H2}K_{H3}K_{H4}} \tag{9}$$

where K_{Ca2} is the Ca^{2+} association constant for the ligand in the protonated form HR. This equation provides some basics of the relationships for a single ligand. However, more complicated solutions have multiple equilibria, which cannot readily be solved simultaneously in an analytical manner.

Note, however, that proceeding from free $[Ca^{2+}]$ to $[Ca_t^{2+}]$ is simpler, especially with no Ca^{2+} competitors, because all the ligands that might bind Ca^{2+} will be in equilibrium with the same free $[Ca^{2+}]$. Thus, one simply could use a series of equations like Eq. 7 for different ligands if the values on the right hand side are known. Then the free $[Ca^{2+}]$ and the $[CaR]$ values from the chelators are added to obtain the $[Ca_t^{2+}]$. If free $[Ca^{2+}]$ is not known (or chosen), multiple versions of equations like Eq. 8 must be solved simultaneously. Thus, iterative computer programs are useful (see Section VII).

IV. Temperature, Ionic Strength, and pH Corrections

Although Section III explains the theoretical basis for calculating the pH effect on K_{Ca}, we should clarify how we normally correct for temperature, ionic strength, and pH for the experimental conditions used (Harrison and Bers, 1989). Thus, the final apparent affinity (or K'_{Ca}) should include correction for temperature and ionic strength as well as for pH. Indeed, both proton affinity (K_{H1}–K_{H4}) and metal affinity constants (e.g., K_{Ca}) should be adjusted for the experimental temperature and ionic strength before adjusting for pH, as just described. Again, readers not interested in the details can proceed to Section V.

A. Temperature Corrections

The standard method of correcting equilibrium constants for changes in temperature depends on knowledge of the enthalpy (ΔH) of the reaction.

$$\log_{10} K' = \log_{10} K + \Delta H(1/T - 1/T')/(2.303 \cdot R) \qquad (10)$$

where T is in °K, ΔH is in kcal/mol, and R is 1.9872×10^{-3} kcal/mol · °K. Unfortunately, the ΔH values are not known for all the reactions in which we might be interested. For example, for EGTA, ΔH is known for the first two acid association constants (K_{H1} and K_{H2}) and for the higher affinity Ca^{2+} constant (K_{Ca1}). This knowledge is generally sufficient for calculations with EGTA (see Fig. 3A). However, no ΔH values have been reported for individual constants for BAPTA and Br_2-BAPTA. Harrison and Bers (1987) measured the temperature dependence of the apparent K'_{Ca} for BAPTA and Br_2-BAPTA. We have fit their data, varying the value of the ΔH for K_{Ca}, a somewhat arbitrary decision because temperature dependence of K_{H1}–K_{H4} also is likely. However, the data were well described using ΔH values (for K_{Ca}) of 4.7 and 5.53 kcal/mol for BAPTA and Br_2-BAPTA, respectively (see Figs. 4, 5). Also, since BAPTA and Br_2-BAPTA are almost completely unprotonated already at neutral pH (see Fig. 1), the adjustments to K_{H1} and K_{H2} are less important than for EGTA.

However, note that one cannot simply use the ΔH values reported by Harrison and Bers (1987) for the overall K'_{Ca} of BAPTA and Br_2-BAPTA (3.32 and 4.04 kcal/mol), as suggested by Marks and Maxfield (1991), because the intrinsic effect of increasing temperature on the K'_{Ca} (with $\Delta H = 0$) is to reduce the

K'_{Ca} (because of the intrinsic temperature dependence of the ionic strength adjustment; see Fig. 4A and subsequent discussion). Consequently, the apparent overall ΔH for K'_{Ca} (3.32 for BAPTA) is smaller than the actual ΔH for K_{Ca} required (our estimate is 4.7 kcal/mol).

Additionally, Harrison and Bers (1987) found the K'_{Ca} for Br$_2$-BAPTA to be somewhat higher than the value predicted by the initial values reported by Tsien (1980). We find that using a slightly higher K_{Ca} (log K_{Ca} = 5.96 rather than 5.8) allowed a considerably better fit to the array of experimental data shown in Fig. 5.

B. Ionic Strength Corrections

Ionic strength also can alter the K'_{Ca} dramatically (see Figs. 3–5). We use the procedure described by Smith and Miller (1985) with ionic equivalents (I_e) rather than formal ionic strength ($I_e = 0.5 \cdot \Sigma C_i \cdot |z_i|$, where C_i and z_i are the concentration and valence of the ith ion). We will use the terms equivalently here. Then the expression used to adjust for ionic strength is

$$\log_{10} K' = \log_{10} K + 2 \, xy(\log_{10} f_j - \log_{10} f'_j) \tag{11}$$

where K' is the constant after conversion, K is the constant before, and x and y are the valences of cation and anion involved in the reaction. The terms $\log_{10} f_j$ and $\log_{10} f'_j$ are adjustment terms related to the activity coefficients for zero ionic strength and desired ionic strength, respectively. To adjust for ionic strength,

$$\log_{10} f_j = \frac{A \cdot I_e^{1/2}}{1 + I_e^{1/2}} - b \cdot I_e \tag{12}$$

where b is a constant (0.25). A is a constant that depends on temperature and on the dielectric constant of the medium (ϵ)

$$A = \frac{1.8246 \cdot 10^6}{(\epsilon T)^{3/2}} \tag{13}$$

where T is the absolute temperature (°K) and ϵ is the dielectric constant for water. The dielectric constant is temperature dependent and can be found in tables, but that is inconvenient for a computer program so, using a curve-fitting program, the following equation provides an excellent empirical description over the range 0–50°C.

$$\epsilon = 87.7251 - 0.3974762 \cdot T + 0.0008253 \cdot T^2 \tag{14}$$

where T is in °C. Thus, some intrinsic temperature dependence exists in the ionic strength adjustment itself (see Fig. 4A, dashed line). These corrections provide a reasonably good description of the influence of ionic strength on the K'_{Ca} in Figs. 3–5.

C. Activity Coefficient for Protons

The association constants as usually reported (e.g., in Martell and Smith, 1974,1977) are often called stoichiometric (or concentration) constants. These terms are sensible because they imply (correctly) that they are to be used with concentrations or stoichiometric amounts in chemical equilibria (e.g., as in Eq. 1). Although we routinely talk about ion concentrations in "concentration" or "stoichiometric" terms, the usual exception is pH [where pH = $-\log$ (hydrogen ion activity) or $10^{-pH} = a_H = \gamma_H[H^+]$]. However, one can convert pH to $[H^+]$ and continue using the "stoichiometric" constants at face value, that is, then all values are in concentration terms and not activity. We have done this in our programs.

The alternative is to change the stoichiometric constants to "mixed" constants (for proton interactions, or $K_{H1}-K_{H4}$ only). Then pH (or 10^{-pH} rather than $10^{-pH}/\gamma_H$) can be used in the calculations. Thus, acid association constants ($K_{H1}-K_{H4}$) should be divided by the value of γ_H. Then the constant can be multiplied by the proton activity (since they are always of the same order in the equations; see Eq. 5). That is, $[H^+]K_{H1} = ([H^+] \cdot \gamma_H) \cdot (K_{H1}/\gamma_H)$ where $[H^+] \cdot \gamma_H = 10^{-pH}$. This method seems a bit more awkward, but the result is the same.

The proton activity coefficient, γ_H, varies with both temperature and ionic strength. The empirical relationship we devised to describe this relationship is

$$\gamma_H = 0.145045 \cdot \exp(-B \cdot I_e) + 0.063546 \cdot \exp(-43.97704 \cdot I_e) + 0.695634$$

(15)

where

$$B = 0.522932 \cdot \exp(0.0327016 \cdot T) + 4.015942$$

where I_e is ionic strength and T is temperature (in °C). This equation gives very good estimates of γ_H from 0–40°C and from 0–0.5 M ionic strength. This expression also was sent to Alex Fabiato for use in his computer program and was listed in his publication (Fabiato, 1991). The first coefficient (0.145045) was sent to him incorrectly and printed that way (i.e., as 1.45045). Fortunately, the typographical error was caught before disks with his modified program were distributed.

V. Ca^{2+} Measurement and Calibration Solutions

A. Measuring [Ca^{2+}]

Although we can calculate the free $[Ca^{2+}]$ or $[Ca_t^{2+}]$ for our solutions with the computer programs described in the following sections, many potential sources of error still exist (e.g., contaminant Ca^{2+}, systematic errors in pH, impurities in chemicals). Thus, measuring the free $[Ca^{2+}]$ to check that the solutions are as expected (especially for complex solutions) is a good idea. Ca^{2+}-sensitive electrodes are convenient tools for measuring [Ca^{2+}] (see Chapter 2). We normally use Ca^{2+} minielectrodes (as described in Chapter 4) or commercial macroelectrodes. Both can be connected to a standard pH meter, but having a meter that reads in increments of 0.1 mV is best. We have had pretty good luck with Orion (Boston, MA) brand Ca^{2+} electrodes, which can be stable for 6 months or so. However, these electrodes are rarely as good as the minielectrodes we make. These minielectrodes are very easy to make and are sensitive to changes in free [Ca^{2+}] to 1 nM or less. They do not last as long as commercial macroelectrodes, but they are extremely inexpensive to make (per electrode) and can be discarded if they become contaminated with protein or are exposed to radioactive molecules.

B. Spreadsheet for Calibration Calculations

Making up calibration solutions for Ca^{2+} electrodes (or fluorescent indicators) is really a simpler version of the multiple equilibria problem that is discussed in Section VII (MaxChelator), because we really only need to consider the Ca^{2+}–EGTA buffer system. The approach we generally use is based on the discussion by Bers (1982). This method has the following general steps:

1. Calculate how much total Ca^{2+} (or free [Ca^{2+}]) is required for the desired solutions (using known constants, corrected as described). All solutions should have the same dominant ionic constituents as the solutions to be measured (e.g., 140 mM KCl, 10 mM HEPES).

2. Measure the free [Ca^{2+}] with a high quality Ca^{2+} electrode, compared with free [Ca^{2+}] standards without EGTA (at higher [Ca^{2+}]).

3. Believing (for the moment) that the values from the electrode are all correct, bound Ca^{2+} ([CaR]) is calculated from free [Ca^{2+}] and total [Ca^{2+}].

4. Scatchard analysis allows the independent measurement of the apparent K'_{Ca} and total [EGTA] in your solutions under experimental conditions (even with systematic errors). Note that the Scatchard plot is very sensitive and deviates from linearity at very low [Ca^{2+}], at which the Ca^{2+} electrodes become sub-Nernstian in response (see Fig. 7).

5. Using these "updated" values of total [EGTA] and K'_{Ca} one can recalculate the free [Ca^{2+}] in the solutions. Then, use the free [Ca^{2+}] predicted from the electrode directly or recalculate from the total [Ca^{2+}] and updated constants. The latter step is necessary for the lowest free [Ca^{2+}], at which the electrode begins to be nonlinear (~pCa 9).

We use a *Lotus-123* spreadsheet to simplify all these steps greatly (see Fig. 6). Three basic versions of this *Lotus* spreadsheet are available: one for free [Ca^{2+}] as the input (DMB-CAF.WK1), one for pCa as input (DMB-

Ca²⁺Calibration   For Entry of pCa

3/17/93

Solution Conditions

7.2	pH
0.15	M Ionic Equiv (0.5*sum |zi|Ci)
23	degrees C
5	mM EGTA
500	ml bottle

K_d = 1.57E-07 M   or   0.1572 uM

Regression Analysis (see I32..M40)

4.953608 [EGTA]tot (mM)
0.997987 r^2

Range for Linear Regression for Scatchard Must be selected "DR"

6.868389 = Log KCa from Scatchard
1.35E-07 = K'Ca Dissociation from Scatchard

K'Ca Calculation (see A32..G47)

B = 5.125357 Intermed
Gamma H= 0.762958 H activity coefficient
Log [H] = -7.08250
[H] (M) = 8.27E-08
K'Ca Assn= 6362839. Log K'Ca= 6.803650
K'Ca Discn= 1.57E-07

Main Table

Initial pCa	Ca–free (nM)	Ca–total (mM)	ml 100 mM CaCl2	V–Ca (mV)	Ca–free (M)	Ca–free (nM)	Ca–Bound (nM)	B/F	Regresn line B/F	inter mediate	Recalc ulated (nM)	pCa
8.5000	3.162277	0.098621	0.493105	-152	5.38E-09	5.383125	0.098615	18319.40	35857.37	-0.00485	2.750294	8.5606
8.0000	10	0.299110	1.495553	-142.6	1.14E-08	11.35461	0.299099	26341.65	34376.67	-0.00465	8.700730	8.0604
7.5000	31.62277	0.837538	4.187690	-131.5	2.74E-08	27.41138	0.837510	30553.38	30400.14	-0.00411	27.54955	7.5599
7.0000	100	1.944334	9.721672	-117.1	8.60E-08	85.99689	1.944248	22608.35	22226.14	-0.00300	87.47565	7.0581
6.5000	316.2277	3.340243	16.70121	-101.4	2.99E-07	299.1314	3.339944	11165.47	11917.99	-0.00161	280.2487	6.5525
6.0000	1000	4.321778	21.60889	-85	1.10E-06	1099.966	4.320678	3928.010	4674.610	-0.00063	924.5800	6.0341
5.5000	3162.277	4.766282	23.83141	-70.4	3.51E-06	3506.126	4.762775	1358.415	1409.423	-0.00018	3381.533	5.4709
5.0000	10000	4.932479	24.66239	-51.9	1.52E-05	15232.05	4.917247	322.8222	268.5445	-0.00002	17312.43	4.7616
4.5000	31622.77	5.006739	25.03369	-38.1	4.56E-05	45563.87	4.961175	108.8839	-55.8930	0.000052	63647.14	4.1962
3.0000	1000000	5.999057	29.99528	0	9.38E-04	938455.7	5.060601	5.392477	-790.217	0.001045	1046090.	2.9804
3.0000	1000000	5.999057	29.99528	0	9.38E-04	938455.7	5.060601	5.392477	-790.217	0.001045	1046090.	2.9804
3.0000	1000000	5.999057	29.99528	0	9.38E-04	938455.7	5.060601	5.392477	-790.217	0.001045	1046090.	2.9804

10 mM	10 mM
1 mM	1 mM
100 uM	100 uM

V–Ca (mV): 28.2 / 0.8 / -29.6
Avg slope = 28.9
Slope (mV) = 29
mV offset at 1 mM Ca = 0.8

Temperature and Ionic Strength Correction

	Std Cond	Ionic Str incl T eff	Final 23 Deg C	Final 23 Deg C	Delta H kcal/mol	valence 2*x*y
Temp	20					
I-Eq	0.100	0.150	0.150 M			
Stoich Const	Log K	Log K'	Log K'	K' (M)		
K1	9.47	9.357636	9.313797	2.1E+09	-5.8	8
K2	8.85	8.765727	8.721888	5.3E+08	-5.8	6
K3	2.66	2.603818	2.603818	401.6226	0	4
K4	2	1.971909	1.971909	93.73657	0	2
KCa	10.97	10.74527	10.68404	4.8E+10	-8.1	16
KCa2	5.3	5.131454	5.131454	135348.8	0	12
Log f	0.109224	0.123270				
Temp	293	296				
A	0.507424	0.510066				
epsil	80.10569	79.01973				

Regression Output:

Constant	36585.72 Y intercept
Std Err of Y Est	571.3961
R Squared	0.997987
No. of Observations	8
Degrees of Freedom	6
X Coefficient(s)	-7385.67 /mM =-K'Ca (slope)
Std Err of Coef.	135.4058

From Regression:

[EGTA]tot	K–Ca–EGTA
4.953608 mM	7.385671 x 10^6 /M
99.07% pure	6.868389 = log K

PCA.WK1), and one for total Ca^{2+} as input (DMB-CAT.WK1). We are pleased to share copies of these spreadsheets with anyone interested (the programs also should be usable with *Quattro Pro* and *Excel*). Instructions for obtaining copies are given at the end of this chapter. Here we present a step-by-step guide to the use of this spreadsheet in making a series of free $[Ca^{2+}]$ standards.

The fields for the input of data are shaded (or green on the computer screen if you have WYSIWYG). For the pCa version of the spreadsheet in Fig. 6, proceed as follows (the other versions are completely analogous):

1. Enter the solution conditions (upper left, pH, ionic strength, temperature, total [EGTA], and bottle size used). The K'_{Ca} values then are adjusted automatically for the selected temperature, pH, and ionic strength (lower left box).

2. Enter the desired pCa values in the first column. The free $[Ca^{2+}]$, total $[Ca^{2+}]$, and ml of 100 mM Ca^{2+} stock are calculated automatically (next 3 columns) using the adjusted K'_{Ca}.

3. Enter the mV readings from a Ca^{2+} electrode (including values for Ca^{2+} standards lacking EGTA at 100 μM, 1 mM, and 10 mM $[Ca^{2+}]$ and the electrode reading at 1 mM free $[Ca^{2+}]$ as the "offset"). This is the fourth column (V-Ca); choosing the electrode slope (rather than assuming the average) is possible. The free $[Ca^{2+}]$, the Ca^{2+} bound to EGTA, and the bound/free (B/F) ratio then are calculated automatically (based on the electrode response and total $[Ca^{2+}]$).

4. You must choose the range for the linear regression of the Scatchard plot (you must know that the keystrokes "/dr" begin the linear regression in *Lotus*). The output is placed in the lower right box, with the important items listed again at the upper right of the page (see Fig. 7A).

5. The free $[Ca^{2+}]$ and pCa are recalculated automatically using the *measured* K'_{Ca} and total [EGTA] as well as the total Ca^{2+} values (results are placed in the last two columns). Also, "named" graphs are built into the spreadsheet that allow you to look at the Scatchard plot and the electrode calibration curves (see Fig. 7).

We routinely use this program for calibration solutions that we use for both Ca^{2+} electrodes and fluorescent indicators. In addition to improving the reliability of Ca^{2+} calibration solutions, one of the convenient aspects of this spreadsheet is that all the details of the process are observable. For example, one can see that the EGTA is saturated almost completely as free $[Ca^{2+}]$ approaches 10 μM. In this range, we usually believe the electrode, rather than

---

Fig. 6 *Lotus-123* spreadsheet used to prepare Ca^{2+} calibration buffers using a Ca^{2+} electrode. This version, printed with WYSIWYG, is used when the input is the pCa of the calibration solutions (DMB-PCA.WK1; see text for details).

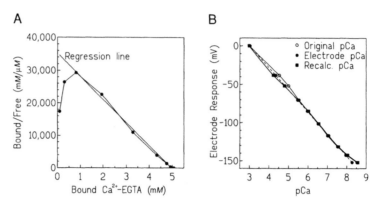

Fig. 7 Scatchard plot (A) and electrode calibration curves (B) for the spreadsheet shown in Fig. 6. The Scatchard plot allows estimation of the total [EGTA] (*x* intercept) and the apparent association constant, K'_{Ca} (−slope). The Scatchard plot is very sensitive to the detection limit of the Ca^{2+} electrode. The two points furthest to the left in A are the lowest free $[Ca^{2+}]$ in the calibration curve in B (and are not included in the regression). The three calibration curves shown are for the original (or planned pCa), the pCa predicted solely by the electrode, and the pCa after recalculation using the values determined in the Scatchard plot as well as the total Ca^{2+} added to the buffers. In this instance, good agreement existed between the three curves, but this is not always the case (see Bers, 1982).

our ability to pipette within 1% of the required volume. On the other hand, as the detection limit of the electrode is approached (e.g., ∼pCa 9), we use the recalculated pCa values. Measured vs predicted K'_{Ca}, EGTA purity, and $[Ca^{2+}]$ also can be useful in identifying potential systematic errors or changes in procedures.

VI. Preparing the Buffer Solution

A. Basic Steps in Solution Preparation

No hard-and-fast rules or special tricks exist in making these buffers, but special care in weighing and pipetting, and common sense, can help avoid some potential problems. The water should be well purified to minimize contamination with Ca^{2+} and other metals. We usually use water that is first distilled and then run through a water purification system containing at least one ion exchange column (e.g., Nanopure; Barnstead, Dubuque, IA). This procedure provides water with resistivity of >15 MΩ · cm as it leaves the column. Starting with good water is probably important for the removal of other metal contaminants as well as Ca^{2+}. Contaminating Ca^{2+} and metals can also be found in the salts and chemicals used to make solutions. In the end, finding 1–3 μ*M* free Ca^{2+} in nominally Ca^{2+}-free solutions is typical, and can be checked with a Ca^{2+} electrode.

Some investigators include 1–2 μ*M* N,N,N',N'-tetrakis (2-pyridylmethyl) ethylene diamine (TPEN), a heavy metal chelator in Ca^{2+} buffer solutions that can chelate submicromolar amounts of heavy metals, which may or may not be chelated by the dominant Ca^{2+} buffer. This step may not be important in routine applications, but may insure that the Ca^{2+}-sensitive process under study will not be altered by trace amounts of other metals. All solutions should be made and stored in clean plasticware (careful washing and extensive rinsing in deionized water is required). Glass containers should be avoided. EGTA can leach Ca^{2+} out of glass, leading to gradual increase in free $[Ca^{2+}]$ in the solutions. We often have been able to store Ca^{2+} calibration solutions for up to 2–3 months in polypropylene bottles (provided no organic substrate to foster bacterial growth is present).

An accurate $[Ca^{2+}]$ standard is important for making Ca^{2+} buffers. Making accurate $[Ca^{2+}]$ using $CaCl_2 \cdot 2H_2O$ typically used to make physiological solutions is difficult because the hydration state varies, making stoichiometric weighing imprecise. $CaCO_3$ can be weighed more accurately, but has the disadvantage that the CO_2 must be driven off with prolonged heating and HCl, unless HCO_3 (which is a weak Ca^{2+} buffer itself) is desired in the solutions. A convenient alternative is buying a $CaCl_2$ standard solution; we use a 100 mM $CaCl_2$ solution from Orion (BDH Pharmaceuticals (London, U.K.) also sells an excellent 1 M $CaCl_2$ standard). To save money, one can titrate a larger volume of $CaCl_2$ to the same free $[Ca^{2+}]$ as the Orion standard using a Ca^{2+} electrode.

Preparing accurate stock solutions of Ca^{2+} chelators is also important. EGTA from different commercial sources differs somewhat in purity (Bers, 1982; Miller and Smith, 1984), but the manufacturers currently provide pretty good estimates. We find about 99% purity with Fluka (Ronkonkoma, New York) Puriss grade, and about 96–97% with Sigma (St. Louis, Missouri) E-4378. BAPTA also has been reported to contain 20% water by weight (Harrison and Bers, 1987), but can be dried at 150°C until the weight is constant to assure removal of water. If the total buffer concentration (as described in Section V) is measured, this problem can be largely obviated. We typically measure the purity of each lot of EGTA or BAPTA that we use, taking this approach. Then we often keep track on the bottle itself, so we can confirm the value on subsequent tests with the same batch. EGTA (in the free acid form) is also not very soluble because of the acid pH. For neutral pH solutions, dissolving EGTA with KOH in a 1:2 stoichiometry is practical since, at neutral pH, two of the four protons on EGTA are dissociated (vs all four for BAPTA).

When Ca^{2+} is added to EGTA solutions, 2 mol H^+ are released for each mol Ca^{2+} bound. Thus, the pH always should be adjusted, as the Ca^{2+} is being added or afterward. The strong pH dependence of the K'_{Ca} of EGTA (Fig. 1) emphasizes the importance of this point. We typically measure $[Ca^{2+}]$ and pH simultaneously just before the solutions are brought up to final volume (for approximate pH adjustment), and after, for final pH adjustment (as close to the third decimal place as possible) and $[Ca^{2+}]$ measurement. The solutions also are checked again later to insure consistency. The rigorous attention to pH adjustment obviously will be less crucial for the BAPTA buffers.

An obvious question is, "Why not use BAPTA rather than EGTA?" The main reason is expense; BAPTA is about 30 times more expensive. The other reason is that EGTA is the "devil we know;" indeed, we do know much about its chemistry (e.g., metal binding constants, ΔH values). For applications with small volumes of solution, however, replacing EGTA with BAPTA may be quite reasonable.

The ionic strength contribution of the pH buffer also should be included in the ionic strength calculation ($I_e = 0.5 \Sigma C_i |z_i|$). This adjustment requires calculation of the fraction of buffer in ionized form (i.e., not protonated).

B. Potential Complications

Not all the desired constants have been determined for the metals and chelators of interest, placing some limitations on how accurately one can predict the free $[Ca^{2+}]$ of a given complex solution or determine how much total Ca^{2+} is required to achieve a desired free $[Ca^{2+}]$. The same is true for other species of interest (e.g., Mg^{2+}, Mg^{2+}–ATP). Some Ca^{2+} buffers also can interact with Ca^{2+} in multiple stoichiometries (e.g., the low affinity Ca^{2+} buffer nitrilotriacetic acid, NTA, can form Ca^{2+}–NTA_2 complexes). Also, systematic errors in pH measurements (Illingworth, 1981) or purity of reagents can occur. Purity can be estimated as described earlier.

The pH problem is actually quite common, especially with combination pH electrodes. Simply put, the reference junctions of some electrodes (particularly with ceramic junctions) can develop junction potentials that are sensitive to ionic strength. This problem can be exacerbated when the ionic strength of the experimental solution differs greatly from that of the pH standards (typically low ionic strength phosphate pH standard buffers). A systematic error in solution pH of about 0.2 pH units is not at all uncommon. As is clear from Fig. 1, this difference could translate into a 0.4 error in log K'_{Ca} and produce a 2- to 3-fold difference in free $[Ca^{2+}]$, even when EGTA is at its best in terms of buffer capacity.

Although measuring the free $[Ca^{2+}]$ with an electrode can be extremely valuable, it is not foolproof. Ca^{2+} electrodes are not perfectly selective for Ca^{2+} (see Chapter 4). For example the selectivity of these electrodes for Ca^{2+} over Mg^{2+} is 30,000–100,000 (Schefer *et al.*, 1986). This range roughly corresponds to the difference in intracellular concentrations. Thus a 100 nM Ca^{2+} solution with 1 mM Mg^{2+} would appear to the electrode as a 110–130 nM Ca^{2+} solution. For the Ca^{2+} electrodes described in Chapter 4 (using the ETH 129 ligand) the interference by Na^+ or K^+ is less. For 140 mM Na^+ or K^+ in a 100 nM Ca^{2+} solution, the apparent $[Ca^{2+}]$ would be only about 101 nM.

Some Ca^{2+} buffers also can interfere with Ca^{2+} electrodes. Citrate, DPA (dipicolinic acid), and ADA (acetamidoiminodiacetic acid), three low affinity

Ca^{2+} buffers, were found to interfere with Ca^{2+} electrode measurements, whereas NTA did not (Bers *et al.,* 1991). Interestingly, citrate, DPA, and ADA (which modified electrode behavior) also modified Ca^{2+} channel characteristics, but NTA did not. When Ca^{2+} electrodes cannot be used practically, optical indicators such as the fluorescent indicators fura-2 or indo-1 still may be used for low [Ca^{2+}] (20 nM–2 μM), or the metallochromic dyes antipyralazo III, murexide, or tetramethylmurexide can be used for higher free [Ca^{2+}] ($K_d \sim$ 200 μM, 3.6 mM, and 2.8 mM, respectively; Ohnishi, 1978,1979; Scarpa *et al.,* 1978). Of course, these indicators require calibration also.

A general potential complication with Ca^{2+} buffers is that they may alter the very processes one is interested in studying with Ca^{2+} buffers. For example, EGTA and other Ca^{2+} chelators have been documented to increase the Ca^{2+} sensitivity of the plasmalemmal and sarcoplasmic reticular (SR) Ca^{2+}–ATPase pumps and also of Na$^+$/Ca^{2+} exchange (Schatzmann, 1973; Sarkadi *et al.,* 1979; Berman, 1982; Trosper and Philipson, 1984). For example, 48 μM EGTA decreased the apparent K_{Ca} of Na$^+$/Ca^{2+} exchange in cardiac sarcolemmal vesicles from 20 to 5 μM Ca^{2+} (Trosper and Philipson, 1984).

These points are not meant to discourage the use of Ca^{2+} buffers, but simply to point out some of the potential problems that might be encountered. Being aware of what might occur can help troubleshoot when results do not make sense. Clearly, the use of Ca^{2+} buffer solutions is essential to the understanding of Ca^{2+}-dependent phenomena. Our aim here is to provide helpful information.

VII. MaxChelator (MAXC v6.50): An Easy-to-Use Computer Program for Ca²⁺ Buffer Preparation

One of the most useful tools in making Ca^{2+} buffer solutions containing multiple metals and/or ligands is a computer program to solve the complicated multiple equilibria. Numerous programs have been developed and numerous descriptions of how to make Ca^{2+} buffers have been published (e.g., Fabiato, 1988,1991; McGuigan *et al.,* 1991; Marks and Maxfield, 1991; Brooks and Storey, 1992; Taylor *et al.,* 1992). Several other programs have been used effectively, although they are not described explicitly in publications (e.g., see Fabiato, 1991). The programs generally have been distributed freely. One of the limitations of some programs used extensively by individual laboratories is that they are generally not very "user friendly," making it difficult to overcome the inertia to use them (sometimes the reason they remain largely in the laboratory that developed them). Until fairly recently, the appropriate constants and methods to adjust them for temperature, ionic strength, and pH also have varied a bit. Our hope is that the program described here is extremely easy to use and provides accurate and useful results as well.

MaxChelator (MAXC) is a program that solves a series of equations to match up free and total chelators and metals in solution. The first version of MaxChelator (v2.7) was released in 1986. The program was written in Pascal for the IBM computer and was based on an early Apple Basic program used in Richard Steinhardt's laboratory, as well as on the program of Fabiato and Fabiato (1979). The program has been improved, in part because of improvements in computers, the Borland Pascal compiler, and the programmer. Feedback from users and a host of experts in metal chelators (e.g., Stephen M. Baylor) also has been essential to these improvements, as has the encouragement and constant support of David Epel.

Under ideal conditions, this program is intended to be a black box. The chemical equations and mathematical expressions that constitute the core of the black box are essentially the same as those described by Fabiato and Fabiato (1979) and are straightforward individual expressions. Much effort has been spent on making the black box easy to use. Simply stated, into the black box go:

1. stability constants (stored in a file)
2. chelator and metal types and concentrations
3. environmental conditions (pH, ionic strength, temperature)

Out comes the answer—the free or total concentration of the chelators or metals that were chosen. To calculate some Ca^{2+} buffers without any further ado, refer to Section VII,A. A brief description of each of the "inputs" follows here:

1. Stability constants—The stability constants are largely from Martell and Smith (1974,1977) and are stored in a default file, MAXC.CCM, which can be used as it stands. We also supply an alternative file (BERS.CCM) which includes a few different chelators and includes the ΔH values for BAPTA and Br_2-BAPTA (and the altered K_{Ca} for the latter), as described earlier. The tables of constants can be modified readily on the screen and saved under a new name. In this way, you can edit existing constants, add new ones as they are reported, or replace some buffers with other ones you may choose.

2. Chelator and metal types and concentrations—You can choose the particular metals and chelators that will be in your solutions from two lists. For example, you may choose to use EGTA and ATP with Ca^{2+} and Mg^{2+}. Be aware of any interactions from other metals and chelators that might affect the result. For instance, ATP and other phosphonucleotides all bind Ca^{2+} and Mg^{2+}, and therefore must be accounted for. Mg^{2+} will bind some chelators in sufficient concentration to affect the calculated free $[Ca^{2+}]$. Other metals ($Zn^{2+/3+}$, Fe^{3+}, Al^{3+}, etc.) will bind more strongly to chelators than Ca^{2+}, if they are present in sufficiently high concentration or are themselves necessary for some enzyme reaction, they can have a profound effect on the result.

3. Environmental conditions (pH, ionic strength, temperature)—The sta-

bility constants are adjusted for ionic strength, temperature, and pH as described in the earlier sections of this chapter.

A. Getting Started Quickly with MAXC

1. Copy the MAXC files into a directory on your hard drive
2. Type MAXC to start the program and be prompted automatically through the following steps
3. Choose a **F**ile of constants: "ALT FILE OPEN" (e.g., MAXC.CCM or BERS.CCM)
4. Choose a **C**alculation: Type "ALT C" then

   "**T**otal" to calculate total [metal]s from free [metal]s and total [chelator]s

   "**F**ree" to calculate free [metal]s from total [metal]s and total [chelator]s

   "**C**helator" to calculate free and total [chelator]s from free and total [metal]s
5. Choose your conditions:

   Array of metals and chelators: "ALT A" (6 metals and 6 chelators is the maximum)

   Input [**M**etal] and [chelator]: "ALT M"

   Choose **E**nvironmental conditions: "ALT E" (temperature, pH, and ionic strength)
6. Type "ENTER" and read the answer from the screen!
7. **F4** prints the answer to a printer or an ASCII file; "ALT F" or **F5** for printer options

This list is superficial and does not show the flexibility and power of MAXC. However, you should calculate some real values with this process within 5 min of putting the floppy disk in your computer.

B. Features of MAXC

The ideal program should be easy to use, accurate, and flexible. Perhaps, it may even trigger new insights.

1. Ease of Use

A great deal of effort was expended to make MAXC as easy to use as possible, as well as to maintain flexibility. Currently, the program is available only in an MS-DOS version (although Macintosh and Windows versions are planned currently). The Macintosh ideal still has had its effect on this version.

MAXC has the pull-down menus, mouse support, and help screens that are expected in modern programs regardless of operating environment. MS-DOS users do not always have a mouse or do not necessarily like to use one, so keyboard equivalents are included also.

2. Accuracy

The programs are not perfect. No program can take into account all the variables encountered in making up a solution (the species of ions, time constants for binding, interactions with proteins and other cell components, purity of reagents used, etc.). In fact, some disagreement exists over the initial values for the constants themselves. Nevertheless, every effort has been made to make MAXC as accurate as possible. Occasional errors in the code are still possible and improvements are suggested by users. Each user is encouraged to register for updates. (Users are free to copy and distribute the program provided they do not change the main program MAXC.EXE).

3. Flexibility MAXC Shines in its Flexibility. The Program Includes the Following Attributes.

1. Three types of calculations (up to 6 chelators and 6 metals per calculation) can be performed:
   a. calculate total metals, given total chelators and free metal concentrations
   b. calculate free metals, given chelator and metal totals (internal checking of results)
   c. calculate the amount of chelator to add, given the free and total metals

   Free chelator and metal–chelator complex concentrations are shown for all three calculation types also

2. Ionic contribution of the chelator/metal is done and shown automatically

3. Calculator for ionic contribution of monovalent buffers and salts is available

4. Multiple files of constants consisting of 12 chelators and 9 metals per file (also may be loaded from the command line for batch file users)

5. You may edit the files of constants at any time

6. You may create as many new files of constants as you need and have space for

7. Constants may be adjusted for ionic strength and temperature changes

8. You may save and load any work in progress (you also may load any previously saved work from the command line for batch file users)

9. Results of multiple calculations can be viewed on the screen simultaneously; EGA/VGA owners can adjust screen size for more room (43/50 line mode)

10. Printing of results and contents of constant files may be done by a printer of your choice or to a file of your choice (e.g., as straight ASCII text); files may be appended to add the results of multiple calculations and/or constants files to the same file for later editing with any word processor

11. Simple four-function calculator available internally

12. Help is specific to location in program or available from a Help menu; the documentation file can be read from within MAXC or by any word processor

13. Screen colors can be set to LCD or monochrome for laptop users, with a command line switch

14. The entire program and files currently fit on one 360K disk and can be run from a one floppy XT clone without a mathchip, even though double precision math is used (of course, the program is faster on bigger machines)

15. GRAPH function does two-dimensional and three-dimensional plots, on almost any monitor (HERC,CGA,EGA,VGA), of your choice of effects of temperature, pH, ionic strength, or metal concentrations (you choose and set upper and lower limits within the bounds of the program)

    a. Results of GRAPH can be saved for later viewing without recalculation

    b. Results can be printed/appended in ASCII text form to printer or file

    c. Results can be printed in graphics form directly to HP Laserjet/ Deskjet compatible printers

    d. Results can be saved in monochrome *.PCX graphics format for printing to other printers or inclusion into desktop publishing documents (sample PCX printer routines are included, with source, for laser and dot matrix printers)

Viewing the results of a GRAPH calculation can be quite useful. At a single glance one can see where the useful range of a chelator is, how the other metals are affected, and what happens when one goes to a higher or lower ionic strength solution. Figure 8 shows examples of graphs produced directly from MAXC, printed as a screen dump to a laser printer (see legend for details).

Donald M. Bers *et al.*

A

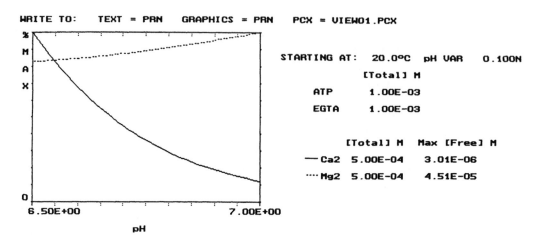

WRITE TO: TEXT = PRN   GRAPHICS = PRN   PCX = VIEW01.PCX

STARTING AT: 20.0°C   pH VAR   0.100N

[Total] M

ATP   1.00E-03

EGTA   1.00E-03

	[Total] M	Max [Free] M
—— Ca2	5.00E-04	3.01E-06
···· Mg2	5.00E-04	4.51E-05

NOT VALID below   0.007 N [ion] (CM Complex conc)

B

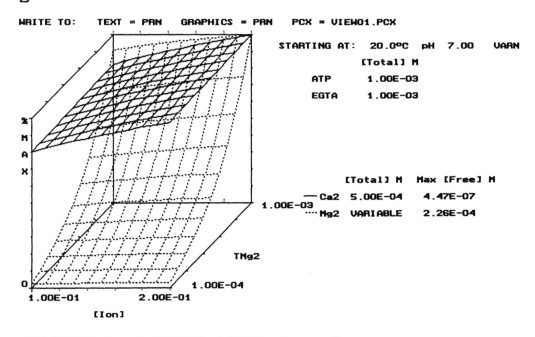

WRITE TO: TEXT = PRN   GRAPHICS = PRN   PCX = VIEW01.PCX

STARTING AT: 20.0°C   pH   7.00   VARN

[Total] M

ATP   1.00E-03

EGTA   1.00E-03

	[Total] M	Max [Free] M
—— Ca2	5.00E-04	4.47E-07
···· Mg2	VARIABLE	2.26E-04

NOT VALID below   0.008 N [ion] (CM Complex conc)

C. Features Not Present in MAXC

MAXC now has a written manual. The program BAD (Brooks and Storey, 1992) produces results for EGTA that agree fairly well with those produced by MAXC. BAD is somewhat better at showing what happens to individual ionic species, such as Cl^- or SO_4^{2-}. Both programs show the chelator–metal complex concentrations. However, BAD requires that you recompile the included partial source code to change constants or to include new chelators and/or metals. No source code is provided with MAXC. Some of the file formats are given for individuals who want to write their own accessories. The source code is omitted so users will be forced to contact the programmer (C. W. Patton) to make changes (including bug fixes) that can then be passed on to other registered users.

D. How to Obtain MAXC and Spreadsheets

To obtain MAXC and the spreadsheets described here, send $10 to cover postage, duplicating, and packaging costs to:

Chris Patton, Stanford University, Hopkins Marine Station, Pacific Grove, California 93950-3094

Please specify disk size and capacity (~400 kB is needed for all the uncompressed files). You will receive, by priority mail, a disk containing all the necessary files. You are free to duplicate the program once you receive the disk, but only registered users will be notified of updates. Some groups are registering one copy for a department and passing around updates as they arrive (saves everyone time and cost). MAXC is continuing to evolve. Hopefully, the program will be much improved by the time this chapter is read and your request is sent.

Fig. 8 Two- and three-dimensional graphs produced directly by MAXC and printed on a laser printer. (A) This two-dimensional example is for solutions containing 1 m*M* EGTA and 1 m*M* ATP, with 0.5 mM total [Ca²⁺] and 0.5 mM total Mg²⁺. The graph illustrates the influence of pH (6.5–7) on the free [Ca²⁺] and [Mg²⁺] (e.g., free [Ca²⁺] declines from 3 μM to 0.3 μM while free [Mg²⁺] increases from 37 to 45 μM). (B) This three-dimensional plot shows the same ionic conditions, except that total [Mg²⁺] is varied from 0.1 to 1 m*M*. Here the effects of changing ionic strength (*x* axis) and total [Mg²⁺] (*z* axis) on free [Ca²⁺] and [Mg²⁺] are shown.

References

Berman, M. C. (1982). Stimulation of calcium transport of sarcoplasmic reticulum vesicles by the calcium complex of ethylene glycol bis(β-aminoethyl ether)-N,N'-tetraacetic acid. *J. Biol. Chem.* **257,** 1953–1957.

Bers, D. M. (1982). A simple method for the accurate determination of free [Ca^{2+}] in Ca–EGTA solutions. *Am. J. Physiol.* **242,** C404–C408.

Bers, D. M., and MacLeod, K. T. (1988). Calcium chelators and calcium ionophores. *In* "Handbook of Experimental Pharmacology" (P. F. Baker, ed.), Vol. 83, pp. 491–507. Berlin: Springer-Verlag.

Bers, D. M., Hryshko, L. V., Harrison, S. M., and Dawson, D. (1991). Citrate decreases contraction and Ca^{2+} current in cardiac muscle independent of its buffering action. *Am. J. Physiol.* **260,** C900–C909.

Brooks, S. P. J., and Storey, K. B. (1992). Bound and determined: A computer program for making buffers of defined ion concentrations. *Anal. Biochem.* **201,** 119–126.

Fabiato, A. (1988). Computer programs for calculating total from specified free or free from specified total ionic concentrations in aqueous solutions containing multiple metals and ligands. *Meth. Enzymol.* **157,** 378–417.

Fabiato, A. (1991). Ca^{2+} buffering: Computer programs and simulations. *In* "Cellular Calcium: A Practical Approach" (J. G. McCormack and P. H. Cobbold, eds.), pp. 159–176. New York: Oxford University Press.

Fabiato, A., and Fabiato, F. (1979). Calculator programs for computing the composition of the solutions containing multiple metals and ligands used for experiments in skinned muscle cells. *J. Physiol. (Paris)* **75,** 463–505.

Good, N. E., Winget, G. D., Winter, W., Connolly, T. N., Izawa, S., and Singh, R. M. M. (1966). Hydrogen on buffers for biological research. *Biochemistry* **5,** 467–477.

Grynkiewicz, G., Poenie, M., and Tsien, R. Y. (1985). A new generation of Ca^{2+} indicators with greatly improved fluorescence properties. *J. Biol. Chem.* **260,** 3440–3450.

Harafuji, H., and Ogawa, Y. (1980). Re-examination of the apparent binding constant of ethylene glycol bis(β-aminoethyl ethylene)-N,N,N′,N′-tetracetic acid with calcium around neutral pH. *J. Biochem.* **87,** 1305–1312.

Harrison, S. M., and Bers, D. M. (1987). The effect of temperature and ionic strength on the apparent Ca-affinity of EGTA and the analogous Ca-chelators BAPTA and dibromo-BAPTA. *Biochim. Biophys. Acta* **925,** 133–143.

Harrison, S. M., and Bers, D. M. (1989). Correction of absolute stability constants of EGTA for temperature and ionic strength. *Am. J. Physiol.* **256,** C1250–C1256.

Hellam, D. C., and Podolsky, R. J. (1969). Force measurements in skinned muscle fibres. *J. Physiol. (London)* **200,** 807–819.

Illingworth, J. A. (1981). A common source of error in pH measurements. *Biochem. J.* **195,** 259–262.

Marks, P. W., and Maxfield, F. R. (1991). Preparation of solutions with free calcium concentration in the nanomolar range using 1,2-bis(*o*-aminophenoxy)ethane-N,N,N′,N′-tetraacetic acid. *Anal. Biochem.* **193,** 61–71.

Martell, A. E., and Smith, R. M. (1974). "Critical Stability Constants," Vol. 1. New York: Plenum Press.

Martell, A. E., and Smith, R. M. (1977). "Critical Stability Constants," Vol. 3. New York: Plenum Press.

McGuigan, J. A. S., Lüthi, D., and Buri, A. (1991). Calcium buffer solutions and how to make them: A do it yourself guide. *Can. J. Physiol. Pharmacol.* **69,** 1733–1749.

Miller, D. J., and Smith, G. L. (1984). EGTA purity and the buffering of calcium ions in physiological solutions. *Am. J. Physiol.* **246,** C160–C166.

Nakon, R. (1979). Free metal ion depletion by Good's buffers. *Anal. Biochem.* **95,** 527–532.

Ohnishi, S. T. (1978). Characterization of the murexide method: Dual wavelength spectrophotometry of cations under physiological conditions. *Anal. Biochem.* **85,** 165–179.

Ohnishi, S. T. (1979). A method of estimating the amount of calcium bound to the metallochromic indicator arsenazo III. *Biochim. Biophys. Acta* **586,** 217–230.

Pethig, R., Kuhn, M., Payne, R., Adler, E., Chen, T.-H., and Jaffe, L. F. (1989). On the dissociation constants of BAPTA-type calcium buffers. *Cell Calcium* **10,** 491–498.

Sarkadi, B., Shubert, A., and Gardos, G. (1979). Effect of Ca-EGTA buffers on active calcium transport in inside-out red cell membrane vesicles. *Experientia* **35,** 1045–1047.

Scarpa, A., Brinley, F. J., and Dubyak, G. (1978). Antipyralazo III, a middle range Ca²⁺ metallochromic indicator. *Biochemistry* **17,** 1378–1386.

Schatzmann, H. J. (1973). Dependence on calcium concentrations and stoichiometry of the calcium pump in human red cells. *J. Physiol.* **235,** 551–569.

Schefer, U., Ammann, D., Pretsch, E., Oesch, U., and Simon, W. (1986). Neutral carrier based Ca²⁺-selective electrode with detection limit in the subnanomolar range. *Anal. Chem.* **58,** 2282–2285.

Smith, G. L., and Miller, D. J. (1985). Potentiometric measurements of stoichiometric and apparent affinity constants of EGTA for protons and divalent ions including calcium. *Biochim. Biophys. Acta* **839,** 287–299.

Smith, P. D., Berger, R. L., and Podolsky, R. J. (1977). Stopped-flow study of the rate of calcium binding by EGTA. *Biophys. J.* **17,** 159a.

Taylor, R. B., Trimble, C., Valdes, J. J., Wayner, M. J., and Chambers, J. P. (1992). Determination of free calcium. *Brain Res. Bull.* **29,** 499–501.

Trosper, T. L., and Philipson, K. D. (1984). Stimulatory effect of calcium chelators on Na⁺–Ca²⁺ exchange in cardiac sarcolemmal vesicles. *Cell Calcium* **5,** 211–222.

Tsien, R. Y. (1980). New calcium indicators and buffers with high selectivity against magnesium and protons: Design, synthesis, and properties of prototype structures. *Biochemistry* **19,** 2396–2404.

Tsien, R. Y., and Zucker, R. S. (1986). Control of cytoplasmic calcium with photolabile tetracarboxylate 2-nitrobenzhydrol chelators. *Biophys. J.* **50,** 843–853.

CHAPTER 2

Photorelease Techniques for Raising or Lowering Intracellular Ca^{2+}

Robert Zucker

Molecular and Cell Biology Department
University of California at Berkeley
Berkeley, California 94720

I. Introduction

Photolabile Ca^{2+} chelators, sometimes called caged Ca^{2+} chelators, are used to control $[Ca^{2+}]_i$ in cells rapidly and quantitatively. A beam of light is aimed at cells filled with a photosensitive substance that changes its affinity for binding Ca^{2+}. In the last few years, several such compounds have been invented that allow the effective manipulation of $[Ca^{2+}]_i$ in cells. These compounds offer tremendous advantages over the alternative methods of microinjecting Ca^{2+} salts, pharmacologically releasing Ca^{2+} from intracellular stores, or increasing cell membrane permeability to Ca^{2+} using ionophores, detergents, electroporation, fusion with micelles, or activation of voltage-dependent channels, in terms of specificity of action, repeatability and reliability of effect, maintenance of cellular integrity, definition of spatial extent, and rapidity of effect, all combined with the ability to maintain the $[Ca^{2+}]_i$ change for sufficient time to measure its biochemical or physiological consequences. Only photosensitive chelators allow the concentration of Ca^{2+} in the cytoplasm of intact cells to be changed rapidly by a predefined amount over a selected region or over the whole cell. Since loading can precede photolysis by a substantial amount of time, cells can recover from the adverse effects of the loading procedure before the experiments begin. The ideal photosensitive Ca^{2+} chelator does not exist, but would have the following properties.

1. The compound could be introduced easily into cells, by microinjection or by loading a membrane-permeating derivative that would be altered enzymatically to an impermeant version trapped in cells.

2. The compound could be loaded with Ca^{2+} to such a level that the unphotolyzed form would buffer the $[Ca^{2+}]_i$ to near the normal resting level, so its introduction into cells would not perturb the resting Ca^{2+} level. Additionally, by adjusting the Ca^{2+} loading or selecting chelator variants, the initial resting Ca^{2+} level could be set to somewhat higher or lower than the normal resting concentration.

3. The chelator should be chemically and photolytically stable.

4. Photolysis by a bright flash of light should allow rapid changes in the free Ca^{2+} level; this characteristic requires rapid photochemical and subsequent dark reactions of the chelator.

5. Photolysis should be achievable with biologically appropriate wavelengths, which requires a high quantum efficiency and absorbance at wavelengths that readily penetrate cytoplasm but cause little biological damage, that is, that are not highly ionizing. For the chelator to be protected from photolysis by light needed to view the preparation would also be useful.

6. The photoproducts, or post-photolysis buffer mixture, should continue to buffer Ca^{2+}, and so hold it at the new level in the face of homeostatic pressure from membrane pumps and transport processes.

7. Neither the unphotolyzed chelator nor its photoproducts should be toxic, but rather should be inert with respect to all ongoing cellular molecular and physiological processes. Three classes of compounds, the nitr series, DM-nitrophen, and the diazo series share enough of these properties to have generated intense interest and widespread popularity, and form the subjects of this review.

Numerous more general reviews of photolabile or caged compounds, which contain some information on photolabile Ca^{2+} chelators, have appeared (Ogden, 1988; Kaplan and Somlyo, 1989; McCray and Trentham, 1989; Walker, 1991; Parker, 1992; Adams and Tsien, 1993; Gurney, 1993; Kao and Adams, 1993). Reviews focused more on photosensitive Ca^{2+} chelators may be consulted also (Kaplan, 1990; Ashley *et al.*, 1991a; Gurney, 1991).

II. Nitr Compounds

A. Chemical Properties

The first useful class of photosensitive Ca^{2+} chelators to be developed was the series of nitr compounds. These compounds rely on the substitution of a photosensitive nitrobenzyl group on one or both of the aromatic rings of the Ca^{2+} chelator 1,2-bis(*o*-aminophenoxy)ethane-*N*,*N*,*N'*,*N'*-tetracetic acid (BAPTA; Tsien and Zucker, 1986; Adams *et al.*, 1988; Adams and Tsien, 1993). Light absorption results in the abstraction of the benzylic hydrogen atom by the excited nitro group and oxidation of the alcohol group to a ketone. The resulting nitrosobenzoyl group is strongly electron withdrawing, reducing the electron density around the metal-coordinating nitrogens and reducing the affinity of the tetracarboxylate chelator for Ca^{2+}. In the first member of this series, nitr-2, methanol is formed as a by-product of photolysis, but in subsequent members (nitr-5, nitr-7, and nitr-8) only water is produced. Photolysis of nitr-2 is also slow (200 msec time constant). For the other nitr chelators, the dominant photolysis pathway is much faster (nitr-7, 1.8 msec; nitr-5, 0.27 msec; nitr-8, not reported). For these reasons, nitr-2 is no longer used. For the three remaining nitr compounds, photolysis is most efficient at the absorbance maximum for the nitrobenzhydrol group, about 360 nm, although light between 330 and 380 nm is nearly as effective. The quantum efficiency of the Ca^{2+}-bound form is about 1/25 (nitr-5, 0.035; nitr-7, 0.042), and is somewhat less in the Ca^{2+}-free form (0.012 and 0.011). The absorbance at this wavelength is 5500 $M^{-1}cm^{-1}$ (decadic molar extinction coefficient) for nitr-5 and nitr-7, and 11,000 $M^{-1}cm^{-1}$ for nitr-8. The structures of the nitr series of compounds are given in Fig. 1; the photochemical reaction of the most popular member of this group, nitr-5, is shown in Fig. 2.

These chelators share the advantages of the parent BAPTA chelator: high specificity for Ca^{2+} over H$^+$ and Mg^{2+} (Mg^{2+} affinities, 5–8 m*M*), lack of

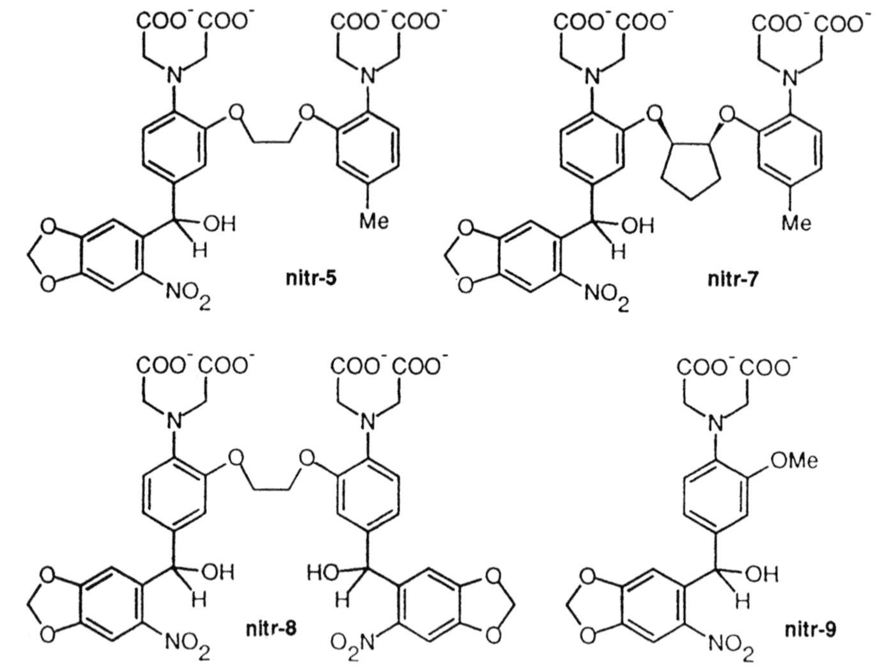

Fig. 1 Structures of the nitr series of photolabile chelators, which release calcium on exposure to light.

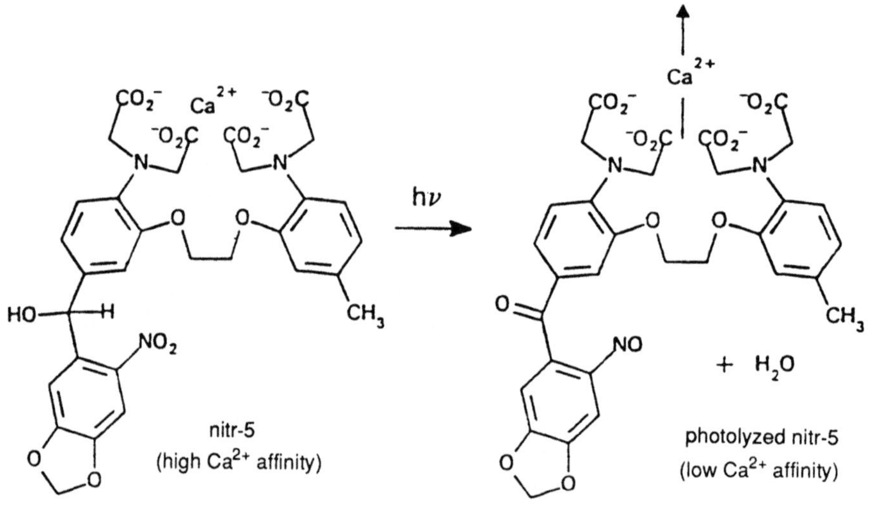

Fig. 2 Reaction scheme for the photolysis of nitr-5.

dependence of Ca^{2+} affinity on pH near pH 7, and fast buffering kinetics. One limitation is that the drop in affinity in the nitr compounds after photolysis is relatively modest, about 40-fold for nitr-5 and nitr-7. The Ca^{2+} affinity of nitr-5 drops from 0.15 to 6 μM at 120 mM ionic strength after complete photolysis. These affinities must be reduced at higher ionic strength, roughly in proportion to the tonicity (Tsien and Zucker, 1986). By incorporating a *cis*-cyclopentane ring into the bridge between the chelating ether oxygens of BAPTA, nitr-7 was created with significantly higher Ca^{2+} affinities (54 nM, decreasing to 3 μM after photolysis at 120 mM ionic strength). To increase the change in Ca^{2+} binding affinity on photolysis, nitr-8 was created with a 2-nitrobenzyl group on each aromatic ring of BAPTA. Photolysis of each group reduces affinity only about 40-fold, as for nitr-5 and nitr-7, but photolysis of both nitrobenzyl groups reduces affinity nearly 3000-fold, to 1.37 mM, with a quantum efficiency of 0.026. Finally, nitr-9 is a dicarboxylate 2-nitrobenzhydrol with a low Ca^{2+} affinity that is unaffected by photolysis; this compound can be used to control for nonspecific effects of the photoproducts.

Nitr-5 is the substance most often applied in biological experiments, largely because it was the first photolabile chelator to have most of the qualities of the ideal substance. The limited affinity for Ca^{2+} of this substance in the unphotolyzed form requires that it be lightly loaded with Ca^{2+} when introduced into cells; otherwise the resting [Ca^{2+}]$_i$ will be too high. However, the compound in a lightly loaded state contains little Ca^{2+} to be released on photolysis. Nitr-7 alleviates this problem with an affinity closer to that of normal resting [Ca^{2+}]$_i$, but its synthesis is more difficult and its photochemical kinetics are significantly slower. Both compounds permit less than two orders of magnitude increase in [Ca^{2+}]$_i$, generally to only the low micromolar range, and then only with very bright flashes or prolonged exposures to steady light to achieve complete photolysis. Nitr-8 permits a very large change in [Ca^{2+}]$_i$, but requires a brighter flash. Photolysis kinetics for this compound have not yet been reported. Neither nitr-8 nor the control compound nitr-9 is presently commercially available; nitr-5 and nitr-7 are supplied by CalBiochem (La Jolla, California).

B. Calculating [Ca^{2+}]$_i$ Changes in Cells

If a nitr compound is photolyzed partially by a flash of light, the reduction in Ca^{2+} affinity of a portion of the nitr requires ~0.3 msec. During this period of photolysis, low affinity buffer is being formed and high affinity buffer is vanishing while the total amount of Ca^{2+} remains unchanged. As the buffer concentrations change, Ca^{2+} ions re-equilibrate among the new buffer concentrations by shifting from the newly formed low affinity nitrosobenzophenone to the remaining unphotolyzed high affinity nitrobenzhydrol. Since the on-rate of binding is close to the diffusion limit (as calculated from Adams *et al.*, 1988; see also Ashley *et al.*, 1991b), this equilibration occurs much faster than photolysis, and Ca^{2+} remains in quasi-equilibrium throughout the photolysis period. The $[Ca^{2+}]_i$ in a cell rises smoothly in a step-like fashion over a period of 0.3 msec from the low level determined by the initial total concentrations of Ca^{2+} and nitrobenzhydrol to a higher level determined by the final concentrations of all the chelator species after partial photolysis. $[Ca^{2+}]_i$ remains under the control of the low and high affinity nitr species, so the elevated Ca^{2+} is removed only gradually by extrusion and uptake into organelles. Thus, the nitr compounds are well suited to producing a modest but quantifiable step-like rise in $[Ca^{2+}]_i$ in response to a partially photolyzing light flash, or a gradually increasing $[Ca^{2+}]_i$ during exposure to steady light. Subsequent flashes cause further increments in $[Ca^{2+}]_i$. These increments actually increase because, with each successive flash, the remaining unphotolyzed nitr is loaded more heavily with Ca^{2+}. Eventually, the unphotolyzed nitr is fully Ca^{2+} bound, and subsequent flashes elevate Ca^{2+} by smaller increments as the amount of unphotolyzed nitr drops.

If a calibrated light source is used that photolyzes a known fraction of nitr in the light path, or in cells filled with nitr and exposed to the light, then the mixture of unphotolyzed nitr and photoproducts may be calculated with each flash (Landò and Zucker, 1989; Lea and Ashley, 1990). The different quantum efficiencies of free and Ca^{2+}-bound chelators must be taken into account. Simultaneous solution of the buffer equations for low and high affinity nitr species and native Ca^{2+} buffers predicts the $[Ca^{2+}]_i$. For sufficiently high nitr concentration (above 5 mM), the native buffers have little effect and usually may be ignored in the calculation. Further, since $[Ca^{2+}]_i$ depends on the ratio of the different nitr species, the exact concentration of total nitr in the cell makes little difference, at least in small cells or cell processes.

If the cell is large, the light intensity will drop as it passes from the front to the rear of the cell. Knowing the absorbance of cytoplasm and nitr species at 360 nm, and the nitr concentrations before a flash, the light intensity and photolysis rate at any point in the cytoplasm may be calculated. A complication in this calculation is that the nitr photoproducts have very high absorbance (Ca^{2+}-free nitr-5 nitrosobenzophenone, 24,000 $M^{-1}cm^{-1}$; Ca^{2+}-bound nitr-5 nitrosobenzophenone, 10,000 $M^{-1}cm^{-1}$; Adams *et al.*, 1988). As photolysis proceeds, the cell darkens and photolysis efficiency is reduced by self-screening. Nevertheless, with estimation of the spatial distribution of light intensity, the spatial concentrations of photolyzed and unphotolyzed chelator can be computed; from this calculation follows the distribution of the rise in $[Ca^{2+}]_i$. The subsequent spatial equilibration of $[Ca^{2+}]_i$ can be calculated by solving diffusion equations, using the initial $[Ca^{2+}]_i$ and chelator distributions as the boundary conditions. Effects of endogenous buffers, uptake, and extrusion mechanisms on the rise in $[Ca^{2+}]_i$ can be included in such models. Simulations of the temporal and spatial distribution of $[Ca^{2+}]_i$ have been devised (Zucker, 1989) and applied to experimental data on physiological effects of $[Ca^{2+}]_i$; the predicted changes in $[Ca^{2+}]_i$ have been confirmed with Ca^{2+}-

sensitive dyes (Landò and Zucker, 1989). Simplified and approximate models using the volume-average light intensity to calculate volume-average photolysis rate and average $[Ca^{2+}]_i$ changes often suffice when the spatial distribution of $[Ca^{2+}]_i$ is not important, for example, when estimating the change in $[Ca^{2+}]_i$ in a cell after diffusional equilibration has occurred.

<hr>

III. DM-Nitrophen

A. Chemical Properties

A different strategy for releasing Ca^{2+} involves attaching a 2-nitrobenzyl group to one of the chelating amines of ethylenediaminetetraacetic acid (EDTA) to form the photosensitive chelator DM-nitrophen (Ellis-Davies and Kaplan, 1988; Kaplan and Ellis-Davies, 1988). Photolysis by UV light in the wavelength range 330–380 nm cleaves the DM-nitrophen through a series of intermediates (McCray *et al.*, 1992) to form iminodiacetic acid and a 2-nitrosoacetophenone derivative within less than 180 μsec. The reaction is shown in Fig. 3. Although DM-nitrophen binds Ca^{2+} with an affinity of 5 nM at pH 7.2, the products bind with affinities of about 3 mM and 0.25 mM at an ionic strength of 150 mM (Kaplan and Ellis-Davies, 1988; Neher and Zucker, 1993). Thus, complete photolysis of DM-nitrophen can elevate Ca^{2+} 50,000-fold, much more than photolysis of the nitr compounds. This significant advantage is counterbalanced to some extent by the facts that the photoproducts buffer Ca^{2+} so weakly that the final $[Ca^{2+}]_i$ will be determined largely by native cytoplasmic buffers, and that the Ca^{2+} liberated by photolysis of DM-nitrophen will be removed more readily by extrusion and uptake pumps. The

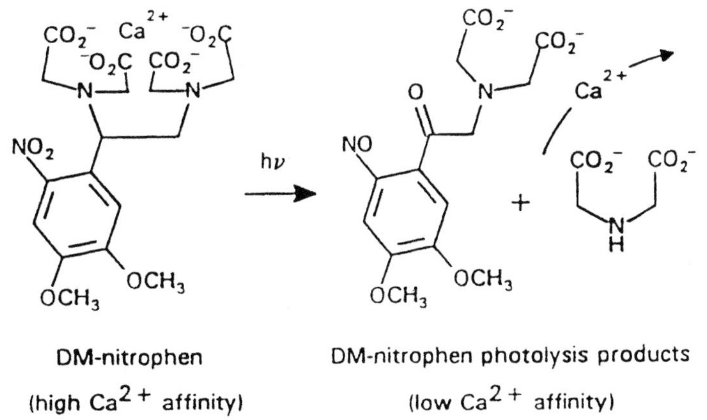

Fig. 3 Structure of and reaction scheme for DM-nitrophen, which releases calcium on exposure to light.

absorbance and quantum efficiency of Ca^{2+}-saturated DM-nitrophen are 4300 $M^{-1}cm^{-1}$ and 0.18. The Ca^{2+}-free and photolyzed forms have nearly the same absorbance, but the quantum efficiency of the unbound chelator is 40% that of the bound form (Zucker, 1993a). Photolysis of Ca^{2+}-loaded DM-nitrophen produces a different spectrum than photolysis of the Ca^{2+}-free form followed by addition of Ca^{2+}. Therefore, multiple photochemical pathways are involved; the photochemistry of DM-nitrophen is still not well understood.

A serious drawback of DM-nitrophen is that it shares the cation-binding properties of its parent molecule EDTA. In particular, H^+ and Mg^{2+} compete for Ca^{2+} at the hexacoordinate binding site. The affinity of DM-nitrophen for Mg^{2+} at pH 7.2 is 0.8 μM, whereas the photoproducts bind Mg^{2+} with affinities of about 3 mM. Further, both the Ca^{2+} and Mg^{2+} affinities of DM-nitrophen are highly pH dependent (Grell et $al.$, 1989), changing by a factor of 2 for 0.3 pH units. Thus, in the presence of typical $[Mg^{2+}]_i$ levels of 1–3 mM, DM-nitrophen that is not already bound to Ca^{2+} will be largely in the Mg^{2+} form. Further, excess DM-nitrophen will pull Mg^{2+} off of ATP, which binds it substantially more weakly, compromising the ability of ATP to serve as an energy source or as a substrate for ATPases. Finally, photolysis of DM-nitrophen will lead to a jump in $[Mg^{2+}]_i$ as well as $[Ca^{2+}]_i$, and to a rise in pH. Unless controlled by native or exogenous pH buffers, this pH change can alter the Ca^{2+} and Mg^{2+} affinities of the remaining DM-nitrophen. In the absence of Ca^{2+}, DM-nitrophen even may be used as a caged Mg^{2+} chelator. Attributing physiological responses to a $[Ca^{2+}]_i$ jump requires control experiments in which DM-nitrophen is not loaded with Ca^{2+}. DM-nitrophen currently is sold by CalBiochem.

B. Calculating Changes in $[Ca^{2+}]_i$

Quantifying changes in $[Ca^{2+}]_i$ caused by photolysis is much more difficult for DM-nitrophen than for the nitr compounds. The initial level of $[Ca^{2+}]_i$ before photolysis is dependent on the total concentrations of Mg^{2+}, Ca^{2+}, DM-nitrophen, ATP, and native Ca^{2+} buffers, because two buffers (DM-nitrophen and native buffer) compete for Ca^{2+}, two (ATP and DM-nitrophen) compete for Mg^{2+}, and, after partial photolysis, both cations also bind to the photoproducts. Calculating equilibrium Ca^{2+} levels involves simultaneous solution of six nonlinear buffer equations (Delaney and Zucker, 1990), which is a tedious chore at best. Also, the various dissociation constants depend on ionic strength, and have been measured only at 150 mM. The high affinity of DM-nitrophen for Ca^{2+} might appear to dominate the buffering of Ca^{2+} in cytoplasm, but this idea is misleading. A solution of DM-nitrophen that is 50% saturated with Ca^{2+} will hold the free $[Ca^{2+}]_i$ at 5 nM at pH 7.2; this action will be independent of the total DM-nitrophen concentration. However, 5 mM DM-nitrophen with 2.5 mM Ca^{2+} and 5 mM Mg^{2+} will buffer free $[Ca^{2+}]_i$ to 2.5 μM; now doubling all concentrations results in a final $[Ca^{2+}]_i$ of 5 μM.

Since the total $[Mg^{2+}]_i$ available, free or weakly bound to ATP, is several millimolar, clearly partially Ca^{2+}-loaded DM-nitrophen may bring the resting Ca^{2+} level to a surprisingly high level. Because the solution is still buffered, this Ca^{2+} may be reduced only gradually by pumps and uptake, but eventually Ca^{2+} will be pumped off the DM-nitrophen until the $[Ca^{2+}]_i$ is restored to its normal level. Then photolysis may lead to rather small jumps in $[Ca^{2+}]_i$. On the other hand, if a large amount of DM-nitrophen is introduced into a cell relative to the total $[Mg^{2+}]_i$, the compound may be loaded heavily with Ca^{2+} and still buffer free $[Ca^{2+}]_i$ to low levels, while releasing a large amount on photolysis. In fact, if DM-nitrophen is introduced into cells with no added Ca^{2+}, it may absorb Ca^{2+} from cytoplasm and intracellular stores gradually, so photolysis still produces a jump in $[Ca^{2+}]_i$. Therefore, both the resting and the postphotolysis levels of Ca^{2+} may vary over very wide ranges, depending on [DM-nitrophen], $[Mg^{2+}]_i$, and cellular $[Ca^{2+}]_i$ control processes, all of which are difficult to estimate or control. Thus, quantification of changes in $[Ca^{2+}]_i$ is not easy to achieve.

The situation may be simplified by perfusing cells with Ca^{2+}–DM-nitrophen solutions while dialyzing out all Mg^{2+} and native buffer (Neher and Zucker, 1993; Thomas et al., 1993). Of course, this procedure will not work in studies of cell processes requiring Mg^{2+}-ATP or if perfusion through whole-cell patch pipettes is not possible.

Another consequence of Mg^{2+} binding by DM-nitrophen is that cytoplasmic Mg^{2+} may displace Ca^{2+} from DM-nitrophen early in the injection or perfusion procedure, leading to a transient rise in $[Ca^{2+}]_i$ before sufficient DM-nitrophen is introduced into the cell. Such a "loading transient" has been calculated from models of changes of the concentrations of total $[Ca^{2+}]_i$, $[Mg^{2+}]_i$, ATP, native buffer, and DM-nitrophen during filling from a whole-cell patch electrode (R. S. Zucker, unpublished), and has been confirmed by measurements with Ca^{2+} indicators introduced into the cell with the DM-nitrophen (Neher and Zucker, 1993). Since this process may have important physiological consequences, controlling it is important. The process may be eliminated largely by separating the Ca^{2+}–DM-nitrophen filling solution in the pipette from the cytoplasm by an intermediate column of neutral solution [such as dilute ethylene glycol bis(β-aminoethylether)-N,N,N',N'-tetraacetic acid (EGTA) or BAPTA] in the tip of the pipette, which allows most of the Mg^{2+} to escape from the cell before the DM-nitrophen begins to enter. Then most of the loading transient occurs within the tip of the pipette.

One method of better controlling the change in $[Ca^{2+}]_i$ in DM-nitrophen experiments is filling cells with a mixture of Ca^{2+}–DM-nitrophen and another weak Ca^{2+} buffer such as N-hydroxyethylethylenediaminetriacetic acid (HEEDTA) or 1,3-diaminopropan-2-ol-tetraacetic acid (DPTA). These tetracarboxylate Ca^{2+} chelators have Ca^{2+} affinities in the micromolar or tens of micromolar range. If cells are filled with such a mixture without Mg^{2+}, the initial Ca^{2+} level can be set by saturating the DM-nitrophen and adding appropriate Ca^{2+} to the other buffer. Then photolysis of DM-nitrophen releases its

Ca^{2+} onto the other buffer; the final Ca^{2+} can be calculated from the final buffer mixture in the same fashion as for the nitr compounds. Since all the constituent affinities are highly pH dependent, a large amount of pH buffer (e.g., 100 mM) should be included in the perfusion solution, and the pH of the final solution adjusted carefully.

The kinetic behavior of DM-nitrophen is even more complex than its equilibrium reactions. Photolysis of DM-nitrophen is as fast as or faster than that of the nitr compounds, but the on-rate of Ca^{2+} binding is much slower, about 1.5 mM^{-1}msec^{-1} (Zucker, 1993a). This characteristic has particularly interesting consequences for partial photolysis of partially Ca^{2+}-loaded DM-nitrophen. A flash of light will release some Ca^{2+}, which initially will be totally free. If the remaining unphotolyzed and unbound DM-nitrophen concentration exceeds that of the released Ca^{2+}, this Ca^{2+} will rebind to the DM-nitrophen by displacing H^+ within milliseconds, producing a brief intense $[Ca^{2+}]_i$ "spike" (Grell *et al.*, 1989; Kaplan, 1990; McCray *et al.*, 1992). If Mg^{2+} is also present, a secondary relaxation of $[Ca^{2+}]_i$ follows because of the slower displacement of Mg^{2+} from DM-nitrophen (Delaney and Zucker, 1990). Moreover, if a steady UV source is used to photolyze the DM-nitrophen, Ca^{2+} rebinding continually lags release, leading to a low (micromolar range) free $[Ca^{2+}]_i$ while the illumination persists. When the light is extinguished, the $[Ca^{2+}]_i$ drops rapidly to a low level under control of the remaining DM-nitrophen buffer. If the remaining DM-nitrophen is bound to Mg^{2+}, achievement of equilibrium is somewhat slower (tens of milliseconds). Thus a reversible "pulse" of $[Ca^{2+}]_i$ is generated, the amplitude of which depends on light intensity and the duration of which is controlled by the length of the illumination. This situation remains so until the remaining DM-nitrophen becomes fully saturated with Ca^{2+}, whereupon $[Ca^{2+}]_i$ escapes from the control of the chelator, imposing a practical limit on the product of $[Ca^{2+}]_i$ and duration of about 0.75 $\mu M \cdot$ sec. Model calculations of the "spike" and "pulse" behavior of DM-nitrophen have been confirmed with experimental measurements (Zucker, 1993a; R. Zucker, unpublished results). Similar kinetic considerations apply when Ca^{2+} is passed by photolysis from DM-nitrophen to another slow buffer such as HEEDTA or DPTA. If this behavior is considered undesirable, it may be avoided by using only fully Ca^{2+}-saturated DM-nitrophen, for which rebinding to unphotolyzed chelator is impossible. Thus, the kinetic complexity of DM-nitrophen can be turned to experimental advantage, increasing the flexibility of control of $[Ca^{2+}]_i$.

IV. Diazo Compounds

A. Chemical Properties

In some experiments, being able to lower the $[Ca^{2+}]_i$ rapidly, rather than raise it, is desirable. For this purpose, the caged calcium chelators were developed. Initial attempts involved attachment of a variety of photosensitive protecting groups to mask one of the carboxyl groups of BAPTA, thus reducing its Ca^{2+} affinity until restored by photolysis. Such compounds displayed low quantum efficiency (Adams *et al.*, 1989; Ferenczi *et al.*, 1989) and their development has not been pursued. A more successful approach (Adams *et al.*, 1989) involved substituting one (diazo-2) or both (diazo-4) of the aromatic rings of BAPTA with an electron-withdrawing diazoketone that reduces Ca^{2+} affinity, much like the photoproducts of the nitr compounds. Figure 4 shows the structures of the diazo series of chelators. Photolysis converts the substituent to an electron-donating carboxymethyl group while releasing a proton; the Ca^{2+} affinity of the photoproduct is, thereby, increased. The reaction is illustrated in Fig. 5.

Diazo-2 absorbs one photon with quantum efficiency 0.03 to increase affinity, in 433 μsec, from 2.2 μM to 73 nM at 120 mM ionic strength (150 nM at 250 mM ionic strength). The absorbance maximum of the photosensitive group is 22,200 $M^{-1}cm^{-1}$ at 370 nm, and drops to negligible levels at this wavelength after photolysis. A small remaining absorbance reflects formation of a side product of unenhanced affinity and unchanged molar extinction coefficient in 10% of the instances of effective photon absorption. This "inactivated" diazo still binds Ca^{2+} (with some reduction in absorbance), but is incapable of further photolysis. The Ca^{2+}-bound form of diazo-2 has about one-tenth the absorbance of the free form, dropping to negligible levels after photolysis, with quantum efficiency of 0.057 and a time constant of 134 μsec. Binding of Ca^{2+} to photolyzed diazo-2 is fast, with an on-rate of 8 × 10^8 $M^{-1}sec^{-1}$. Mg^{2+} binding is weak, dropping from 5.5 to 3.4 mM after photolysis, and pH interference is small with this class of compounds.

One limitation of diazo-2 is that the unphotolyzed chelator has sufficient Ca^{2+} affinity that its incorporation into cytoplasm is likely to reduce resting levels to some degree, and certainly will have some effect on [Ca^{2+}]$_i$ rises that occur physiologically. To obviate this problem, diazo-4 was developed with two photolyzable diazoketones. Absorption of one photon increases the Ca^{2+}

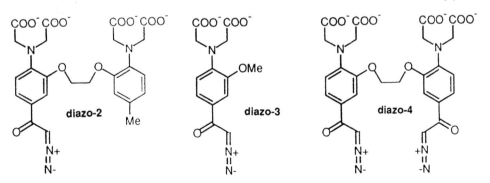

Fig. 4 Structures of the diazo series of photolabile chelators, which take up calcium on exposure to light.

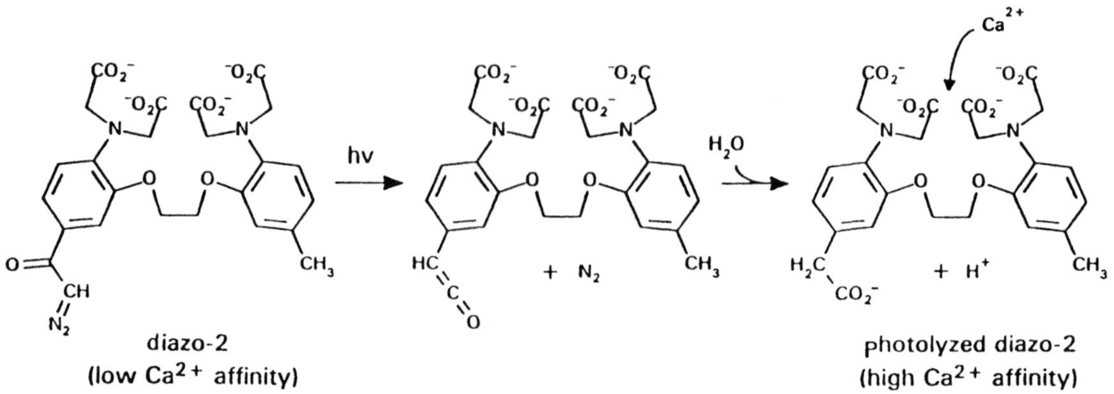

Fig. 5 Reaction scheme for the photolysis of diazo-2.

affinity from 89 μM to 2.2 μM (with a 10% probability of producing a side-product with one inactivated group). Absorption of two photons (with a probability assumed to equal the square of the probability of one group absorbing one photon, and with a measured quantum efficiency of 0.015) results in further increase of the affinity to 55 nM, a total increase of 1,600-fold. This large increase in affinity is, to some extent, offset by the small fraction of diazo-4 that can be doubly photolyzed readily. Thus, a flash of light produces a variety of species: unphotolyzed, singly photolyzed, doubly photolyzed, singly inactivated, doubly inactivated, and singly photolyzed–singly inactivated, with a variety of transition probabilities among species (Fryer and Zucker, 1993). Unphotolyzed diazo-4 is highly absorbant (46,000 $M^{-1}cm^{-1}$ at 371 nm for the free form; about 4,600 $M^{-1}cm^{-1}$ for the Ca^{2+}-bound form). The singly photolyzed species have absorbances of half these values; doubly photolyzed diazo-4 has negligible absorbance at this wavelength. Inactivation causes little change in absorbance.

A third member of this series, diazo-3, has a diazoketone attached to half the cation-coordinating structure of BAPTA, and has negligible Ca^{2+} affinity. On photolysis, diazo-3 produces the photochemical intermediates of diazo-2 plus a proton, and may be used to control for these effects of photolysis of the diazo series. At this time, only diazo-2 and diazo-3 are commercially available (Molecular Probes, Eugene, Oregon).

B. Calculating Effects of Photolysis

As for the nitr compounds, equilibration is faster than photolysis, so a flash of light leads to a smooth step transition in the concentration of Ca^{2+} chelator species. If the percentage of photolysis caused by a light flash is known, the proportions of photolyzed and inactivated diazo-2, or of the six species of diazo-4, can be calculated. Usually, diazo is injected without any added Ca^{2+}, so the effect of photoreleased buffers is to reduce the [Ca^{2+}]$_i$ from its resting value. This change can be calculated only if the total Ca^{2+} bound to the native buffer in cytoplasm as well as the characteristics of that buffer are known. These characteristics often can be inferred from available measurements on cytoplasmic Ca^{2+} buffer ratio and the normal resting [Ca^{2+}]$_i$ level. The more usual application of these substances is to reduce the effect of a physiologically imposed rise in [Ca^{2+}]$_i$. In many cases, the magnitude of the source of this Ca^{2+} is known, as in the case of a Ca^{2+} influx measured as a Ca^{2+} current under voltage clamp or the influx through single channels estimated from single channel conductances. Also, the magnitude of the total Ca^{2+} increase in a response can be estimated from measured increases in [Ca^{2+}]$_i$ and estimates of cytoplasmic buffering. With this information, the expected effect of newly formed diazo photoproducts on a physiological rise in [Ca^{2+}]$_i$ can be calculated by solving diffusion equations that are appropriate for the distribution of Ca^{2+} sources before and after changing the composition of the mixture of buffers in the cytoplasm. Examples of such solutions of the diffusion equation exist for spherical diffusion inward from the cell surface (Sala and Hernandez-Cruz, 1990; Nowycky and Pinter, 1993), cylindrical diffusion inward from membranes of nerve processes (Zucker and Stockbridge, 1983; Stockbridge and Moore, 1984), diffusion from a point source (Stern, 1992; Fryer and Zucker, 1993), and diffusion from arrays of point sources (Fogelson and Zucker, 1985; Simon and Llinás, 1985). For large cells, the spatial nonuniformity of light intensity and photolysis rate also must be considered, taking into account the absorbances of all the species of diazo and the changes in their concentration with photolysis. Unlike the nitr chelators, the self-screening imposed by diazo chelators is reduced with photolysis, so successive flashes (or prolonged illumination) are progressively more effective.

V. Introduction into Cells

The photolabile chelators are introduced into cells by pressure injection from micropipettes, perfusion from whole-cell patch pipettes, or permeabilization of the cell membrane. Iontophoresis is also suitable for diazo compounds, since this procedure inserts only the Ca^{2+}-free form. For the caged Ca^{2+} substances, this method of introduction requires that the chelator load itself with Ca^{2+} by absorbing it from cytoplasm or intracellular stores. Filling cells from a patch pipette has the special property that, if the photolysis light is confined to the cell and excludes all but the tip of the pipette, the pipette acts as an infinite reservoir of unphotolyzed chelator. Then the initial conditions of solutions in the pipette can be restored within minutes after photolysis of the chelator in the cell. The nitr and diazo compounds are soluble at concentrations over 100 mM and DM-nitrophen is soluble at 75 mM, so levels in cytoplasm exceeding 10 mM can be achieved relatively easily, even by microinjection, making the exogenous chelator compound the dominant Ca^{2+} buffer.

Nitr and diazo chelators also have been produced as membrane-permeant acetoxymethyl (AM) esters (Kao *et al.*, 1989). Exposure of intact cells to medium containing these esters (available from CalBiochem and Molecular Probes, respectively) might result in the loading of cells with up to millimolar concentrations, if sufficient activity of intracellular esterase is present to liberate the membrane-impermeant chelator. However, nitr-5 or nitr-7 introduced in this manner is not bound to Ca^{2+}, so it must sequester Ca^{2+} from cytoplasm, from intracellular stores, or after Ca^{2+} influx is enhanced, for example, by depolarizing excitable cells. The final concentration, level of Ca^{2+} loading, and localization of the chelator are uncertain, so this method of incorporation does not lend itself to quantification of effects of photolysis.

During loading and other preparatory procedures, the photolabile chelators may be protected from photolysis with low pass UV-blocking filters in the light path of the tungsten or quartz halide beams used for viewing. For more detail on these filling procedures, the reader is referred to Gurney (1991). Other methods of loading cells, used primarily with other sorts of caged compounds, are discussed by Adams and Tsien (1993).

VI. Light Sources

Photolysis of caged Ca^{2+} chelators requires a bright source of near UV light. If time resolution is unimportant, an ordinary mercury or xenon arc lamp may be used. Mercury lamps have a convenient emission line at 366 nm. Exposure can be controlled with a shutter, using MgF-coated Teflon blades for particularly bright sources. Lamps of 100–150 W power with collimating quartz lenses provide sufficient energy to photolyze ~25% of caged Ca^{2+} compounds in ~2 sec. Bulbs of larger power only generate bigger arcs, with more energy in a larger spot of similar intensity. With additional focusing, photolysis can be achieved in one-tenth the time or even less. These light sources are the appropriate choice in applications using reversible $[Ca^{2+}]_i$ elevation with DM-nitrophen.

Fast events require the use of a laser or xenon arc flashlamp. The xenon lamps are less expensive and cumbersome; convenient commercial systems are available from Chadwick Helmuth (El Monte, California) and Rapp Opto-elektronik (Hamburg, Germany). Both flashlamps discharge up to 200 J electrical energy across the bulb to provide a pulse of ~1-msec duration with up to 300 mJ energy in the 330- to 380-nm band. The Chadwick-Helmuth unit includes only a power supply and lamp socket, so a housing with focusing optics must be constructed (see Rapp and Guth, 1988). Focusing can be accomplished with a UV-optimized elliptical reflector or with quartz refractive optics. The reflector can be designed to capture more light (i.e., have a larger effective numerical aperture), but reflectors have greater physical distortion than well-made lenses. In practice, the reflector generates a larger spot with more total energy, but somewhat less intensity, than refractive methods. One advantage of reflectors is that they are not subject to chromatic aberration—focusing is independent of wavelength—so the UV will be focused in the same spot as visual light. This is not true of refractive lenses. To focus and aim them accurately at the sample, a UV filter must be used to block visual light and the beam must be focused on a fluorescent surface. Both types of housing are available from Rapp Optoelektronik. Using either system, photolysis rates approaching 80–90% in one flash are achievable. This rate may be reduced by imposing neutral density filters or reducing discharge energy, but the relationship between electrical and light energy is not linear and should be measured with a photometer. Flashlamps can be reactivated only after their storage capacitors have recharged, setting the minimal interval between successive flashes at 10 sec or more.

Flashlamps are prone to generating a number of artifacts. The discharge causes electrical artifacts that can burn out semiconductors and op amps, and reset or clear digital memory in other nearby equipment. Careful electrostatic shielding, wrapping inductors with paramagnetic metal, power source isolation, and using isolation circuits in trigger pulse connections to other equipment prevent most problems. The discharge generates a mechanical thump at the coil used to shape the current pulse through the bulb; this thump can dislodge electrodes from cells or otherwise damage the sample. Mechanical isolation of the offending coil solves the problem. The light pulse also generates a movement artifact at electrodes, which can be seen to oscillate violently for a fraction of a second when videotaped during a flash. This movement can damage cells severely, especially those impaled with multiple electrodes. Small cells sealed to the end of a patch pipette often fare better against such mistreatment. To reduce this source of injury, the light can be filtered to eliminate all but the near UV. Commercial Schott filters (UG-1, UG-11), coated to reflect infrared (IR) light, serve well for this purpose, but can cut the 330- to 380-nm energy to 30% or less. Liquid filters to remove IR and far UV also have been described (Tsien and Zucker, 1986). Removing IR reduces temperature changes, which otherwise can exceed 1°C, whereas removing far UV prevents the damaging effects of this ionizing radiation. Chlorided silver pellets and wires often used in electrophysiological recording constitute a final source of artifact. These components must be shielded from the light source or they will generate large photochemical signals.

To simply aim and focus the light beam directly onto the preparation is easiest. If isolating the lamp from the preparation is necessary, the light beam may be transmitted by a fiber optic or liquid light guide, with some loss of intensity. If a microscope is being used already, the photolysis beam may be directed through the epifluorescence port of the microscope. The lamp itself, or a light guide, may be mounted onto this port. Microscope objectives having high numerical aperture and good UV transmission will focus the light quite effectively onto a small area, which can be delimited further by a field stop aperture. With the right choice of objectives and direct coupling of the lamp to the microscope port, light intensities similar to those obtained by simply aiming the focused steady lamp or flashlamp can be achieved. Half reflective mirrors can be used to combine the photolysis beam with other light sources, such as those used for $[Ca^{2+}]_i$ measurement. However, as the optical arrangement becomes more complex, photolysis intensity inevitably decreases.

Lasers provide an alternative source of light, with the advantages of a coherent collimated beam that is focused easily to a very small spot. Pulsed lasers such the frequency-doubled ruby laser or the XeF excimer laser provide at least 200 mJ energy at 347 or 351 nm in 50 and 10 nsec, with possible repetition rates of 1 and 80 Hz, respectively. Liquid coumarin-dye lasers, with up to 100 mJ tunable energy in the UV and 1-μsec pulse duration, are also available. New and inexpensive nitrogen lasers providing lower pulse energies (0.25 mJ) in 3-nsec pulses at 337 nm also have been developed (Laser Science, Cambridge, Massachusetts) and, with appropriate focusing, might be useful. To date, lasers have found their widest application in studies of muscle contraction. More information on these laser options is contained in discussions by Goldman *et al.* (1984) and McCray and Trentham (1989).

An adaptation of laser photolysis is the two-photon absorption technique (Denk *et al.*, 1990). A colliding-pulse mode-locked laser generating 100-fsec pulses of 630-nm light at 80 MHz is focused through a confocal scanning microscope. Photolysis of UV-sensitive caged compounds requires simultaneous absorption of two red photons, so photolysis occurs only in the focal plane of the scanning beam. This behavior restricts photolysis in three dimensions, but the photolysis rate is so slow that several minutes of exposure are required with currently available equipment. This technique is expensive and specialized, and is still under development, but may have practical application after undergoing further refinement.

Near UV light alone seems to have little effect on most biological tissues, with the obvious exception of photoreceptors and the less obvious case of smooth muscle (Gurney, 1993). Control experiments on the effects of light on unloaded cells, and on the normal physiological response under study, can be used to ascertain the absence of photic effects.

VII. Calibration

When designing a new optical system or trying a new caged compound, being able to estimate the rate of photolysis of the apparatus used is important. This information is necessary to adjust the light intensity or duration for the desired degree of photolysis, and to insure that photolysis is occurring at all.

In principle, the fraction (F) of a substance photolyzed by a light exposure of energy J can be computed from the formula $e^{-(J - J')} = (1 - F)/0.1$, where J' is the energy needed to photolyze 90% of the substance and is given by $J' = hcA/Q\varepsilon\lambda$, where h is Planck's constant, c is the speed of light, A is Avogardro's number, Q is the quantum efficiency, ε is the decadic molar excinction coefficient, and λ is the wavelength of the light. In practice, however, this equation is rarely useful, for the following reasons.

1. Measuring the energy of the incident light on a cell accurately is difficult, especially for light of broad bandwidth with varying intensity at different wavelengths.

2. The quantum efficiency, although provided for all the photolabile Ca^{2+} chelators, is not such a well-defined quantity. The value depends critically on how it is measured, which is not always reported. In particular, the effective quantum efficiency for a pulse of light of moderate duration (e.g., from a flashlamp) is often greater than that of either weak steady illumination or a very brief pulse (e.g., from a laser), because of the possibility of multiple photon absorptions of higher efficiency by photochemical intermediates. This phenomenon has been noted to play a particularly strong role in nitr-5 photolysis (McCray and Trentham, 1989). Thus, apparent differences in quantum efficiencies between different classes of chelators may be mainly the results of different measurement procedures.

3. Finally, the quantum efficiency is a function of wavelength, which is rarely given.

A more practical and commonly adopted approach is mixing a partially Ca^{2+}-loaded photolabile chelator with a Ca^{2+} indicator in a solution with appropriate ionic strength and pH buffering, and measuring the [Ca^{2+}] change in a small volume of this solution, the net absorbance of which is sufficiently small to minimize inner filtering of the photolyzing radiation. Suitable indicators include fura-2, indo-1 (Grynkiewicz et al., 1985), furaptra (Konishi et al., 1991), fluo-3, rhod-2 (Minta et al., 1989), Calcium Green™, Orange™, and Crimson™ (Eberhard and Erne, 1991), arsenazo III (Scarpa et al., 1978), and fura-red (Kurebayashi et al., 1993). The choice depends largely on available equipment. Fura-2, indo-1, and furaptra are dual-excitation or -emission wavelength fluorescent dyes, allowing more accurate ratiometric measurement of [Ca^{2+}], but they require excitation at wavelengths that photolyze the photolabile Ca^{2+} chelators and are subject to bleaching by the photolysis light. The former problem may be minimized by using low intensity measuring light with a high sensitivity detection system. Furaptra is especially useful for DM-nitrophen, because of its lower Ca^{2+} affinity. Fluo-3 and rhod-2 were designed specifically for use with photolabile chelators (Kao et al., 1989), being excited at wavelengths different from those used to photolyze the chelators, but they are not ratiometric dyes and are difficult to calibrate accurately. Calcium Green, Orange, and Crimson suffer the same limitation. Arsenazo and antipyralazo are metallochromic dyes that change absorbance on binding Ca^{2+}, fortunately at wavelengths different from those at which the photolabile chelators show any significant absorbance. However, these dyes are also difficult to calibrate for absolute levels of $[Ca^{2+}]$, although changes in $[Ca^{2+}]$ may be determined fairly accurately. Fura-red is a new ratiometric dye that is excited by visible light, so it might have some application in calibrating photolysis. A problem common to all the fluorescent indicators is that their fluorescent properties may be altered by the presence of photolabile chelators, which generally are used at millimolar levels whereas the indicators are present at 100 μM or less. The photolabile chelators often produce contaminating fluorescence, which also may be Ca^{2+}-dependent and may partially quench the fluorescence of the indicators (Zucker, 1992a). Thus, the indicators must be calibrated in the presence of photolabile chelator at three well-controlled $[Ca^{2+}]$ levels, preferably before and after exposure to the photolysis flash, before they can be used to measure the effects of photolysis on $[Ca^{2+}]$ (Neher and Zucker, 1993). The low and high $[Ca^{2+}]$ calibrating solutions may be made with excess Ca^{2+} or another buffer such as EGTA or BAPTA, but the intermediate $[Ca^{2+}]$ solution is more difficult to generate, since photolysis of the chelator will release some Ca^{2+} and change the $[Ca^{2+}]_i$ and pH in this solution unless it contains a very high concentration of controlling chelator and pH buffer.

The calibration procedure is generally the same for any combination of chelator and indicator. A small sample of the mixture is placed in a 1-mm length of microcuvette with a 20- to 100-μm pathlength (In Vitro Dynamics, Rockaway, New Jersey) under mineral oil to prevent evaporation. This cuvette is exposed repeatedly to the photolysis light or to flashes, which should illuminate the whole cuvette uniformly, and the $[Ca^{2+}]$ after each flash or exposure is measured using a microscope-based fluorescence or absorbance photometer. A small droplet of solution under mineral oil alone would work, and may be necessary if the photolysis beam is directed through the microscope and illuminates a very small area, but sometimes the fluorescent properties of the indicators are affected by the mineral oil. This effect would be detected in the procedure for calibrating the chelator–indicator mixture, but is best avoided using the microcuvettes, in which contact with oil is only at the edges, the fluorescence or absorbance change of which need not be measured. In some applications, such as whole-cell patch clamping of cultured cells, using the cell as a calibration chamber can be easier than any other procedure.

The expected changes in $[Ca^{2+}]$ depend on the chelator used. The nitr and diazo chelators should lead to a stepwise rise or fall in $[Ca^{2+}]$ after each exposure; the results can be fit to models of the chelators and their photoproducts, using their affinities and the relative quantum efficiencies of free and bound chelators (Landò and Zucker, 1989; Fryer and Zucker, 1993). The percentage photolysis of the chelator in response to each light exposure is the only free parameter, and is varied until the model fits the results. In the case of the high affinity DM-nitrophen, little rise in $[Ca^{2+}]$ will occur until the total amount of remaining unphotolyzed chelator equals the total amount of Ca^{2+} in the solution, whereupon the $[Ca^{2+}]$ will increase suddenly. Equations relating initial and final concentrations of DM-nitrophen, total $[Ca^{2+}]$, and photolysis rate (Zucker, 1993a) then may be used to calculate photolysis rate per flash or per second of steady light exposure.

The photosensitive compounds also undergo substantial absorbance changes after photolysis. These changes can be monitored during repeated exposure to the light source without a Ca^{2+} indicator; the number of flashes or the duration of light exposure required to reach a given percentage photolysis then can be determined. Realizing that photolysis proceeds exponentially to completion (Zucker, 1993a), these data can be used to determine the photolysis rate directly. Ideally, both methods should be used to check for consistent results. A final method for determining photolysis rate is using high pressure liquid chromatography (HPLC) to separate and quantify parent chelators and photoproducts in the reaction solution after partial photolysis (Walker, 1991).

VIII. Purity and Toxicity

When experiments do not work as planned, the first suspected source of error is the integrity of the photolabile chelator. Different procedures have proved most useful for testing the different classes of compounds. The nitr and diazo compounds undergo large absorbance changes on binding calcium and on photolysis. A 100 μM solution (nominally) of the chelator is mixed with 50 μM Ca^{2+} in 100 mM chelexed HEPES solution (pH 7.2), and 0.3 ml is scanned in a 1-mm pathlength spectrometer. Then 1 μl 1 M K_2EGTA is added to bring the $[Ca^{2+}]$ to 0, and the sample is scanned again. Finally, 1 μl 5 M $CaCl_2$ is added to provide excess Ca^{2+}, and a third scan is recorded. The first scan should be midway between the other two. If the first scan is closer to the excess Ca^{2+} scan, it is indicative of a lower than expected concentration of the chelator, probably because of an impurity. Alternatively, Ca^{2+} may have been present with the chelator, which may be checked by running a scan on the chelator with no added Ca^{2+} and comparing the result with a scan with added EGTA; they should be identical. Ca^{2+}-free and Ca^{2+}-saturated chelator solutions also are scanned before and after exposure to UV light sufficient to cause complete photolysis; the spectra are compared with published figures (Adams et al., 1988,1989; Kaplan and Ellis-Davies, 1988) to determine whether the sample was partially photolyzed at the outset. The Ca^{2+} affinities of unphotolyzed and photolyzed chelators can be checked by measuring $[Ca^{2+}]$ of 50%-loaded chelators with a Ca^{2+}-selective electrode.

The absorbance of DM-nitrophen is almost Ca^{2+} independent, so these procedures are not effective for this chelator. A solution of DM-nitrophen nominally of 2 mM concentration is titrated with concentrated CaCl$_2$ until the [Ca^{2+}] measured with ion-selective electrodes suddenly increases; this change indicates the actual concentration of the chelator and gives an estimate of purity. The affinity of the photolysis products can be measured as for the other chelators; spectra before and after photolysis indicate whether the sample was already partially photolyzed.

Purities of 80–90% are typical for commercial samples of all the chelators, but occasional batches of 60% purity or less have been seen; these also sometimes show high degrees of toxicity. Whether such low purity is the result of poor synthesis or storage is unclear. Nitr compounds decompose detectably after only 1 day at room temperature, and exposure to ambient fluorescent lighting for 1 day causes detectable photolysis. Chelators should be shipped on dry ice and stored at −80°C in the dark; even under these conditions they might not last forever. Repeated thawing and freezing also may degrade the compounds.

Some of the photolabile Ca^{2+} chelators display a degree of biological toxicity in some preparations. Commercial samples of nitr-5 have been seen to lyse sea urchin eggs (R. S. Zucker and L. F. Jaffe, unpublished results) and leech blastomeres (K. R. Delaney and B. Nelson, unpublished results) within minutes. Zucker and Haydon (1988) found that nitr-5 blocked transmitter release within 10 min of perfusion in snail neurons, whereas DM-nitrophen has no similar effect (P. Haydon, unpublished results). These effects are not caused by the photoproducts, since photolysis is not necessary for the problems to occur. DM-nitrophen has been observed to reduce secretion in chromaffin cells; higher chelator concentrations, photolyzed to give the same final [Ca^{2+}]$_i$ level, caused less secretion (C. Heinemann and E. Neher, unpublished results). The effect was overcome partially by inclusion of glutathione in the perfusion solution, as reported for the photoproducts of other 2-nitrobenzhydrol-based caged compounds (Kaplan *et al.*, 1978). These signs of toxicity have been observed sporadically; whether they are properties of the chelators themselves or of impurities in the samples used is unclear. The chelators have been applied successfully to a wide range of preparations without obvious deleterious results, although subtle effects may have been missed.

IX. Biological Applications

A brief synopsis of the biological applications of the caged Ca^{2+} chelators follows, and is included in this chapter because many of the original papers include a wealth of detail about methodology and interpretation of these new techniques.

A. Ion Channel Modulation

1. Potassium and Nonspecific Cation Channels

The first and still one of the major applications of photosensitive Ca^{2+} chelators is studying the properties of Ca^{2+}-dependent ion channels in excitable cells. In 1987, Gurney and colleagues used nitr-2, -5, and -7 to activate Ca^{2+}-dependent K^+ current in rat sympathetic neurons. These researchers found that a single Ca^{2+} ion binds to the channel with rapid kinetics and 350 nM affinity.

The next application of the nitr chelators was in an analysis of Ca^{2+}-activated currents in *Aplysia* neurons (Landò and Zucker, 1989). These investigators found that Ca^{2+}-activated K^+ and nonspecific cation currents in bursting neurons were linearly dependent on $[Ca^{2+}]_i$ jumps in the micromolar range, as measured by arsenazo spectrophotometry and modeling studies. Both currents relaxed at similar rates after photolysis of nitr-5 or nitr-7, reflecting diffusional equilibration of $[Ca^{2+}]_i$ near the front membrane surface facing the light source. Potassium current relaxed more quickly than nonspecific cation curent, after activation by Ca^{2+} entry during a depolarizing pulse, because of the additional voltage sensitivity of the K^+ channels. This difference was responsible for the more rapid decay of hyperpolarizing after-potentials than of depolarizing after-potentials.

The role of Ca^{2+}-activated K^+ current in shaping plateau potentials in gastric smooth muscle was explored by Carl *et al.* (1990). In fibers loaded with nitr-5/AM, Ca^{2+} photorelease accelerated repolarization during plateau potentials and delayed the time to subsequent plateau potentials, suggesting a role for changes in $[Ca^{2+}]_i$ and Ca^{2+}-activated K^+ current in slow wave generation.

Another current modulated by $[Ca^{2+}]_i$ is the so-called M current, a muscarine-blocked K^+ current in frog sympathetic neurons. Although inhibition is mediated by an as yet unidentified second messenger other than Ca^{2+}, resting M current is enhanced by modest elevation of $[Ca^{2+}]_i$ (some tens of nanomolar) and reduced by greater elevation of $[Ca^{2+}]_i$, which also suppresses the response to muscarine (Marrion *et al.*, 1991). As for ventricular I_{Ca} (see subsequent discussion), apparently several sites of modulation of M current by $[Ca^{2+}]_i$ exist. In these experiments, $[Ca^{2+}]_i$ was elevated by photorelease from nitr-5 and simultaneously measured with fura-2.

The after-hyperpolarization that follows spikes in rat hippocampal pyramidal neurons is caused by a class of Ca^{2+}-dependent K^+ channels called I_{AHP} channels. This after-hyperpolarization and the current underlying it rise slowly to a peak 0.5 sec after the end of a brief burst of spikes. Photorelease of Ca^{2+} from either nitr-5 or DM-nitrophen activates this current without delay; the current may be terminated rapidly by photolysis of diazo-4 (Lancaster and Zucker, 1991; Zucker, 1992b), suggesting that the delay in its activation following action potentials is caused by a diffusion delay between points of Ca^{2+} entry and the I_{AHP} channels.

The Ca^{2+} sensitivity of the mechanoelectrical transduction current in chick cochlear hair cells was studied using nitr-5 introduced by hydrolysis of the AM form (Kimitsuki and Ohmori, 1992). Elevation of $[Ca^{2+}]_i$ to 0.5 μM (measured with fluo-3) diminished responses to displacement of the hair bundle, and accelerated adaptation during displacement when Ca^{2+} entry occurred. Preventing Ca^{2+} influx blocked adaptation. Evidently, adaptation of this current was the result of an action of Ca^{2+} ions entering through the transduction channels.

In guinea pig hepatocytes, noradrenaline evokes a rise in K^+ conductance after a seconds-long delay. Photorelease of Ca^{2+} from nitr-5 and use of caged inositol 1,4,5-trisphosphate (IP_3) show that this delay arises from steps prior to or during generation of IP_3 (Ogden *et al.*, 1990), which releases Ca^{2+} from intracellular stores to activate K^+ current.

2. Ca^{2+} Channels

The first application of DM-nitrophen was in a study of Ca^{2+} channels in chick dorsal root ganglion neurons (Morad *et al.*, 1988). With divalent charge carriers, inactivation by photorelease of intracellular Ca^{2+} occurred within 7 msec, whereas with monovalent charge carriers a nearly instantaneous block occurred, especially when Ca^{2+} was released extracellularly. A similar rapid block of monovalent current through Ca^{2+} channels was observed in response to photorelease of extracellular Ca^{2+} in frog ventricular cells (Näbauer *et al.*, 1989). Different Ca^{2+} binding sites may be exposed if altered conformational states are induced in the channels by the presence of different permeant ions.

The regulation of Ca^{2+} current (I_{Ca}) in frog atrial cells by $[Ca^{2+}]_i$ also has been studied with nitr-5 (Gurney *et al.*, 1989; Charnet *et al.*, 1991). Rapid elevation of $[Ca^{2+}]_i$ potentiated high-voltage-activated or L-type I_{Ca} and slowed its deactivation rate when Ba^{2+} was the charge carrier, after a delay of several seconds. Inclusion of BAPTA in the patch pipette solution blocked the effect of nitr-5 photolysis. The similarity of effect of Ca^{2+} and cAMP and their mutual occlusion suggest a common phosphorylation mechanism.

Regulation of I_{Ca} in guinea pig ventricular cells appears to be more complex (Hadley and Lederer, 1991; Bates and Gurney, 1993). A fast phase of inactivation seems to be the result of direct action of $[Ca^{2+}]_i$ on Ca^{2+} channels, since I_{Ca} inactivation caused by photorelease of Ca^{2+} from nitr-5 is independent of the phosphorylation state of the channels. Ca^{2+} inactivates the current through the channels without affecting the gating current, perhaps indicating an effect on permeation rather than on voltage-dependent gating. A slower phase of potentiation is also present, the magnitude of which depends on the flash intensity delivered during a depolarizing pulse, but not on the initial $[Ca^{2+}]_i$ level, the degree of loading of nitr-5, or the presence of BAPTA in the patch pipette. This result suggests that, during a depolarization, nitr-5 becomes locally loaded by Ca^{2+} entering through Ca^{2+} channels, and that the

Ca^{2+} binding site regulating potentiation is near the channel mouth. Larger $[Ca^{2+}]_i$ jumps elicited by photolysis of DM-nitrophen evoke greater I_{Ca} inactivation, but no potentiation, perhaps because of the more transient rise in $[Ca^{2+}]_i$ when DM-nitrophen is photolyzed. Differences in Ca^{2+} sensitivity of the effects are also likely. An effect of resting $[Ca^{2+}]_i$ on the kinetics of potentiation is explained in terms of a third Ca^{2+} binding site that inhibits potentiation. As in atrial cells, potentiation is suppressed by isoprenaline, which leads to cAMP-dependent channel phosphorylation. However, kinase inhibitors block the response to isoprenaline but not Ca^{2+}-dependent potentiation, suggesting the involvement of different pathways.

DM-nitrophen loaded with magnesium in the absence of Ca^{2+} was used to study the magnesium-nucleotide regulation of L-type I_{Ca} in guinea pig cardiac cells (Backx *et al.*, 1991; O'Rourke *et al.*, 1992). In the presence of ATP, a rise in $[Mg^{2+}]_i$ to 50–200 μM led to a near doubling of the magnitude of I_{Ca} in a few seconds. Omitting ATP prevented this effect, although the rise in $[Mg^{2+}]_i$ still blocked inwardly rectifying K^+ channels. Release of caged ATP also increased I_{Ca}. Therefore, the effect on Ca^{2+} channels was caused by a rise in Mg^{2+}–ATP. Nonhydrolyzable ATP analogs worked as well as ATP and kinase inhibitors failed to block the potentiation, so Mg^{2+}–ATP seems to modulate Ca^{2+} channels directly.

In another study, microinjection of nitr-5, DM-nitrophen, and diazo-4 was used to characterize Ca^{2+}-dependent inactivation of Ca^{2+} current in *Aplysia* central neurons (Fryer and Zucker, 1993). Elevation of $[Ca^{2+}]_i$ to the low micromolar range with nitr-5 caused little inactivation, but photolysis of DM-nitrophen rapidly inactivated half the I_{Ca}, presumably that in the half of the cell facing the light source. Thus, inactivation requires high $[Ca^{2+}]_i$ levels and occurs rapidly in all channels, even if they are closed. Experiments with diazo-4 showed that an increase in buffering power reduced the rate of inactivation of I_{Ca} only modestly. Simulations of the effects of the calculated change in Ca^{2+} buffering on diffusion of Ca^{2+} ions from the channel mouth indicated that Ca^{2+} appeared to act at a site with a mean free path 25 μm from the channel mouth to cause I_{Ca} inactivation (see also Johnson and Byerly, 1993).

B. Muscle Contraction

One of the earliest applications of photolabile Ca^{2+} chelators was initiating muscle contraction in frog cardiac ventricular cells by photorelease of extracellular Ca^{2+} from DM-nitrophen (Näbauer *et al.*, 1989). The strength of contraction elicited by a stepwise rise in $[Ca^{2+}]_e$ showed a membrane potential dependence that was indicative of entry through voltage-dependent Ca^{2+} channels rather than of transport by Na^+–Ca^{2+} exchange.

Several laboratories have used caged Ca^{2+} chelators to study Ca^{2+}-dependent Ca^{2+} release from the sarcoplasmic reticulum in rat ventricular myocytes. Valdeolmillos *et al.* (1989) loaded intact cells with the AM form of nitr-5, Kentish *et al.* (1990) subjected saponin-skinned fibers to solutions containing Ca^{2+}-loaded nitr-5, and Näbauer and Morad (1990) perfused single myocytes with DM-nitrophen loaded with Ca^{2+}. Photolysis of the chelator elicited a contraction that was blocked by ryanodine or caffeine pretreatment, procedures that prevent release of Ca^{2+} from the sarcoplasmic reticulum, so the investigators concluded that the contractions resulted from Ca^{2+}-induced Ca^{2+} release. When Ca^{2+} release was confined to a portion of a fiber (O'Neill *et al.*, 1990), contraction remained localized, indicating that Ca^{2+}-induced Ca^{2+} release does not invariably lead to propagation of and rise in $[Ca^{2+}]_i$ throughout myocytes.

Recently, Györke and Fill (1993) used Ca^{2+}-DM-nitrophen to show that the ryanodine receptor in cardiac muscle adapts to a maintained elevation of $[Ca^{2+}]_i$ in the micromolar range, remaining sensitive to larger $[Ca^{2+}]$ changes and responding by releasing still more Ca^{2+}. In smooth muscle from guinea-pig portal vein, the IP_3-dependent release of Ca^{2+} was also found to depend on $[Ca^{2+}]_i$ (Iino and Endo, 1992). Elevation of $[Ca^{2+}]_i$ by photolysis of Ca^{2+}-DM-nitrophen and detected with fluo-3 accelerated the release of Ca^{2+} from a ryanodine-insensitive, $InsP_3$-activated store.

Ca^{2+}-loaded nitr-5 was used in skinned muscle fibers of the scallop and the frog to show that, for both myosin- and actin-regulated muscles, the rate-limiting step in contraction is not the time course of the rise in $[Ca^{2+}]_i$ but the response time of the contractile machinery (Lea *et al.*, 1990; Ashley *et al.*, 1991b). Using isolated myofibrillar bundles from barnacle muscle, Lea and Ashley (1990) showed that photorelease of 0.2–1.0 μM Ca^{2+} from Ca^{2+}-loaded nitr-5 not only activated contraction directly and rapidly (within ~200 msec) but also evoked a slower phase of contraction (~2-sec rising half-time) that was dependent on Ca^{2+}-induced Ca^{2+} release from the sarcoplasmic reticulum.

The first biological application of the caged chelator diazo-2 was in the study of muscle relaxation. Mulligan and Ashley (1989) showed that rapid reduction in $[Ca^{2+}]_i$ in skinned frog semitendinosus muscle fibers resulted in a relaxation similar to that occurring normally in intact muscle, indicating that mechanochemical events subsequent to the fall in $[Ca^{2+}]_i$ were rate limiting. However, Lännergren and Arner (1992) reported some speeding of isometric relaxation after photolysis of diazo-2, loaded in the AM form into frog lumbrical fibers. Lowered pH slowed relaxation to a step reduction in $[Ca^{2+}]_i$ (Palmer *et al.*, 1991), perhaps accounting for a contribution of low pH to the sluggish relaxation of fatigued muscle. In contrast to frog muscle, photorelease of Ca^{2+} chelator caused a much faster relaxation in skinned scallop muscle than in intact fibers (Palmer *et al.*, 1990), suggesting that, in these cells, relaxation is rate limited primarily by $[Ca^{2+}]_i$ homeostatic processes.

C. Synaptic Function

Action potentials evoke transmitter release in neurons by admitting Ca^{2+} through Ca^{2+} channels. Because of the usual coupling between depolarization and Ca^{2+} entry, assessing the possibility of an additional direct action of membrane potential on the secretory apparatus has been difficult. Photolytic release of presynaptic Ca^{2+} by nitr-5 perfused into a presynaptic snail neuron in culture was combined with voltage clamp of the presynaptic membrane potential to distinguish the roles of [Ca^{2+}]$_i$ and potential in neurosecretion (Zucker and Haydon, 1988). These researchers found no direct effect of membrane potential on the rate of transmitter release triggered by a rise in [Ca^{2+}]$_i$.

Hochner *et al.* (1989) injected Ca^{2+}-loaded nitr-5 into the preterminal axon of a crayfish motor neuron, and used a low [Ca^{2+}] medium to block normal synaptic transmission. These investigators found that action potentials transiently accelerated the transmitter release that was evoked at a low level by photolysis of the nitr-5. However, Mulkey and Zucker (1991) used fura-2 to show that the extracellular solutions used by Hochner *et al.* (1989) failed to block Ca^{2+} influx through voltage-dependent Ca^{2+} channels. When external Ca^{2+} chelators or more effective channel blockers were used to eliminate Ca^{2+} influx completely, spikes failed to have any influence on transmitter release, even when it was activated strongly by photolysis of intracellularly injected Ca^{2+}-loaded DM-nitrophen.

Delaney and Zucker (1990) confirmed that action potentials at the squid giant synapse have no effect on transmitter release triggered by a rise in [Ca^{2+}]$_i$ caused by photolysis of presynaptically injected Ca^{2+}–DM-nitrophen. Using a flashlamp to photolyze the DM-nitrophen rapidly, a transient postsynaptic response occurred that resembled the response normally caused by an action potential. The early intense phase of transmitter release probably was caused by the brief spike in [Ca^{2+}]$_i$ that followed partial flash photolysis of partially Ca^{2+}-loaded DM-nitrophen (Zucker *et al.*, 1991; Zucker, 1993a). This response began a fraction of a millisecond after the rise in [Ca^{2+}]$_i$, a delay similar to the usual synaptic delay following Ca^{2+} influx during an action potential; both delays had the same temperature dependence. Thus, under the conditions of these experiments, photolysis of DM-nitrophen caused a [Ca^{2+}]$_i$ transient resembling that occurring normally at transmitter release sites in the vicinity of Ca^{2+} channels that open briefly during an action potential. After the 2- to 3-msec intense phase of secretion, a moderate phase of transmitter release persisted for ~15 msec, corresponding to a 60-msec relaxation in [Ca^{2+}]$_i$ measured with fura-2 that probably reflected displacement of Mg^{2+} bound to unphotolyzed DM-nitrophen by the photolytically liberated Ca^{2+} ions; the difference in time constants reflects the cooperativity of Ca^{2+} action in evoking neurosecretion. A small persistent phase of secretion, lasting a few seconds, was likely to be the result of the rise in resting [Ca^{2+}]$_i$ after partial DM-nitrophen photolysis and of the restoring effects of Ca^{2+} sequestering and extrusion mechanisms.

Similar responses to partial flash photolysis of partially Ca^{2+}-loaded DM-nitrophen have been obtained at crayfish neuromuscular junctions (Zucker, 1993b). Transmitter release evoked by slow photolysis of Ca^{2+}–DM-nitrophen using steady illumination also has been studied at this junction (Mulkey and Zucker, 1993). The rate of quantal transmitter release, measured as the frequency of miniature excitatory junctional potentials (MEJPs), was increased ~1000-fold during the illumination. Brief illuminations (0.3–2 sec) evoked a rise in MEJP frequency that dropped abruptly back to normal when the light was extinguished, as would be expected from the reversible rise in $[Ca^{2+}]_i$ that should be evoked by such illumination, which leaves most of the DM-nitrophen unphotolyzed (Zucker, 1993a). Longer light exposures caused an increase in MEJP frequency that outlasted the light signal, as would be expected from the rise in resting $[Ca^{2+}]_i$ after photolysis of most of the DM-nitrophen. These experiments illustrate the utility of slow photolysis of partially Ca^{2+}-loaded DM-nitrophen in generating reversible changes in $[Ca^{2+}]_i$ in cells.

Caged Ca^{2+} has been used to study the modulation of transmitter release at cultured snail synapses by the neuropeptide FMRFamide (Man-Son-Hing *et al.*, 1989). This peptide had a dual effect—reducing the Ca^{2+} current during depolarization and inhibiting neurosecretion, measured as the increase in miniature inhibitory postsynaptic currents evoked by a rise in $[Ca^{2+}]_i$ from photolysis of Ca^{2+}–nitr-5 perfused into the presynaptic neuron. FMRFamide also blocked the phasic release of transmitter caused by partial flash photolysis of partially Ca^{2+}-loaded DM-nitrophen (Haydon *et al.*, 1991). As in crayfish and squid synapses, these flash-evoked postsynaptic responses resembled the spike-evoked responses and were triggered by the spike in $[Ca^{2+}]_i$ that results when DM-nitrophen is used in this fashion.

At central serotonergic synapses in the leech, a presynaptic uptake system participates in the recovery of released transmitter and the termination of postsynaptic responses. The kinetics of this process were studied (Bruns *et al.*, 1993) by recording a presynaptic serotonin transport current following the phasic activation of transmitter release by photolysis of presynaptically injected Ca^{2+}-loaded DM-nitrophen. Blocking serotonin uptake by Na^+ removal or by zimelidine, eliminated the transport current and greatly prolonged the postsynaptic response.

DM-nitrophen has been used to probe the steps involved in exocytosis in endocrine cells. Measuring $[Ca^{2+}]_i$ changes with furaptra, Neher and Zucker (1993), working with bovine chromaffin cells, and Thomas *et al.* (1993), working with rat melanotrophs, found three distinct kinetic phases in secretion in response to steps of $[Ca^{2+}]_i$ to ~100 μM. These phases were interpreted to reflect vesicles released from different pools that were more or less accessible to the release machinery. In chromaffin cells, investigators also showed that prior exposure to a modest (micromolar) rise in $[Ca^{2+}]_i$ primed the response to a subsequent step in $[Ca^{2+}]_i$ released from DM-nitrophen, indicating that $[Ca^{2+}]_i$ not only triggers exocytosis but also mobilizes vesicles into a docked or releasable position. After exocytosis, another $[Ca^{2+}]_i$ stimulus often evoked a rapid reduction in membrane capacitance, interpreted as a phase of $[Ca^{2+}]_i$-dependent endocytosis. A similar effect was observed in rat neuronal synaptosomes from the neurohypophysis (Chernevskaya *et al.*, 1993).

Zoran *et al.* (1991) used photorelease of Ca^{2+} in a study of the developmental sequence of synapse maturation. When cultured snail neurons are brought into contact with a postsynaptic target, spike-evoked transmitter release begins only after several hours. Photorelease of Ca^{2+} from DM-nitrophen was used to show that this developmental change is the result of the delayed appearance of sensitivity of the secretory machinery to a rise in $[Ca^{2+}]_i$.

Long-term potentiation (LTP) is a complex form of synaptic plasticity thought to be involved in cognitive processes such as memory consolidation and spatial learning in the mammalian brain. Blocking a $[Ca^{2+}]_i$ rise in postsynaptic pyramidal cells in area CA1 of rat hippocampus during afferent stimulation is known to prevent the establishment of LTP. Malenka *et al.* (1988) microinjected Ca^{2+}–nitr-5 into these cells and showed that a rise in $[Ca^{2+}]_i$ was sufficient to trigger LTP in the injected neuron but in none of its neighbors. The duration of postsynaptic $[Ca^{2+}]_i$ increase necessary to induce LTP was explored by terminating the $[Ca^{2+}]_i$ rise caused by a brief afferent tetanus with photorelease of Ca^{2+} chelator from diazo-4 (Malenka *et al.*, 1992). $[Ca^{2+}]_i$ had to remain elevated for ~2 sec to generate LTP; shorter or smaller increases led only to a slowly decrementing or short-term potentiation, whereas an abbreviated $[Ca^{2+}]_i$ elevation of less than 1 sec was ineffective in generating synaptic plasticity.

D. Other Applications

In addition to the three major fields of application of caged Ca^{2+} chelators that were just described, the technique of $[Ca^{2+}]_i$ regulation has been used to address an increasingly diverse range of biological problems. Nitr and diazo compounds were inserted by AM loading into fibroblasts that were activated by mitogenic stimulation to produce $[Ca^{2+}]_i$ oscillations monitored using fluo-3 (Harootunian *et al.*, 1988). Photorelease of Ca^{2+} from nitr-5 enhanced and accelerated the oscillations, whereas release of caged chelator by photolysis of diazo-2 inhibited them. Nitr-7 photolysis caused not only an immediate rise in $[Ca^{2+}]_i$ liberated from the photolyzed chelator, but also elicited a later rise in $[Ca^{2+}]_i$ (Harootunian *et al.*, 1991). This effect was shown, pharmacologically, to be caused by IP$_3$-sensitive stores, suggesting that an interaction between $[Ca^{2+}]_i$ and these stores underlies the $[Ca^{2+}]_i$ oscillations.

Photorelease of Ca^{2+} from DM-nitrophen has been used to study the binding kinetics of Ca^{2+} to the Ca^{2+}–ATPase of sarcoplasmic reticulum vesicles (De-Long *et al.*, 1990). The relaxation of the $[Ca^{2+}]$ step, measured by arsenazo spectrophotometry after photolysis, revealed the kinetics of binding to the ATPase. Changes in the Fourier transform infrared spectrum consequent to photorelease of Ca^{2+} from nitr-5 provided information on structural changes in the ATPase after binding Ca^{2+} (Buchet *et al.*, 1991, 1992). In a final application to the study of enzyme conformational changes, photolysis of Mg^{2+}-loaded DM-nitrophen was used to form Mg^{2+}–ATP rapidly to activate the Na^+/K^+ pump, the state of which was monitored by fluorescence of aminostyrylpyridinium dyes (Forbush and Klodos, 1991). Rate-limiting steps were measured at 45 sec^{-1} by this method.

In other applications, Gilroy *et al.* (1990) and Fricker *et al.* (1991) microinjected Ca^{2+}-loaded nitr-5 into guard cells of lily leaves and showed that photorelease of about 600 nM intracellular Ca^{2+} (measured with fluo-3) initiated stomatal pore closure. Kao *et al.* (1990) loaded Swiss 3T3 fibroblasts with nitr-5/AM, and showed that photolysis that elevated $[Ca^{2+}]_i$ by hundreds of nanomolar (measured by fluo-3) triggered nuclear envelope breakdown, an early step in mitosis, while having little effect on the metaphase to anaphase transition. Control experiments using nitr-9 showed no effect of reactive photochemical intermediates or products. Tisa and Adler (1992) used electroporation to introduce Ca^{2+}-loaded nitr-5 or DM-nitrophen into *Escherichia coli* bacteria, and showed that elevation of $[Ca^{2+}]_i$ enhanced tumbling behavior characteristic of chemotaxis whereas photorelease of caged chelator from diazo-2 decreased tumbling. Photolysis of diazo-3, which reduces pH without affecting $[Ca^{2+}]_i$, caused only a small increase in tumbling. Mutants with methyl-accepting chemotaxis receptor proteins still responded to Ca^{2+}, whereas mutants of specific Che proteins did not, indicating that the action of these proteins lay downstream of the Ca^{2+} signal.

X. Conclusions

Interest in photolabile Ca^{2+} chelators is burgeoning. Their range of application is broadening beyond the original nerve, muscle, and fibroblast preparations. The kinetic properties of the chelators are beginning to be recognized and exploited (Mulkey and Zucker, 1993; Zucker, 1993a). New techniques of photolysis, such as two-photon absorption (Denk *et al.*, 1990), are under development. New classes of chelators are being designed (Adams and Tsien, 1993). These probes offer promise of a more detailed understanding of the function of Ca^{2+} as the most common and versatile cellular second messenger.

Acknowledgments

I am grateful to Steve Adams for valuable discussion and to Dr. Joseph Kao for drawings of chelator structures. The research done in my laboratory is supported by National Institutes of Health Grant NS 15114.

References

Adams, S. R., and Tsien, R. Y. (1993). Controlling cell chemistry with caged compounds. *Ann. Rev. Physiol.* **55,** 755–784.

Adams, S. R., Kao, J. P. Y., Grynkiewicz, G., Minta, A., and Tsien, R. Y. (1988). Biologically useful chelators that release Ca^{2+} upon illumination. *J. Am. Chem. Soc.* **110,** 3212–3220.

Adams, S. R., Kao, J. P. Y., and Tsien, R. Y. (1989). Biologically useful chelators that take up Ca^{2+} upon illumination. *J. Am. Chem. Soc.* **111,** 7957–7968.

Ashley, C. C., Griffiths, P. J., Lea, T. J., Mulligan, I. P., Palmer, R. E., and Simnett, S. J. (1991a). Use of fluorescent TnC derivatives and "caged" compounds to study cellular Ca^{2+} phenomena. *In* "Cellular Calcium: A Practical Approach" (J. G. McCormack and P. H. Cobbold, eds.), pp. 177–203. New York: Oxford University Press.

Ashley, C. C., Mulligan, I. P., and Lea, T. J. (1991b). Ca^{2+} and activation mechanisms in skeletal muscle. *Q. Rev. Biophys.* **24,** 1–73.

Backx, P. H., O'Rourke, B., and Marban, E. (1991). Flash photolysis of magnesium–DM-nitrophen in heart cells. A novel approach to probe magnesium- and ATP-dependent regulation of calcium channels. *Am. J. Hypertension (Suppl.)* **4,** 416–421.

Bates, S. E., and Gurney, A. M. (1993). Ca^{2+}-dependent block and potentiation of L-type calcium current in guinea-pig ventricular myocytes. *J. Physiol. (London)* **466,** 345–365.

Bruns, D., Engert, F., and Lux, H. D. (1993). A fast activating presynaptic reuptake current during serotonergic transmission in identified neurons of Hirudo. *Neuron* **10,** 559–72.

Buchet, R., Jona, I., and Martonosi, A. (1991). Ca^{2+} release from caged-Ca^{2+} alters the FTIR spectrum of sarcoplasmic reticulum. *Biochim. Biophys. Acta* **1069,** 209–217.

Buchet, R., Jona, I., and Martonosi, A. (1992). The effect of dicyclohexylcarbodiimide and cyclopiazonic acid on the difference FTIR spectra of sarcoplasmic reticulum induced by photolysis of caged-ATP and caged-Ca^{2+}. *Biochim. Biophys. Acta* **1104,** 207–14.

Carl, A., McHale, N. G., Publicover, N. G., and Sanders, K. M. (1990). Participation of Ca^{2+}-activated K^{+} channels in electrical activity of canine gastric smooth muscle. *J. Physiol. (London)* **429,** 205–221.

Charnet, P., Richard, S., Gurney, A. M., Ouadid, H., Tiaho, F., and Nargeot, J. (1991). Modulation of Ca^{2+} currents in isolated frog atreal cells studied with photosensitive probes. Regulation by cAMP and Ca^{2+}: A common pathway? *J. Mol. Cell. Cardiol.* **23,** 343–356.

Chernevskaya, N. J., Zucker, R. S., and Nowycky, M. C. (1993). Capacitance changes associated with exocytosis from mammalian peptidergic nerve terminals produced by release of caged calcium. *Biophys. J.* **64,** A317.

Delaney, K. R., and Zucker, R. S. (1990). Calcium released by photolysis of DM-nitrophen stimulates transmitter release at squid giant synapse. *J. Physiol.* **426,** 473–498.

DeLong, L. J., Phillips, C. M., Kaplan, J. H., Scarpa, A., and Blasie, J. K. (1990). A new method for monitoring the kinetics of calcium binding to the sarcoplasmic reticulum Ca^{2+}–ATPase employing the flash-photolysis of caged-calcium. *J. Biochem. Biophys. Meth.* **21,** 333–339.

Denk, W., Strickler, J. H., and Webb, W. W. (1990). Two-photon laser scanning fluorescence microscopy. *Science* **248,** 73–76.

Eberhard, M., and Erne, P. (1991). Calcium binding to fluorescent calcium indicators: Calcium green, calcium orange, and calcium crimson. *Biochem. Biophys. Res. Commun.* **180,** 209–215.

Ellis-Davies, G. C. R., and Kaplan, J. H. (1988). A new class of photolabile chelators for the rapid release of divalent cations: Generation of caged Ca and caged Mg. *J. Org. Chem.* **53,** 1966–1969.

Ferenczi, M. A., Goldman, Y. E., and Trentham, D. R. (1989). Relaxation of permeabilized, isolated muscle fibres of the rabbit by rapid chelation of Ca^{2+}-ions through laser pulse photolysis of "caged-BAPTA." *J. Physiol. (London)* **418,** 155P.

Fogelson, A. L., and Zucker, R. S. (1985). Presynaptic calcium diffusion from various arrays of single channels: Implications for transmitter release and synaptic facilitation. *Biophys. J.* **48,** 1003–1017.

Forbush, B., III, and Klodos, I. (1991). Rate-limiting steps in Na translocation by the Na/K pump. *Soc. Gen. Physiol. Ser.* **46,** 210–225.

Fricker, M. D., Gilroy, S., Read, N. D., and Trewavas, A. J. (1991). Visualisation and measurement of the calcium message in guard cells. *Symp. Soc. Exp. Biol.* **45,** 177–90.

Fryer, M. W., and Zucker, R. S. (1993). Ca^{2+}-dependent inactivation of Ca^{2+} current in *Aplysia* neurons: Kinetic studies using photolabile Ca^{2+} chelators. *J. Physiol. (London)* **464,** 501–528.

Gilroy, S., Read, N. D., and Trewavas, A. J. (1990). Elevation of cytoplasmic calcium by caged calcium or caged inositol trisphosphate initiates stomatal closure. *Nature (London)* **346,** 769–771.

Goldman, Y. E., Hibberd, M. G., and Trentham, D. R. (1984). Relaxation of rabbit psoas muscle fibres from rigor by photochemical generation of adenosine-5′-triphosphate. *J. Physiol. (London)* **354,** 577–604.

Grell, E., Lewitzki, E., Ruf, H., Bamberg, E., Ellis-Davies, G. C. R., Kaplan, J. H., and De-Weer, P. (1989). Caged-Ca^{2+}: A new agent allowing liberation of free Ca^{2+} in biological systems by photolysis. *Cell. Mol. Biol.* **35,** 515–522.

Grynkiewicz, G., Poenie, M., and Tsien, R. Y. (1985). A new generation of Ca^{2+} indicators with greatly improved fluorescence properties. *J. Biol. Chem.* **260,** 3440–3450.

Gurney, A. (1991). Photolabile calcium buffers to selectively activate calcium-dependent processes. *In* "Cellular Neurobiology: A Practical Approach" (J. Chad and H. Wheal, eds.), pp. 153–177. New York: IRL Press.

Gurney, A. (1993). Photolabile caged compounds. *In* "Fluorescent Probes for Biological Function of Living Cells—A Practical Guide" (W. T. Mason, ed.), pp. 335–348. New York: Academic Press.

Gurney, A. M., Tsien, R. Y., and Lester, H. A. (1987). Activation of a potassium current by rapid photochemically generated step increases of intracellular calcium in rat sympathetic neurons. *Proc. Natl. Acad. Sci. U.S.A.* **84,** 3496–3500.

Gurney, A. M., Charnet, P., Pye, J. M., and Nargeot, J. (1989). Augmentation of cardiac calcium current by flash photolysis of intracellular caged-Ca^{2+} molecules. *Nature (London)* **341,** 65–68.

Györke, S., and Fill, M. (1993). Ryanodine receptor adaptation: control mechanism of Ca^{2+}-induced Ca^{2+} release in heart. *Science* **260,** 807–9.

Hadley, R. W., and Lederer, W. J. (1991). Ca^{2+} and voltage inactivate Ca^{2+} channels in guinea-pig ventricular myocytes through independent mechanisms. *J. Physiol. (London)* **444,** 257–268.

Harootunian, A. T., Kao, J. P. Y., and Tsien, R. Y. (1988). Agonist-induced calcium oscillations in depolarized fibroblasts and their manipulation by photoreleased Ins(1,4,5)P_3, Ca^{2+}, and Ca^{2+} buffer. *Cold Spring Harbor Symp. Quant. Biol.* **53,** 935–943.

Harootunian, A. T., Kao, J. P., Paranjape, S., Adams, S. R., Potter, B. V. L., and Tsien, R. Y. (1991). Cytosolic Ca^{2+} oscillations in REF52 fibroblasts: Ca^{2+}-stimulated IP_3 production or voltage-dependent Ca^{2+} channels as key positive feedback elements. *Cell Calcium* **12,** 153–164.

Haydon, P. G., Man-Son-Hing, H., Doyle, R. T., and Zoran, M. (1991). FMRFamide modulation of secretory machinery underlying presynaptic inhibition of synaptic transmission requires a pertussis toxin-sensitive G-protein. *J. Neurosci.* **11,** 3851–3860.

Hochner, B., Parans, H., and Parnas, I. (1989). Membrane depolarization evokes neurotransmitter release in the absence of calcium entry. *Nature (London)* **342,** 433–435.

Iino, M., Endo, M. (1992). Calcium-dependent immediate feedback control of inositol 1,4,5-triphosphate-induced Ca^{2+} release. *Nature (London)* **360,** 76–8.

Johnson, B. D., and Byerly, L. (1993). Photo-released intracellular Ca^{2+} rapidly blocks Ba^{2+} current in *Lymnaea* neurons. *J. Physiol. (London)* **464,** 501–528.

Kao, J. P. Y., and Adams, S. R. (1993). Photosensitive caged compounds: Design, properties, and biological applications. *In* "Optical Microscopy: New Technologies and Applications" (B. Herman and J. J. Lemasters, eds.), pp. 27–85. New York: Academic Press.

Kao, J. P. Y., Harootunian, A. T., and Tsien, R. Y. (1989). Photochemically generated cytosolic calcium pulses and their detection by fluo-3. *J. Biol. Chem.* **264,** 8179–8184.

Kao, J. P. Y., Alderton, J. M., Tsien, R. Y., and Steinhardt, R. A. (1990). Active involvement of Ca^{2+} in mitotic progression of Swiss 3T3 fibroblasts. *J. Cell Biol.* **111,** 183–196.

Kaplan, J. H. (1990). Photochemical manipulation of divalent cation levels. *Annu. Rev. Physiol.* **52,** 897–914.

Kaplan, J. H., and Ellis-Davies, G. C. R. (1988). Photolabile chelators for rapid photolytic release of divalent cations. *Proc. Natl. Acad. Sci. U.S.A.* **85,** 6571–6575.

Kaplan, J. H., and Somlyo, A. P. (1989). Flash photolysis of caged compounds: New tools for cellular physiology. *Trends Neurosci.* **12,** 54–59.

Kaplan, J. H., Forbush, B., III, and Hoffman, J. F. (1978). Rapid photolytic release of adenosine 5′-triphosphate from a protected analogue: Utilization by the Na:K pump of human red blood cell ghosts. *Biochemistry* **17,** 1929–1935.

Kentish, J. C., Barsotti, R. J., Lea, T. J., Mulligan, I. P., Patel, J. R., and Ferenczi, M. A. (1990). Calcium release from cardiac sarcoplasmic reticulum induced by photorelease of calcium or Ins(1,4,5)P$_3$. *Am. J. Physiol.* **258,** H610–H615.

Kimitsuki, T., and Ohmori, H. (1992). The effect of caged calcium release on the adaptation of the transduction current in chick hair cells. *J. Physiol. (London)* **458,** 27–40.

Konishi, M., Hollingworth, S., Harkins, A. B., and Baylor, S. M. (1991). Myoplasmic calcium transients in intact frog skeletal muscle fibers monitored with the fluorescent indicator furaptra. *J. Gen. Physiol.* **97,** 271–301.

Kurebayashi, N., Harkins, A. B., and Baylor, S. M. (1993). Use of fura red as an intracellular calcium indicator in frog skeletal muscle fibers. *Biophys. J.* **64,** 1934–1960.

Lännergren, J., and Arner, A. (1992). Relaxation rate of intact striated muscle fibres after flash photolysis of a caged calcium chelator (diazo-2). *J. Muscle Res. Cell Motil.* **13,** 630–634.

Lancaster, B., and Zucker, R. S. (1991). Photolytic manipulation of [Ca^{2+}]$_i$ controls hyperpolarization in hippocampal pyramidal cells. *Soc. Neurosci. Abstr.* **17,** 1114.

Landò, L., and Zucker, R. S. (1989). "Caged calcium" in *Aplysia* pacemaker neurons. Characterization of calcium-activated potassium and nonspecific cation currents. *J. Gen. Physiol.* **93,** 1017–1060.

Lea, T. J., and Ashley, C. C. (1990). Ca^{2+} release from the sarcoplasmic reticulum of barnacle myofibrillar bundles initiated by photolysis of caged Ca^{2+}. *J. Physiol. (London)* **427,** 435–453.

Lea, T. J., Fenton, M. J., Potter, J. D., and Ashley, C. C. (1990). Rapid activation by photolysis of nitr-5 in skinned fibres of the striated adductor muscle from the scallop. *Biochim. Biophys. Acta* **1034,** 186–194.

Malenka, R. C., Kauer, J. A., Zucker, R. S., and Nicoll, R. A. (1988). Postsynaptic calcium is sufficient for potentiation of hippocampal synaptic transmission. *Science* **242,** 81–84.

Malenka, R. C., Lancaster, B., and Zucker, R. S. (1992). Temporal limits on the rise in postsynaptic calcium required for the induction of long-term potentiation. *Neuron* **9,** 121–128.

Man-Son-Hing, H., Zoran, M. J., Lukowiak, K., and Haydon, P. G. (1989). A neuromodulator of synaptic transmission acts on the secretory apparatus as well as on ion channels. *Nature (London)* **341,** 237–239.

Marrion, N. V., Zucker, R. S., Marsh, S. J., and Adams, P. R. (1991). Modulation of M-current by intracellular Ca^{2+}. *Neuron* **6,** 533–545.

McCray, J. A., and Trentham, D. R. (1989). Properties and uses of photoreactive caged compounds. *Annu. Rev. Biophys. Biophys. Chem.* **18,** 239–270.

McCray, J. A., Fidler-Lim, N., Ellis-Davies, G. C. R., and Kaplan, J. H. (1992). Rate of release of Ca^{2+} following laser photolysis of the DM-nitrophen-Ca^{2+} complex. *Biochemistry* **31**, 8856–8861.

Minta, A., Kao, J. P. Y., and Tsien, R. Y. (1989). Fluorescent indicators for cytosolic calcium based on rhodamine and fluorescein chromophores. *J. Biol. Chem.* **264**, 8171–8178.

Morad, M., Davies, N. W., Kaplan, J. H., and Lux, H. D. (1988). Inactivation and block of calcium channels by photo-released Ca^{2+} in dorsal root ganglion neurons. *Science* **241**, 842–844.

Mulkey, R. M., and Zucker, R. S. (1991). Action potentials must admit calcium to evoke transmitter release. *Nature (London)* **350**, 153–155.

Mulkey, R. M., and Zucker, R. S. (1993). Calcium released from DM-nitrophen photolysis triggers transmitter release at the crayfish neuromuscular junction. *J. Physiol. (London)* **462**, 243–260.

Mulligan, I. P., and Ashley, C. C. (1989). Rapid relaxation of single frog skeletal muscle fibres following laser flash photolysis of the caged calcium chelator, diazo-2. *FEBS Lett.* **255**, 196–200.

Näbauer, M., and Morad, M. (1990). Ca^{2+}-induced Ca^{2+} release as examined by photolysis of caged Ca^{2+} in single ventricular myocytes. *Am. J. Physiol.* **258**, C189–C193.

Näbauer, M., Ellis-Davies, G. C. R., Kaplan, J. H., and Morad, M. (1989). Modulation of Ca^{2+} channel selectivity and cardiac contraction by photorelease of Ca^{2+}. *Am. J. Physiol.* **256**, H916–H920.

Neher, E., and Zucker, R. S. (1993). Multiple calcium-dependent processes related to secretion in bovine chromaffin cell. *Neuron* **10**, 21–30.

Nowycky, M. C., and Pinter, M. J. (1993). Time courses of calcium and calcium-bound buffers following calcium influx in a model cell. *Biophys. J.* **64**, 77–91.

O'Neill, S. C., Mill, J. G., and Eisner, D. A. (1990). Local activation of contraction in isolated rat ventricular myocytes. *Am. J. Physiol.* **258**, C1165–C1168.

O'Rourke, B., Backx, P. H., and Marban, E. (1992). Phosphorylation-independent modulation of L-type calcium channels by magnesium-nucleotide complexes. *Science* **257**, 245–248.

Ogden, D. (1988). Answer in a flash. *Nature (London)* **336**, 16–17.

Ogden, D. C., Capiod, T., Walker, J. W., and Trentham, D. R. (1990). Kinetics of the conductance evoked by noradrenaline, inositol trisphosphate or Ca^{2+} in guinea-pig isolated hepatocytes. *J. Physiol. (London)* **422**, 585–602.

Palmer, R. E., Mulligan, I. P., Nunn, C., and Ashley, C. C. (1990). Striated scallop muscle relaxation: Fast force transients produced by photolysis of diazo-2. *Biochem. Biophys. Res. Commun.* **168**, 295–300.

Palmer, R. E., Simnett, S. J., Mulligan, I. P., and Ashley, C. C. (1991). Skeletal muscle relaxation with diazo-2: the effect of altered pH. *Biochem. Biophys. Res. Commun.* **181**, 1337–1342.

Parker, I. (1992). Caged intracellular messengers and the inositol phosphate signaling pathway. *In* "Neuromethods: Intracellular Messengers" (A. Boulton, G. Baker, and C. Taylor, eds.), Vol. 20, pp. 369–396. Totowa, New Jersey: Humana Press.

Rapp, G., and Guth, K. (1988). A low cost high intensity flash device for photolysis experiments. *Pflügers Arch.* **411**, 200–203.

Sala, F., and Hernádez-Cruz, A. (1990). Calcium diffusion modeling in a spherical neuron. Relevance of buffering properties. *Biophys. J.* **57**, 313–324.

Scarpa, A., Brinley, F. J., Tiffert, T., and Dubyak, G. R. (1978). Metallochromic indicators of ionized calcium. *Ann. N.Y. Acad. Sci.* **307**, 86–112.

Simon, S. M., and Llinás, R. R. (1985). Compartmentalization of the submembrane calcium activity during calcium influx and its significance in transmitter release. *Biophys. J.* **48**, 485–498.

Stern, M. D. (1992). Buffering of calcium in the vicinity of a channel pore. *Cell Calcium* **13**, 183–192.

Stockbridge, N., and Moore, J. W. (1984). Dynamics of intracellular calcium and its possible relationship to phasic transmitter release and facilitation at the frog neuromuscular junction. *J. Neurosci.* **4**, 803–811.

Thomas, P., Wong, J. G., and Almers, W. (1993). Millisecond studies of secretion in single rat pituitary cells stimulated by flash photolysis of caged Ca^{2+}. *EMBO J.* **12**, 303–306.

Tisa, L. S., and Adler, J. (1992). Calcium ions are involved in *Escherichia coli* chemotaxis. *Proc. Natl. Acad. Sci. U.S.A.* **89**, 11804–11808.

Tsien, R., and Zucker, R. S. (1986). Control of cytoplasmic calcium with photolabile 2-nitrobenzhydrol tetracarboxylate chelators. *Biophys. J.* **50**, 843–853.

Valdeolmillos, M., O'Neill, S. C., Smith, G. L., and Eisner, D. A. (1989). Calcium-induced calcium release activates contraction in intact cardiac cells. *Pflügers Arch.* **413**, 676–678.

Walker, J. W. (1991). Caged molecules activated by light. *In* "Cellular Neurobiology: A Practical Approach" (J. Chad and H. Wheal, eds.), pp. 179–203. New York: IRL Press.

Zoran, M. J., Doyle, R. T., and Haydon, P. G. (1991). Target contact regulates the calcium responsiveness of the secretory machinery during synaptogenesis. *Nature* (*London*) **6**, 145–151.

Zucker, R. S. (1989). Models of calcium regulation in neurons. *In* "Neural Models of Plasticity: Experimental and Theoretical Approaches" (J. H. Byrne and W. O. Berry, eds.), pp. 403–422. Orlando, Florida: Academic Press.

Zucker, R. S. (1992a). Effects of photolabile calcium chelators on fluorescent calcium indicators. *Cell Calcium* **13**, 29–40.

Zucker, R. S. (1992b). Calcium regulation of ion channels in neurons. *In* "Intracellular Regulation of Ion Channels" (M. Morad and Z. Agus, eds.), pp. 191–201. New York: Springer-Verlag.

Zucker, R. S. (1993a). The calcium concentration clamp: Spikes and reversible pulses using the photolabile chelator DM-nitrophen. *Cell Calcium* **14**, 87–100.

Zucker, R. S. (1993b). Calcium and transmitter release. *J. Physiol.* (*Paris*) **87**, 25–36.

Zucker, R. S., and Haydon, P. G. (1988). Membrane potential plays no direct role in evoking neurotransmitter release. *Nature* (*London*) **335**, 360–362.

Zucker, R. S., and Stockbridge, N. (1983). Presynaptic calcium diffusion and the time courses of transmitter release and synaptic facilitation at the squid giant synapse. *J. Neurosci.* **3**, 1263–1269.

Zucker, R. S., Delaney, K. R., Mulkey, R., and Tank, D. W. (1991). Presynaptic calcium in transmitter release and post-tetanic potentiation. *Ann. N.Y. Acad. Sci.* **635**, 191–207.

CHAPTER 3

Pharmacological Tools for Perturbing Intracellular Calcium Storage

David Thomas and Michael R. Hanley

Department of Biological Chemistry
School of Medicine
University of California at Davis
Davis, California 95616

I. Introduction

The regulation of calcium discharge by surface receptors has emphasized the roles of the gene products that control release of calcium into the cytosol. However, comparatively little is known about the calcium storage compartment and how it is loaded with calcium physiologically. Currently, the best evidence suggests that calcium is stored in the endoplasmic reticulum

(ER) in nonmuscle cells (Tsien and Tsien, 1990) in a fashion directly analogous to calcium storage in sarcoplasmic reticulum (SR). Indeed, knowledge of calcium sequestration has been advanced greatly by the recognition that many of the specialized properties of SR have functional counterparts in nonmuscle ER. Significantly, whether the store accessible to receptor action is bulk ER or a specialized domain, which has been termed the "calciosome" (Rossier and Putney, 1991), remains an important debate.

Although pharmacological tools have contributed much to our understanding of calcium signaling, many key Ca^{2+} regulatory proteins in the cell have no means of pharmacological manipulation. Table I lists several reagents that have been applied to studies of intracellular Ca^{2+} regulation, although the level of practical experience and confidence that can be placed in the use of such reagents varies widely. This table clarifies the continuing need to expand the repertoire of reagents for intervention at discrete points in calcium homeostasis. The focus of this chapter is discussing the background and application of a newly described panel of structurally distinct compounds that perturb intracellular Ca^{2+} storage by rapid and potent inhibition of the intracellular Ca^{2+} pumps. These reagents have come into widespread and routine use because of their unique property of bypassing complex surface-receptor-initiated second messenger events and triggering cytosolic Ca^{2+} elevation directly by an action on the intracellular store. Moreover, long-term application of these reagents leads to persistent depletion of Ca^{2+} from the store, activating an unexpected spectrum of cellular changes. The molecular targets of these reagents, the intracellular calcium pumps, are therefore important new sites for experimental intervention in calcium homeostasis. The reagents described here may prove to be the first generation of a potentially much larger class of calcium signaling tools.

II. SERCA Family of Intracellular Ca^{2+} Pumps

A. Distribution and Functional Properties of SERCA Isoforms

In both SR and ER, calcium is loaded from the cytosol into the store by an ion-motive ATPase, the intracellular "calcium pump." Five sarcoplasmic/endoplasmic reticulum Ca^{2+}–ATPase (SERCA) isoforms have been identified that are derived from three genes: SERCA 1, 2, and 3. Alternative splicing of SERCA 1 and 2 generates additional isoform diversity (Brandl *et al.*, 1986; Burk *et al.*, 1989; Table II).

SERCA 1a is the major adult fast-twitch skeletal isoform and is expressed at high levels in membranes of striated muscle sarcoplasmic reticulum, constituting up to 90% of the membrane protein (Carafoli, 1987). Rabbit SERCA 1a was the first SERCA isoform to be cloned and represents the most extensively studied isoform in terms of enzyme mechanism, primarily because of the com-

Table I
Selected List of Pharmacological Compounds Used to Study Intracellular Ca^{2+} Homeostasis

Compound	Putative target	Effects	Comments	References
CICR[a]/Ryanodine receptor				
Dantrolene	Intracellular Ca^{2+} release channel	Blocks Ca^{2+} release from skeletal muscle	May also act on neural cells and exert neuroprotective effect	Frandsen et al. (1992)
Ryanodine	Intracellular Ca^{2+} release channel (CICR) site	High concentrations block Ca^{2+} release; low concentrations activate release	Locks SR[b]/ER[c] CA^{2+} channel in low conductance state	Fill and Coronado (1988)
Caffeine	Intracellular Ca^{2+} release channel (CICR) site	Potentiates Ca^{2+}-induced Ca^{2+} release	Low potency; used at millimolar concentrations	Zachetti et al. (1991)
Ruthenium red	Intracellular Ca^{2+} release channel (CICR) site	Inhibits Ca^{2+}-induced Ca^{2+} release	May have multiple nonspecific polycation effects; blocks other ion channels	Imagawa et al. (1987)
SERCA Pumps				
Ochratoxin A	SR Ca^{2+} pump	Inhibits Ca^{2+} pump; causes Ca^{2+} discharge	Inhibition may be due to membrane perturbation	Khan et al. (1989)
Nafenopin	SR/ER Ca^{2+} pump	Inhibits Ca^{2+} pump; causes Ca^{2+} discharge	Peroxisome proliferator	Ochsner et al. (1990)
Thapsigargin	SR/ER Ca^{2+} pump	Inhibits Ca^{2+} pump; causes Ca^{2+} discharge	Specific and potent inhibition; forms dead-end complex	Thastrup et al. (1990)
Cyclopiazonic acid	SR/ER Ca^{2+} pump	Inhibits Ca^{2+} pump; causes Ca^{2+} discharge	Competitive at ATP binding site	Seidler et al. (1989)
2,5-Di(tert-butyl)hydroquinone	SR/ER Ca^{2+} pump	Inhibits Ca^{2+} pump; causes Ca^{2+} discharge	May be nonspecific; blocks receptor-activated Ca^{2+} entry and ER leak	Foskett and Wong (1992)
Vanadate	SR/ER Ca^{2+} pump	Inhibits Ca^{2+} pump	Nonspecific; high concentrations (millimolar) required	Missiaen et al. (1991)

[a] Calcium-induced calcium release.
[b] Sarcoplasmic reticulum.
[c] Endoplasmic reticulum.

Table II
Intracellular Ca^{2+}–ATPase Gene Family and Tissue Distributions

Gene	Splice variant	Length (residues)	Species	Tissue distribution
SERCA 1	a	942	Rabbit	Adult fast-twitch muscle
SERCA 1	b	1001	Rabbit	Neonatal fast-twitch muscle
SERCA 2	a	997	Rabbit	Cardiac/slow-twitch muscle
SERCA 2	b	1042	Rat	Ubiquitous
SERCA 3	—	999	Rat	Neurons, gut epithelium, limited other sites

parative simplicity of purification of SERCA 1a in its native environment by preparation of rabbit fast-twitch muscle SR (Chu *et al.*, 1988). SERCA 1b is regulated developmentally and is expressed exclusively in neonatal fast-twitch muscle tissues. The SERCA 2 gene also gives rise to two forms: SERCA 2a, found in slow-twitch and cardiac muscle, and SERCA 2b, which has a broader tissue distribution than other isoforms and is likely to be ubiquitous in non-muscle tissues (Lytton *et al.*, 1992). SERCA 2b presently is considered the housekeeping ER-type Ca^{2+} pump of nonmuscle cells. The SERCA 3 gene encodes a single transcript which, like SERCA 2b, has a largely nonmuscle localization, with highest levels of expression in intestine and brain (Lytton *et al.*, 1991).

At the amino acid level, a large degree of sequence conservation exists among the SERCA isoforms (75–85% sequence identity); thus, all are predicted to have very similar transmembrane topology (Fig. 1; see Lytton *et al.*, 1992). All the SERCA pump isoforms have at least 10 predicted transmembrane domains, with evidence for an additional transmembrane domain at the extended C terminus of SERCA 2b. The first five transmembrane helices extend into the cytoplasm to form a pentahelical "stalk." The ATP binding site and an aspartyl residue that is essential to the catalytic cycle are localized to an extensive globular domain between the fourth and fifth transmembrane helices (McClennan *et al.*, 1992). This predicted structure is conserved among the SERCA isoforms and strongly suggests functional conservation of enzyme mechanism and, possibly, many forms of regulation.

The ion transport cycle is achieved by a mechanism shared with the other ion-motive ATPases (Inesi *et al.*, 1990), in which the enzyme undergoes a conformational change that transforms high-affinity cytosol-facing ion binding sites (E1 state) to low-affinity lumen-facing binding sites (E2 state). The enzymatic mechanism of calcium-activated ATP hydrolysis and its coupling to calcium translocation have been discussed in detail (Jencks, 1989; Inesi *et al.*, 1990).

A prominent difference in sequence among the SERCA enzymes is the unique replacement of the otherwise invariant amino acids at the C terminus

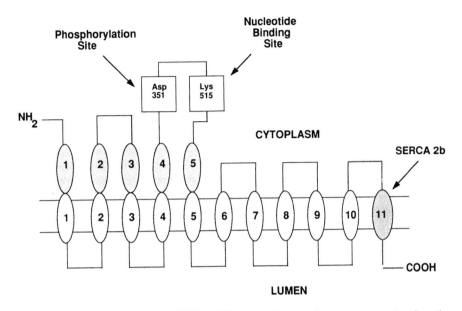

Fig. 1 Proposed topology of the SERCA Ca^{2+} pumps in the endoplasmic/sarcoplasmic reticulum (ER/SR) membrane. Amino acid residues from transmembrane helices 4, 5, 6, and 8 are thought to be critical for Ca^{2+} binding (Clarke *et al.*, 1989). The SERCA 2b isoform may form an additional transmembrane pass at the C terminus (Malcolm *et al.*, 1992).

with an additional 49 amino acids in SERCA 2b (Lytton *et al.*, 1992), sufficient to form an additional transmembrane segment (Campbell *et al.*, 1992; see Fig. 1). Indeed, this tail extension has been proposed to interact with parts of the luminal Ca^{2+} binding domain to account for the observed higher affinity for Ca^{2+} exhibited by SERCA 2b in comparison with SERCA 2a (Verboomen *et al.*, 1992). SERCA 3, despite the high degree of conservation at the amino acid level, appears to be, functionally, significantly different from the other isoforms. For example, SERCA 3 has been observed, to have a lower Ca^{2+} affinity, a more alkaline pH optimum, and a higher vanadate sensitivity than the other SERCA isoforms (Lytton *et al.*, 1992). All these characteristics are consistent with SERCA 3 favoring the E2 state, the conformation in which the enzyme is least sensitive to calcium, over E1 in the E1–E2 equilibrium.

To determine the enzymatic parameters of individual SERCA isoforms, an essential technique has been preparing microsomes from COS cells that overexpress SERCA pumps (Maruyama and MacLennan, 1988; Lytton *et al.*, 1992; Toyofuku *et al.*, 1992; Verboomen *et al.*, 1992). This strategy also is used routinely to study the effects of site-directed mutagenesis of SERCA isoforms. Ectopic expression in COS cells of SERCA 1a has been investigated in the greatest detail. Limited information is available on ectopic expression of other SERCA isoforms or on the use of heterologous cell expression systems such as *Xenopus laevis* oocytes (Camacho and Lechleiter, 1993) or stably transfected mammalian cell lines (Hussain *et al.*, 1992). Clearly, this is a fertile area for future work.

B. Specific Inhibition of SERCA Pumps by Pharmacological Reagents

The study of the sodium–potassium activated ATPase (the "sodium pump") has been aided by the use of potent cardiac glycoside inhibitors such as ouabain. Accordingly, the discovery of compounds that have a comparable inhibitory action on SERCA pumps has been a powerful addition to the tools used for the study of the mechanism of the SERCA pumps (Inesi and Sagara, 1992). Three structurally unrelated SERCA pump inhibitors, shown in Fig. 2, are the most commonly used: thapsigargin, cyclopiazonic acid, and 2,5-di(*tert*-butyl)hydroquinone (tBHQ). However, other compounds have been suggested to have a similar blocking effect on intracellular calcium pumps (see Table I), notably the hepatotoxin ochratoxin A (Khan *et al.*, 1989), the peroxisome proliferator nafenopin (Ochsner *et al.*, 1990), and the plasma membrane calcium pump inhibitor vanadate (Missiaen *et al.*, 1991). These reagents are not considered further here because knowledge of their specificity, mechanism, and cellular applicability is not well developed. However, one or more of these reagents may prove useful in the future for specific applications.

1. Thapsigargin

Thapsigargin is a sesquiterpene lactone isolated from the umbelliferous plant, *Thapsia garganica*. This compound has powerful irritant properties that may arise from its activation of a number of immune cells, including mast cells (Ali *et al.*, 1985). Thapsigargin is also a weak second stage tumor promoter in mouse skin (Hakii *et al.*, 1986), although it does not activate protein kinase C, as do the majority of identified tumor promoters (Jackson *et al.*, 1988). For

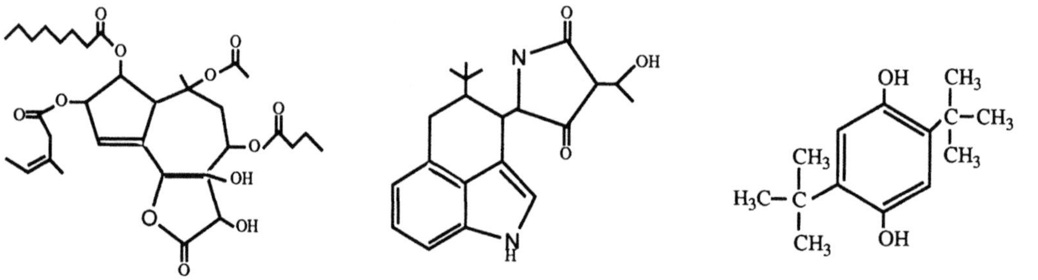

Thapsigargin   Cyclopiazonic acid   2,5-Di(*tert* -butyl)hydroquinone

Fig. 2 Structures of the three most frequently used SERCA pump inhibitors.

this reason, this tumor promoting activity is sometimes termed "non-TPA type" (Hakii *et al.*, 1986).

Thapsigargin has been shown to have the highest inhibitory potency of the known SERCA inhibitors, blocking the known SERCA isoforms with an apparent half-maximal inhibitory potency of 10–20 nM (Lytton *et al.*, 1991). This compound has no detectable action on any other ion-motive ATPase, including the plasma membrane calcium pump. The potency of thapsigargin in inhibiting SERCA pumps can be estimated by its inhibition of calcium-activated ATP hydrolysis in coupled enzyme assays (Chu *et al.*, 1988). A representative curve of the inhibitory potency of thapsigargin using rabbit SR containing SERCA 1a is shown in Fig. 3. In Fig. 4, examples of primary kinetic data obtained using this approach are shown. This measured inhibitory potency may be an underestimation of the true potency of thapsigargin, since thapsigargin can titrate the SERCA activity in SR preparations stoichiometrically (Lytton *et al.*, 1991). Thus, thapsigargin initially was concluded incorrectly to be inactive on SR calcium pumps, because of the very high level of pump expression (Thastrup *et al.*, 1990). Thapsigargin appears to form an irreversible one-to-one "dead end complex" with SERCA 1a (Sagara *et al.*, 1992). The principal use of this compound with isolated pumps has been in the study of enzyme mechanism, since it disrupts both Ca^{2+}-linked and Ca^{2+}-independent enzyme states, unlike any previous reagent (Inesi and Sagara, 1992). However, the initial interaction of thapsigargin may be with a precise enzyme state, despite the evidence for a global disruption of SERCA structure and activity. In this regard, thapsigargin, as well as tBHQ and vanadate, has

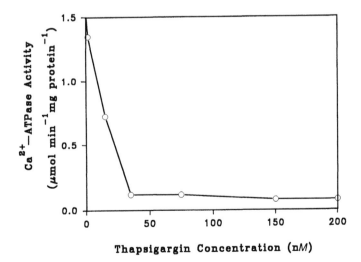

Fig. 3 Dose-potency curve for thapsigargin inhibition of the ATPase activity of the rabbit SR enzyme (SERCA 1a).

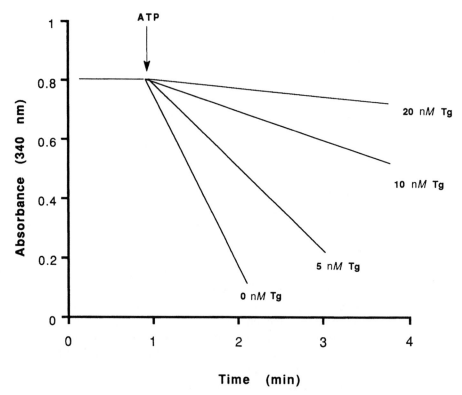

Fig. 4 The coupled enzyme assay using changes in NADH absorbance to monitor ATP hydrolysis (Chu *et al.*, 1988). The figure shows the pronounced inhibition of ATP hydrolysis with increasing thapsigargin (Tg) concentration.

been reported to stabilize the enzyme in the E2 conformation (Wictome *et al.*, 1992). The E2 state of calcium pumps represents a structural conformation of the enzyme in which Ca^{2+} sensitivity is diminished greatly. The recognition site at which thapsigargin interacts is not yet known, but in chimeric E1E2 ATPases of calcium and sodium pumps, thapsigargin sensitivity can be conferred by limited portions of the calcium pump including the N-terminal region that contains the first two transmembrane segments (Ishii *et al.*, 1993; Sumbilla *et al.*, 1993).

Structural features of thapsigargin that are required for SERCA blockade have not been established clearly because no complete chemical synthesis of thapsigargin has been achieved. However, chemical modifications of native thapsigargin (Norup *et al.*, 1986) were investigated using their histamine-releasing action. This study indicated that the *trans*-dihydroxyglycol moiety was critical because its acetylation abolished thapsigargin activity (Fig. 5). In contrast, modification of the ester acyl groups of thapsigargin, or of the lactone carbonyl, reduced but did not eliminate activity.

These structural considerations are important in at least three contexts concerning the use of thapsigargin. The first is that the apparent irreversibility of thapsigargin has been interpreted to arise from formation of a covalent linkage with the pump, perhaps through the active lactone ring (Christensen *et al.*, 1992). The second is that thapsigargin is quite stable in conventional physiological media at neutral pH. However, extremes of pH, particularly alkaline pH, have the potential to open the lactone ring with subsequent loss of activity. The third is that production of a biologically active radioactively labeled version of thapsigargin has proven remarkably difficult. Although a ^{3}H-labeled version of thapsigargin has been tested for its potential in a radioligand binding assay (Christensen, 1985), the high lipophilicity of thapsigargin results in enormous levels of nonspecific binding that compromise the usefulness of the compound. An analog of thapsigargin, comparable in hydrophobic properties to the tumor promoter phorbol-12,13-dibutyrate that is used conventionally to identify protein kinase C, is likely to be needed.

Thapsigargin is commercially available from several suppliers, including LC Services (Woburn, Massachusetts), CalBiochem (LaJolla, California), Gibco (Grand Island, New York), and Sigma (St. Louis, Missouri). The compound is provided as a dried film in sealed ampules, and should be dissolved for storage in anhydrous dimethylsulfoxide (DMSO). A 1 mg/ml stock corresponds to 1.53 mM and can be maintained indefinitely at $-20°$C, but should not be subjected to repeated freezing and thawing because of progressive loss of thapsigargin by its gradually reduced solubility. Thapsigargin should be handled cautiously because of its powerful irritant action and its greatly enhanced penetrating potential when dissolved in vehicles such as DMSO.

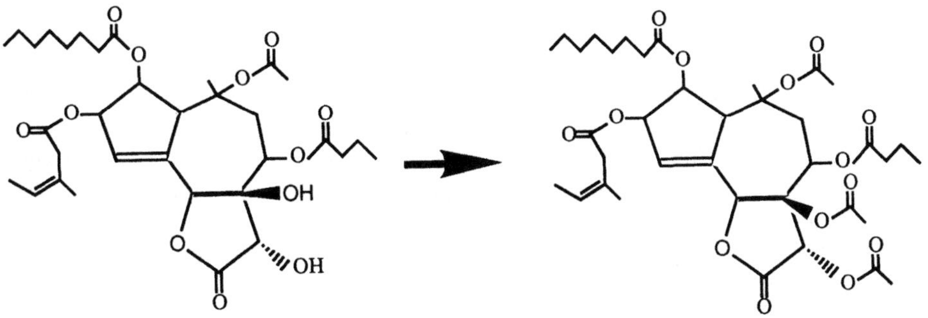

Fig. 5 Modification of the *trans*-dihydroxyglycol moiety of thapsigargin (*left*) to the diacetate derivative (*right*), which lacks activity.

2. Cyclopiazonic Acid

The mycotoxin cyclopiazonic acid (Fig. 2) is an indole tetramic acid derivative produced by certain species of the common fungal genera *Aspergillus* and *Penicillium* (Goeger and Riley, 1989). Adverse effects on the musculature arising from cyclopiazonic acid toxicity initially suggested that the compound had a muscle-specific target. Subsequently, using rat skeletal muscle SR researchers showed that cyclopiazonic acid was a potent inhibitor of the Ca^{2+}-activated ATPase activity of the calcium pump (Goeger and Riley, 1989). In addition, the inhibitor was shown to be largely specific since no inhibition of the Na^+/K^+–ATPase, H^+/K^+–ATPase, mitochondrial F_1 ATPase or plasma membrane Ca^{2+}–ATPase was detected (Seidler *et al.*, 1989). An important feature of the inhibitory action of cyclopiazonic acid is that its action appears to be as a competitive inhibitor of ATP binding to SERCA isoforms (Seidler *et al.*, 1989). As a competitive inhibitor of ATP binding, cyclopiazonic acid provides a complementary tool in circumstances in which rapid reversible inhibition of the Ca^{2+} pump is desired. Cyclopiazonic acid potency is substantially less than that of thapsigargin, since complete inhibition of the SR Ca^{2+}–ATPase requires 1–2 μM (IC_{50} for SR calcium pump $\sim$ 300–500 nM). Cyclopiazonic acid also inhibits the ER Ca^{2+}-activated ATPase, which is likely to correspond to SERCA 2b. However, investigators have reported that cyclopiazonic acid may not be as selective a probe as thapsigargin because it can inhibit passive Ca^{2+} efflux from SR and ER (Missiaen *et al.*, 1992). Cyclopiazonic acid can be purchased from the suppliers that distribute thapsigargin. This compound is dissolved in DMSO and does not appear to be labile.

3. 2,5-Di(*tert*-butyl)hydroquinone

tBHQ is a synthetic compound (Fig. 2) that is less potent and selective than either thapsigargin or cyclopiazonic acid. For example, tBHQ may interfere with additional aspects of Ca^{2+} signaling dynamics, such as inhibition of receptor-activated plasma membrane Ca^{2+} influx (Foskett and Wong, 1992) or inhibition of the passive Ca^{2+} leak from internal stores (Missiaen *et al.*, 1992). Moreover, tBHQ has not been tested rigorously for its specificity in inhibition of other ion-motive ATPases. Nonetheless, tBHQ is used frequently as an intracellular Ca^{2+} pump inhibitor and exhibits an inhibitory potency similar to that of cyclopiazonic acid, requiring a half-maximal inhibitory concentration of about 0.4 μM for the rabbit SERCA 1a isoform (Wictome *et al.*, 1992). In terms of mechanism of pump inhibition, tBHQ appears to act like thapsigargin by stabilizing the E2 state of the enzyme, reducing the conversion rate to E1 (Wictome *et al.*, 1992). The compound is available from Fluka (Ronkonkoma, New York) and should be prepared as a stock solution in DMSO.

III. Acute Effects of Ca^{2+} Pump Inhibitors on Intact Cells

Thapsigargin has proven very useful for intact cell experiments because of its high membrane permeability. Indeed, the results from its application to living cells have proven quite unexpected and are highly informative about calcium storage dynamics. In virtually every eukaryotic cell tested to date, the application of thapsigargin gives a rapid and remarkably similar response (Table III): an initial elevation of cytosolic $[Ca^{2+}]_i$ within 15 sec to 2 min, followed by a sustained elevation for many minutes. This pattern is shared with cyclopiazonic acid, underscoring that intracellular pump inhibition is responsible for this receptor-independent calcium mobilization. Numerous control experiments have established that thapsigargin-induced intracellular $[Ca^{2+}]_i$ elevation is not the result of ionophoric or detergent activity, nor is it the result of the production of calcium-releasing signals such as inositol 1,4,5-trisphosphate (IP_3) (Jackson et al., 1988). Thus, SERCA-type pump arrest leads paradoxically to a dramatic and long-lasting calcium elevation in the cytosol. Ca^{2+} release induced by thapsigargin treatment in fura-2-loaded cells is initiated roughly 15–20 sec after drug application, a considerably slower latency than hormone-induced responses (1–5 sec). Also, the Ca^{2+}-release profile given by thapsigargin treatment differs from hormone-induced responses because Ca^{2+} may elevate more slowly. Thus, the surprising result that elevation of Ca^{2+} results from intracellular calcium pump inhibition reveals an intriguing picture of Ca^{2+} homeostasis in the resting cell. These results suggest the existence of a dynamic state that balances Ca^{2+} uptake with a constant passive Ca^{2+} leak from internal stores (i.e., the Ca^{2+} pump must be working continuously against a native high Ca^{2+} permeability of the SR/ER membrane). This "pump-leak" concept is depicted in Fig. 6, showing the calcium leak pathway to be discrete from the pump. Thus, although static calcium loading without a leak would appear energetically more efficient, cells maintain their releasable Ca^{2+} stores only by constant energy expenditure. This conclusion is supported by observations that enzymatic depletion of medium ATP by hexokinase–glucose reveals a rapid monotonic calcium leak from microsomes (Meyer and Stryer, 1990; Thastrup et al., 1990). At present, the physical entity responsible for Ca^{2+} efflux from the ER/SR stores remains undefined, but note that the calcium pump itself can exhibit leak properties when arrested and thus can support measurable Ca^{2+} efflux (de Meis and Inesi, 1992). The pump-leak hypothesis leads to the prediction that thapsigargin-arrested Ca^{2+} pumps, on their own, will not generate a positive cytosolic Ca^{2+} signal, but that a separate Ca^{2+} leak site is required also. On this basis, one might expect a dissociation of the coexpression of these gene products. One situation in which this dissociation might occur is in the sea urchin egg, which shows thapsigargin-blocked calcium transport but does not respond to thapsigargin treatment with a calcium elevation (Buck et al., 1992). Similarly, preliminary evidence sug-

Table III
Examples of Cells that Discharge Intracellular Ca^{2+} in Response to Thapsigargin

Cell type	EC_{50}^a (nM)	Ca^{2+} Influx[b]	Reference
Blood and Immune Cells			
Neutrophils	10	Yes	Foder et al. (1989)
Lymphocytes	1–10	Yes	Scharff et al. (1988)
Macrophages	15	Yes	Ohuchi et al. (1988)
Platelets	60	Yes	Thastrup et al. (1987)
HL-60	20	Yes	Demaurex et al. (1992)
Neuronal Cells			
PC12	—	Yes	Clementi et al. (1992)
NG115-401L	20	No	Jackson et al. (1988)
N1E-115	—	No	Ghosh et al. (1991)
Endocrine Cells			
Adrenal chromaffin cells	—	Yes	Cheek et al. (1988)
Adrenal glomerulosa cells	50	Yes	Ely et al. (1991)
Exocrine Cells			
Parotid acinar cells	—	Yes	Takemura et al. (1989)
Lacrimal gland cells	—	Yes	Kwan et al. (1990)
Germ Cells			
Mouse eggs	—	Yes	Kline and Kline (1992)
Sperm	—	?	Blackmore (1993)
Smooth Muscle Cells			
DDT_1MF-2	10	No	Bian et al. (1991)
Vascular smooth muscle	—	Yes	Xuan et al. (1992)
Connective Tissue Cells			
NIH 3T3	—	Yes	Polverino et al. (1991)
Chondrocytes	—	Yes	Benton et al. (1989)
Liver Cells			
Hepatocytes	80	Yes	Thastrup et al. (1990)
Epithelial Cells			
Colonic adenocarcinoma	—	Yes	Brayden et al. (1989)
A431 epidermoid cells	—	Yes	Hughes et al. (1991)

[a] EC_{50} refers to thapsigargin concentration at which 50% of maximal Ca^{2+} release is observed.

[b] Influx refers to whether initial thapsigargin release induces Ca^{2+} entry through plasma membrane.

gests that thapsigargin does not induce calcium release from *Xenopus* oocyte stores (Lechleiter and Clapham, 1992).

Generally, to achieve measurable Ca^{2+} discharge in fura-2- or indo-1-loaded cells, thapsigargin should be used at concentrations ranging from 20 to 100 nM, but frequently is used at higher concentrations (1 µM) to insure maximal Ca^{2+} release. However, a growing, and unwise, tendency to use thapsigargin at levels above 1 µM has emerged. Nonspecific effects, including

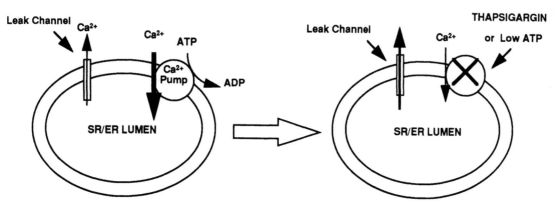

Fig. 6 Schematic diagram illustrating the pump-leak concept; the leak site is represented as a separate channel.

membrane disruption, are likely to be a risk at such heroic concentrations (Tao and Haynes, 1992). In many cells, a significant mismatch exists between SERCA inhibitor concentrations required to elicit calcium discharge rather than pump arrest, because of the efficiency and kinetics of cell penetration by thapsigargin. Thus, calcium discharge from intact cells requires higher concentrations of the reagent. This problem is exacerbated for the more polar reagent cyclopiazonic acid, which can show a 100- to 1000-fold lower potency in stimulating calcium discharge relative to its direct inhibitory activity on SERCA pumps. In fura-2-loaded rat thymocytes, cyclopiazonic acid appears to induce maximal Ca^{2+} discharge at $\sim$30 μM; tBHQ also requires a concentration of 30 μM for maximal Ca^{2+} release (Mason *et al.*, 1991).

The sustained elevation of cytosolic $[Ca^{2+}]_i$ after pump inhibition might be expected to be a direct consequence of blockage of internal calcium sequestration. However, this long-lived elevation is entirely dependent on extracellular calcium, is voltage insensitive, and can be blocked by inorganic ions such as Ni^{2+} or by a new channel blocker specific for this site, SK&F 96365 (Merritt *et al.*, 1990; Demaurex *et al.*, 1992). Thus, the initial transient $[Ca^{2+}]_i$ peak elicited by thapsigargin is caused by release from internal stores whereas the sustained calcium elevation is caused by calcium entry through a novel calcium channel. The results shown in Fig. 7A show representative calcium traces illustrating these two phases. As described earlier, these experiments were performed on fura-2-loaded cells in which the Ca^{2+}-discharge component

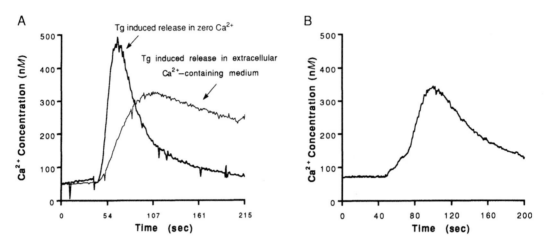

Fig. 7 Response of fura-2 loaded cells to 1 μM thapsigargin. (A) Response of NIH 3T3 cells to 1 μM thapsigargin in the presence and absence of extracellular calcium. (B) Response of NG115-401L cells to 1 μM thapsigargin in presence of extracellular calcium. These experiments were performed using standard fura-2 fluorescence methodology for cell populations (Thomas and Delaville, 1991).

from internal stores was distinguished from influx by repeating the experiments in the presence of 1 mM ethylene glycol bis(β-aminoethylether)-N,N,N',N'-tetraacetic acid (EGTA) to remove extracellular calcium (Fig. 7a). Alternatively, the influx component can be observed separately from Ca^{2+} discharge by inducing Ca^{2+} release with thapsigargin (or other pump inhibitors) in Ca^{2+}-free medium and, subsequently, exposing the enhanced Ca^{2+} permeability of the plasma membrane by adding back 1 mM Ca^{2+}. Another approach relies on the permeability of this capacitative entry pathway in some cells to Mn^{2+}. This technique exploits the ability of Mn^{2+} to quench fura-2 fluorescent signals. Cells loaded with fura-2 are excited at the fura-2 isosbestic wavelength (360 nm) to monitor Ca^{2+}-independent changes in fluorescence. Activated divalent cation entry then can be observed by adding 1 mM Mn^{2+} to the cells while measuring the rate of decrease in fura-2 fluorescence (Demaurex *et al.*, 1992; Glennon *et al.*, 1992).

This $[Ca^{2+}]_i$ response pattern is very similar to that stimulated by calcium-mobilizing surface stimuli, leading to the proposal that calcium entry of this type is gated by depletion of luminal calcium stores. This proposal, termed the "capacitative entry" model of receptor activated calcium influx (Putney, 1986), suggests that reduction of the hormone-releasable store activates the calcium entry channel by direct coupling between the ER and the plasma membrane or by an unknown signal, perhaps a novel diffusible messenger. This voltage-insensitive calcium channel may correspond to I_{CRAC}, a highly selective Ca^{2+} current identified by patch clamping in nonexcitable cells (Hoth and Penner, 1992). Capacitative Ca^{2+} entry now is identified routinely by the appearance of sustained calcium entry after thapsigargin treatment, as suggested earlier (Hanley et al., 1988). In the vast majority of cell types, this biphasic response is observed with SERCA inhibition (Table III). However, Table III also emphasizes that a significant subset of tested cell types does not show such capacitative calcium entry. Figure 7B illustrates one such cell type, the NG115-401L neuronal cell line, in which thapsigargin stimulated a transient $[Ca^{2+}]_i$ peak that returned rapidly to baseline levels and was unaffected by extracellular calcium removal, suggesting no activation of capacitative calcium influx (Jackson et al., 1988). However, these cells do exhibit receptor-coupled activation of calcium entry, as shown by stimulation with a calcium-mobilizing hormone such as bradykinin (Jackson et al., 1987). Consequently, other mechanisms must be invoked to explain receptor-operated calcium entry under these circumstances, for example, gating by a receptor-generated signal. The inositol metabolite (1,3,4,5)P_4 has been proposed as such a gating signal (Irvine, 1991). Indeed, a distinct calcium current that requires IP_4 for gating has been identified by patch clamping (Luckoff and Clapham, 1992), but this channel has distinctive cation selectivity properties and can carry Mn^{2+}, unlike the I_{CRAC} current described earlier. The existence of multiple calcium entry pathways that can be distinguished by their Mn^{2+} permeability, as well as their gating sensitivities (Neher, 1992), has been suggested. This consideration is crucial when using Mn^{2+} quenching to detect capacitative entry, and suggests that multiple experimental designs may be needed to establish the origin of the sustained calcium elevation that follows receptor or thapsigargin activation.

An additional complication in using the SERCA inhibitors interchangeably is that only thapsigargin has a single molecular target in this sequence of events. Cyclopiazonic acid, as noted earlier, interferes to some degree with the ER leak (Missiaen et al., 1992), whereas tBHQ has more pronounced interference with the ER calcium leak (Missiaen et al., 1992) and blocks the capacitative calcium entry channel as well (Foskett and Wong, 1992). Ironically, tBHQ thereby generates a simpler monophasic $[Ca^{2+}]_i$ response that is exclusively attributable to intracellular discharge in cells such as hepatocytes (Kass et al., 1989), in which thapsigargin gives a biphasic response (Thastrup et al., 1990; Mason et al., 1991). This potential for interference by cyclopiazonic acid and tBHQ with more than one site in the calcium mobilization sequence should be considered in the design of experiments on new cell types. However, under some circumstances, generating a single $[Ca^{2+}]_i$ transient may be advantageous in which case tBHQ may prove useful.

===== **IV. Chronic Effects of Ca^{2+} Pump Inhibitors**

A. Growth Modulation

Use of SERCA inhibitors, particularly thapsigargin, is increasing in long-term experimental designs. Note that thapsigargin shows a cell-specific variation in its effects with chronic treatment. In cells of the myeloid cell lineage, such as basophils and lymphocytes, protracted thapsigargin administration elicits a mitogenic response (Ali et al., 1985; Norup et al., 1986; Scharff et al., 1988). On the other hand, central nervous system neurons are exquisitely sensitive to thapsigargin as a neurotoxin (Silverstein and Nelson, 1992). Paradoxically, thapsigargin treatment can rescue sympathetic neurons from some forms of induced cell death (Lampe et al., 1992), suggesting that neurons may be heterogeneous in terms of their response to depletion of calcium stores or sustained cytosolic $[Ca^{2+}]_i$ elevation.

However, in other cells, such as DDT_1MF-2 smooth muscle cells (Ghosh et al., 1991), thapsigargin induces cytostatic growth arrest that can last for days but is fully reversible. Remarkably, only a brief exposure to thapsigargin (30 min) elicited growth arrest for up to 7 days, correlated with a sustained depletion of the IP_3-sensitive pool as judged by the failure of these cells to respond to epinephrine (Ghosh et al., 1991). These quiescent DDT_1MF-2 cells otherwise appeared normal in morphology and exhibited several aspects of normal biochemical integrity, for example, functional mitochondria and resting Ca^{2+} levels.

The complexity of cellular growth responses to thapsigargin predicts a comparable complexity in altered gene expression. Thapsigargin has not been used extensively to study transcriptional control, but the reagent has many advantages over calcium ionophores in generating a controlled calcium signal. For example, thapsigargin has been shown to induce expression of the immediate early genes c-*fos* and c-*jun* in NIH 3T3 cells. The thapsigargin stimulation of an immediate early gene expression response is abolished by clamping cytosolic $[Ca^{2+}]_i$ levels using 1,2-bis(o-aminophenoxy)ethane-N,N,N',N'-tetraacetic acid/acetoxymethyl ester (BAPTA/AM) loading, suggesting that cytosolic $[Ca^{2+}]_i$ elevation, not luminal Ca^{2+} depletion, activates growth-promoting transcription factors (Schonthal et al., 1991).

B. Induction of Stress Responses

Treatment of cells with calcium ionophores has long been appreciated to be a cellular stress (Drummond et al., 1987) to which cells respond with a well-characterized enhancement of expression of a limited number of "stress proteins" against a background of reduced protein synthesis (Lee, 1992). Various perturbations of the physiological state, such as nutrient deprivation, wounding, viral infection, and loss of Ca^{2+} from internal stores, induce the stress response (Lee, 1992). The onset of the cellular stress response is marked by elevated transcription, followed by subsequent enhanced translation, of a characteristic group of glucose-regulated proteins (GRPs) that share considerable sequence and functional homology with the heat-shock proteins. The major GRPs are luminal ER proteins that may act, normally, as chaperones to assist protein folding or, in the stressed state, as detectors for malformed proteins. In addition, these proteins share a common property of low-affinity, high-capacity Ca^{2+} binding, suggesting that they also may act as calcium buffers in the lumen (Koch, 1990).

Thapsigargin treatment first was shown to generate a stress response in chondrocytes (Benton et al., 1989), but a number of subsequent reports noted reduced protein synthesis or other changes that directly or indirectly supported the notion of thapsigargin-induced stress responses (Hanley and Benton, 1990; Preston and Berlin, 1992). This complication is important in the chronic use of thapsigargin, that is, whether induction of a stress response may be the underlying causative factor in cytostatic growth arrest is unclear. Thus, a crucial control in any experiment in which a SERCA inhibitor is administered on a long-term basis is testing for a global reduction in protein synthesis with the parallel induction of, for example, GRP78 and GRP94, using [^{35}S]methionine-labeling of new protein synthesis.

Unexpectedly, the signal for induction of the stress response after Ca^{2+} perturbation is depletion of the Ca^{2+} store, not Ca^{2+} elevation in the cytosol, since neither BAPTA nor EGTA buffering of cytosolic Ca^{2+} prevents the onset of the stress response (Drummond et al., 1987; Hanley and Benton, 1990). Moreover, Ca^{2+} store depletion exerts a distinct regulatory effect on the initiation of protein synthesis, in which the key event is loss of stored Ca^{2+}, not merely elevated cytosolic Ca^{2+} (Brostrom and Brostrom, 1990). In HeLa cells, cytosolic buffering of $[Ca^{2+}]_i$ did not block the thapsigargin-induced inhibition of protein synthesis (Preston and Berlin, 1992). These observations collectively reinforce the concept that at least two sites, the plasma membrane capacitative entry calcium channel and the nuclear stress genes, exist at which levels of calcium in the stores, not levels in the cytosol, constitute the primary signal.

C. Protein Folding and Trafficking

Long-term depletion of luminal Ca^{2+} content by thapsigargin has been shown to influence protein processing and folding events. The ER plays a critical role in protein sorting and assembly, as well as translation; these pathways are sensitive to stored Ca^{2+} levels, as was indicated by the failure of the asialoglycoprotein receptor to undergo proper maturation in the presence of ionophore- or thapsigargin-depleted Ca^{2+} stores (Lodish et al., 1992). This observation may be an indication of a requirement for Ca^{2+} for proper chaperone function in the ER (Suzuki et al., 1991). Thapsigargin treatment also perturbs normal oligosaccharide processing of glycoproteins in the ER (Kuznetsov *et al.*, 1993). Note that some of these effects may be epiphenomenal to distortion or rearrangement of the ER, since thapsigargin treatment has been shown to induce major physical disruption of the ER in some cell types, fragmenting cisternal ER into vesicles (Booth and Koch, 1989). The roles of calcium in the luminal ER compartment in protein targeting and maturation is not well defined, so SERCA inhibitors may be applied increasingly to studies of problems of this type. The initial results indicate that thapsigargin may be highly complementary to other pharmacological tools, such as brefeldin A, used to study protein folding and processing in the ER.

V. Potential Applications of Intracellular Ca^{2+} Pump Inhibitors in Calcium Cell Biology

The major points that influence the experimental use of thapsigargin and other SERCA inhibitors as well as the interpretation of results are (1) the generation of a pure calcium signal that is triggered by discharge from internal stores, but can be extended and amplified by subsequent capacitative calcium entry, and (2) depletion of Ca^{2+} stores, constituting a distinct signaling system from cytosolic elevation. These points are key to understanding the uniqueness, but also the complexities, of SERCA inhibition. In this section, three examples are given briefly to illustrate how SERCA inhibition has been applied to problems in calcium cell biology.

A. Relationships between Calcium Pools

In addition to studies on calcium dynamics in intact cells, thapsigargin has been used with permeabilized cells to provide additional insights into Ca^{2+} store properties, particularly the functional definition of distinct storage compartments. In the smooth muscle cell line DDT_1MF-2, thapsigargin treatment defined at least three distinct storage compartments: (1) an IP_3-sensitive/thapsigargin-insensitive pool, (2) an IP_3-insensitive/thapsigargin-sensitive pool, and (3) a pool insensitive to both IP_3 and thapsigargin, dischargeable only by Ca^{2+} ionophore (Bian *et al.*, 1991). Thapsigargin has been applied in conjunction with other reagents such as caffeine and ryanodine, to functional studies of Ca^{2+} pools. Although, not every cell possesses a ryanodine-sensitive Ca^{2+} store, virtually every cell exposed to thapsigargin initiates Ca^{2+} discharge because of the presence of at least one thapsigargin-sensitive store. Thapsigargin and other SERCA inhibitors have indicated repeatedly that the IP_3-sensitive store constitutes a subset of total thapsigargin-sensitive Ca^{2+} storage compartments (Verma *et al.*, 1990; Bian *et al.*, 1991; Ely *et al.*, 1991; Koshiyama and Tashjian, 1991; Menniti *et al.*, 1991). A frequent and important misconception, however, is that thapsigargin exclusively targets the IP_3-

sensitive store and that a one-to-one correspondence exists between IP_3- and thapsigargin-sensitive storage compartments. In fact, partial, complete, or virtually no overlap between the hormone-sensitive and thapsigargin-sensitive stores has been observed (Bian *et al.,* 1991; Zacchetti *et al.,* 1991; Ely *et al.,* 1991; Koshiyama and Tashjian, 1991; Menniti *et al.,* 1991; Foskett and Wong, 1992).

In practice, experiments on calcium storage pools are performed simply by the sequential addition of multiple hormones or drugs to intact cells to determine whether Ca^{2+} discharge induced by one agent can abolish the effect of the other. In general, thapsigargin pretreatment abolishes the IP_3-releasable pool in the majority of cell types (for examples, see Ely *et al.,* 1991; Menniti *et al.,* 1991). The other commonly used SERCA-type Ca^{2+} pump inhibitors, cyclopiazonic acid and tBHQ, also release Ca^{2+} from IP_3-sensitive stores. No distinction has been found between the stores defined by the use of different SERCA inhibitors, that is, SERCA inhibition defines the same functional store regardless of the inhibitor used.

B. Testing Models of Calcium Oscillation

In many cells, cytoplasmic Ca^{2+} oscillations are initiated after hormonally induced Ca^{2+} discharge; such repetitive Ca^{2+} changes may be a major basis for calcium signaling (Berridge, 1993). Proposed models for Ca^{2+} oscillations and wave propagation (Berridge, 1993) have a universal feature in the necessity to move Ca^{2+} between the cytosol and storage sites by sequestration, a process that is principally, perhaps exclusively, attributable to the action of SERCA pumps (Missiaen *et al.,* 1991). Accordingly, SERCA inhibitors should be powerfully disruptive to induced oscillation, an effect that has been observed (for example, see Kline and Kline, 1992). In rat parotid acinar cells, however, thapsigargin itself can initiate sustained oscillations that are amplified by caffeine and blocked by ryanodine (Foskett *et al.,* 1991). Such observations challenge current notions of calcium movement between pools and may indicate novel mechanisms for ATP-dependent calcium sequestration or, alternatively, noncytosolic communication influencing calcium signaling. One indication that the role of SERCA pumps in the spatial and temporal patterning of calcium signals may have been underestimated comes from results demonstrating that overexpression of SERCA 1a in *Xenopus* oocytes decreased the refractory period, narrowed the width, and increased the frequency of IP_3-induced calcium waves (Camacho and Lechleiter, 1993).

C. Physiological Regulation of SERCA Pumps

The existence of multiple, structurally distinct inhibitors of the SERCA family raises the issue of whether an endogenous counterpart to this pharmacological regulation exists. Specifically, is calcium pump arrest used physiologically to generate or modulate calcium signals? Very little is known about physiological regulation of the SERCA pumps but, in cardiac muscle cells, a specific interaction between a negatively acting regulatory protein, phospholamban, and the SERCA 2a isoform has many parallels with SERCA inhibition by exogenous reagents. Phospholamban is a small (6 kDa) membrane protein that forms a homopentamer in the cardiac SR (Tada and Kadoma, 1989). In the native unphosphorylated state, phospholamban acts as an endogenous pump inhibitor through as yet poorly defined mechanisms. Phospholamban has been reported to exert two inhibitory effects on the heart muscle pump: a lowering of enzyme turnover rate (V_{max}) and a lowering of the Ca^{2+} affinity of the pump (K_{Ca}) (Sasaki *et al.*, 1992). Note, in light of the previous discussion, that phospholamban inhibits the pump by a selective interaction with the E2 state (James *et al.*, 1989), as has been seen with SERCA inhibitors. The inhibition is relieved when phospholamban is phosphorylated in response to surface-receptor-generated signals. Phospholamban contains, for example, a consensus phosphorylation site for cAMP-dependent kinase, which is used physiologically (Tada and Kadoma, 1989). Although phospholamban is an endogenous SERCA inhibitor, is it mechanistically like the inhibitory reagents discussed in this chapter? At this time, such a conclusion seems unlikely for two reasons: (1) phospholamban inhibition has been mapped to a region of the SERCA 2b structure different from the one at which known SERCA inhibitors would be expected to act (James *et al.*, 1989) and (2) SERCA 3 is completely resistant to phospholamban (Toyofuku *et al.*, 1993). Moreover, no counterpart to phospholamban has been found in nonmuscle tissues although, in COS cell expression studies, both SERCA 1a and SERCA 2b were susceptible to regulatory control by phospholamban (Verboomen *et al.*, 1992).

Nevertheless, one of the exciting possibilities arising from the extensive use of SERCA inhibitors is that pump regulation may provide an unsuspected site of physiological control in initiating, extinguishing, or modulating calcium signaling.

Acknowledgments

David Thomas was supported by a National Institutes of Health training grant in molecular and cellular biology.

References

Ali, H., Christensen, S. B., Foreman, J. C., Pearce, F. L., Piotrowski, N., and Thastrup, O. (1985). The ability of thapsigargin and thapsigargicin to activate cells involved in the inflammatory response. *Br. J. Pharmacol.* **85,** 705–712.

Benton, H. P., Jackson, T. R., and Hanley, M. R. (1989). Identification of a novel inflammatory stimulant of chondrocytes: Early events in cell activation by bradykinin receptors on pig articular chondrocytes. *Biochem. J.* **258,** 861–867.

Berridge, M. J. (1993). Inositol trisphosphate and calcium signalling. *Nature (London)* **361**, 315–325.

Bian, J., Ghosh, T., Wang, J. C., and Gill, D. L. (1991). Identification of intracellular calcium pools: Selective modification by thapsigargin. *J. Biol. Chem.* **266**, 8801–8806.

Blackmore, P. F. (1993). Thapsigargin elevates and potentiates the ability of progesterone to increase intracellular free calcium in human sperm. *Cell Calcium* **14**, 53–60.

Booth, C., and Koch, G. (1989). Perturbation of cellular calcium induces secretion of luminal ER proteins. *Cell* **59**, 729–737.

Brandl, C. J., Green, N. M., Korczak, B., and MacLennan, D. H. (1986). Two Ca^{2+} ATPase genes: Homologies and mechanistic implications of deduced amino acid sequences. *Cell* **44**, 597–607.

Brayden, D. J., Hanley, M. R., Thastrup, O., and Cuthbert, A. W. (1989). Thapsigargin, a new calcium dependent epithelial anion secretogogue. *Br. J. Pharmacol.* **98**, 809–816.

Brostrom, C. O., and Brostrom, M. A. (1990). Calcium-dependent regulation of protein synthesis in intact mammalian cells. *Annu. Rev. Physiol.* **52**, 577–590.

Buck, W. R., Rakow, T. L., and Shen, S. S. (1992). Synergistic release of calcium in sea urchin eggs by caffeine and ryanodine. *Exp. Cell Res.* **202**, 59–66.

Burk, S. E., Lytton, J., MacLennan, D. H., and Shull, G. E. (1989). cDNA cloning, functional expression, and mRNA tissue distribution of a third organellar Ca^{2+} pump. *J. Biol. Chem.* **264**, 18,561–18,568.

Camacho, P., and Lechleiter, J. D. (1993). Increased frequency of calcium waves in *Xenopus laevis* oocytes that express a calcium-ATPase. *Science* **260**, 226–229.

Campbell, M. A., Kessler, P. D., and Fambrough, D. M. (1992). The alternative carboxyl termini of avian cardiac and brain sarcoplasmic reticulum/endoplasmic reticulum Ca^{2+}-ATPases are on opposite sides of the membrane. *J. Biol. Chem.* **267**, 9321–9325.

Carafoli, E. (1987). Intracellular calcium homeostasis. *Ann. Rev. Biochem.* **56**, 395–433.

Cheek, T. R., Moreton, R. B., Berridge, M. J., and Thastrup, O. (1988). Effects of the Ca^{2+}-mobilizing tumor promoter thapsigargin and an inositol trisphosphate mobilizing agonist on cytosolic Ca^{2+} in bovine adrenal chromaffin cells. *Biochem. Soc. Trans.* **17**, 94–95.

Christensen, S. B. (1985). Radiolabelling of the histamine liberating sesquiterpene lactone, thapsigargin. *J. Label. Comp. Radiopharm.* **22**, 71–77.

Christensen, S. B., Andersen, A., Lauridsen, A., Moldt, P., Smitt, S. W. and Thastrup, O. (1992). Thapsigargin: A lead to design of drugs with the calcium pump as target. *In* "New Leads and Targets in Drug Research" (P. Krogsgaard-Larsen, S. B. Christensen, and H. Kofod, eds.), pp. 243–252. Copenhagen: Munksgaard.

Chu, A., Dixon, M. C., Saito, A., Seiler, S., and Fleischer, S. (1988). Isolation of sarcoplasmic reticulum fractions referable to longitudinal tubules and junctional terminal cisternae from rabbit skeletal muscle. *Meth. Enzymol.* **157**, 36–46.

Clarke, D. M., Loo, T. W., Inesi, G., and MacLennan, D. H. (1989). Location of high affinity Ca^{2+}-binding sites within the predicted transmembrane domain of the sarcoplasmic reticulum Ca^{2+}-ATPase. *Nature (London)* **339**, 476–478.

Clementi, E., Scheer, H., Zacchetti, D., Fasolato, C., Pozzan, T., and Meldolesi, J. (1992). Receptor-activated influx. *J. Biol. Chem.* **267**, 2164–2172.

Demaurex, N., Lew, D. P., and Krause, K. H. (1992). Cyclopiazonic acid depletes intracellular Ca^{2+} stores and activates an influx pathway for divalent cations in HL-60 cells. *J. Biol. Chem.* **267**, 2318–2324.

de Meis, L., and Inesi, G. (1992). Functional evidence of a transmembrane channel within the Ca^{2+} transport ATPase of sarcoplasmic reticulum. *FEBS Lett.* **299**, 33–35.

Drummond, I. A., Lee, A. S., Resendez, E., Jr., and Steinhardt, R. A. (1987). Depletion of intracellular calcium stores by calcium ionophore A23187 induces the genes for glucose-regulated proteins in hamster fibroblasts. *J. Biol. Chem.* **262**, 12,801–12,805.

Ely, J. A., Ambroz, C., Baukal, A. J., Christensen, S. B., Balla, T., and Catt, K. (1991). Relationship between agonist- and thapsigargin-sensitive calcium pools in adrenal glomerulosa cells. *J. Biol. Chem.* **266**, 18,635–18,641.

Fill, M., and Coronado, R. (1988). Ryanodine receptor channel of sarcoplasmic reticulum. *Trends Neurosci.* **11**, 453–457.

Foder, B., Scharff, O., and Thastrup, O. (1989). Ca^{2+} transients and Mn^{2+} entry in human neutrophils induced by thapsigargin. *Cell Calcium* **10**, 477–490.

Foskett, J. K., and Wong, D. (1992). Calcium oscillations in parotid acinar cells induced by microsomal Ca^{2+}-ATPase inhibition. *Am. J. Physiol.* **262**, C656–C663.

Foskett, J. K., Roifman, C. M., and Wong, D. (1991). Activation of calcium oscillations by thapsigargin in parotid acinar cells. *J. Biol. Chem.* **266**, 2778–2782.

Frandsen, A., and Schousboe, A. (1992). Mobilization of dantrolene-sensitive intracellular calcium pools is involved in the cytotoxicity induced by quisqualate and *N*-methyl-D-aspartate but not by 2-amino-3-(3-hydroxy-5-methylisooxazol-4-yl)propionate and kainate in cultured cerebral cortical neurons. *Proc. Natl. Acad. Sci. U.S.A.* **89**, 2590–2594.

Ghosh, T. K., Bian, J., Short, A. D., Rybak, S. L., and Gill, D. L. (1991). Persistent intracellular calcium pool depletion by thapsigargin and its influence on cell growth. *J. Biol. Chem.* **266**, 24,690–24,697.

Glennon, M. C., Bird, G. St. J., Kwan, C.-Y., and Putney, J. W., Jr. (1992). Actions of vasopressin and the Ca^{2+}-ATPase inhibitor, thapsigargin, on Ca^{2+} signaling in hepatocytes. *J. Biol. Chem.* **267**, 8230–8233.

Goeger, D. E., and Riley, R. T. (1989). Interaction of cyclopiazonic acid with rat skeletal muscle sarcoplasmic reticulum vesicles. *Biochem. Pharmacol.* **38**, 3995–4003.

Hakii, H., Fujiki, H., Suganuma, M., Nakayasu, M., Tahira, T., Sugimura, T., Scheuer, P. J., and Christensen, S. B. (1986). Thapsigargin, a histamine secretagogue, is a non-12-*O*-tetradecanoylphorbol-13-acetate (TPA) type tumor promoter in two-stage mouse skin carcinogenesis. *Cancer Res. Clin. Oncol.* **111**, 177–181.

Hanley, M. R., and Benton, H. P. (1990). Proto-oncogenes, stress proteins, and signalling mechanisms in neural injury and recovery. *In* "Advances in Neural Regeneration Research" (F. J. Seil, ed.), pp. 277–289. New York: Wiley-Liss.

Hanley, M. R., Jackson, T. R., Vallejo, M., Patterson, S. I., Thastrup, O., Lightman, S., Rogers, J., Henderson, G., and Pini, A. (1988). Neural function: Metabolism and actions of inositol metabolites in mammalian brain. *In* "Inositol Lipids and Transmembrane Signalling" (M. J. Berridge and R. H. Michell, eds.), pp. 145–161. London: The Royal Society.

Hoth, M., and Penner, R. (1992). Depletion of intracellular calcium stores activates a calcium current in mast cells. *Nature (London)* **355**, 353–355.

Hughes, A. R., Bird, G. S., Obie, J. F., Thastrup, O., and Putney, J. W., Jr. (1991). Role of inositol (1,4,5) trisphosphate in epidermal growth factor-induced Ca^{2+} signaling in A431 cells. *Mol. Pharmacol.* **40**, 254–262.

Hussain, A., Lewis, D., Sumbilla, C., Lai, L. C., Malera, P. W., and Inesi, G. (1992). Coupled expression of Ca^{2+} transport ATPase and a dihydrofolate reductase selectable marker in a mammalian cell system. *Arch. Biochem. Biophys.* **296**, 539–546.

Imagawa, T., Smith, J. S., Coronado, R., and Campbell, K. P. (1987). Purified ryanodine receptor from skeletal muscle sarcoplasmic reticulum is the Ca^{2+}-permeable pore of the calcium release channel. *J. Biol. Chem.* **262**, 16,636–16,643.

Inesi, G., and Sagara, Y. (1992). Thapsigargin, a high affinity and global inhibitor of intracellular Ca^{2+} transport ATPases. *Arch. Biochem. Biophys.* **298**, 313–317.

Inesi, G., Sumbilla, C., and Kirtley, M. E. (1990). Relationships of molecular structure and function in Ca^{2+}-transport ATPase. *Physiol. Rev.* **70**, 749–760.

Irvine, R. F. (1991). Inositol tetrakisphosphate as a second messenger: Confusions, contradictions, and a potential resolution. *BioEssays* **13**, 419–427.

Ishii, T., Ganjeizadeh, M., and Takeyasu, K. (1993). Construction and expression of thapsigargin-, ouabain-, and Ca^{2+}-sensitive chimeric E1E2-ATPases. *Biophys. J. Abs.* **64**, A332.

Jackson, T. R., Hallam, T. J., Downes, C. P., and Hanley, M. R. (1987). Receptor-coupled events in bradykinin action: Rapid production of inositol phosphates and regulation of cytosolic free Ca^{2+} in a neural cell line. *EMBO J.* **6**, 49–54.

Jackson, T. R., Patterson, S. I., Thastrup, O., and Hanley, M. R. (1988). A novel tumor promoter, thapsigargin, transiently increases cytoplasmic free Ca^{2+} without generation of inositol phosphates in NG115-401L neuronal cells. *Biochem. J.* **253**, 81–86.

James, P., Inui, M., Tada, M., Chiesi, M., and Carafoli, E. (1989). Nature and site of phospholamban regulation of the Ca^{2+} pump of sarcoplasmic reticulum. *Nature (London)* **342**, 90–92.

Jencks, W. P. (1989). How does a calcium pump pump calcium? *J. Biol. Chem.* **264**, 18,855–18,858.

Kass, G. E., Duddy, S. K., Moore, G. A., and Orrenius, S. (1989). 2,5-Di-(tert-butyl)-1,4-benzohydroquinone rapidly elevates cytosolic Ca^{2+} concentration by mobilizing the inositol 1,4,5-trisphosphate-sensitive Ca^{2+} pool. *J. Biol. Chem.* **264**, 15,192–15,198.

Khan, S., Martin, M., Bartsch, H., and Rahimtula, A. D. (1989). Perturbation of liver microsomal calcium homeostasis by ochratoxin A. *Biochem. Pharmacol.* **38**, 67–72.

Kline, D., and Kline, J. T. (1992). Thapsigargin activates a calcium influx pathway in the unfertilized mouse egg and suppresses repetitive calcium transients in the fertilized egg. *J. Biol. Chem.* **267**, 17,264–17,630.

Koch, G. L. E. (1990). The endoplasmic reticulum and calcium storage. *BioEssays* **12**, 527–531.

Koshiyama, H., and Tashjian, A. (1991). Evidence for multiple intracellular calcium pools in GH4C1 cells: Investigations using thapsigargin. *Biochem. Biophys. Res. Commun.* **177**, 551–558.

Kuznetsov, G., Brostrom, M. A., and Brostrom, C. O. (1993). Role of endoplasmic reticulum calcium in oligosaccharide processing of alpha 1-antitrypsin. *J. Biol. Chem.* **268**, 2001–2008.

Kwan, C. Y., Takemura, H., Obie, J. F., Thastrup, O., and Putney, J. W., Jr. (1990). Effects of MeCh, thapsigargin, and La^{3+} on plasmalemmal and intracellular Ca^{2+} transport in lacrimal acinar cells. *Am. J. Physiol.* **258**(6), C1006–C1015.

Lampe, P., Cornbrooks, E., Juhasz, A., Franklin, J., and Johnson, E. (1992). Thapsigargin enhances survival of sympathetic neurons by elevating intracellular calcium concentration ($[Ca^{2+}]_i$). *Soc. Neurosci. Abstr.* **18**, 50.

Lechleiter, J. D., and Clapham, D. E. (1992). Molecular mechanisms of intracellular calcium excitability in *X. laevis* oocytes. *Cell* **69**, 283–294.

Lee, A. S. (1992). Mammalian stress response: Induction of the glucose-regulated protein family. *Curr. Opin. Cell Biol.* **4**, 267–273.

Lodish, H. F., Kong, N., and Wikstrom, L. (1992). Calcium is required for folding of newly made subunits of the asialoglycoprotein receptor within the endoplasmic reticulum. *J. Biol. Chem.* **267**, 12,753–12,760.

Luckoff, A., and Clapham, D. E. (1992). Inositol 1,3,4,5-tetrakisphosphate activates an endothelial Ca^{2+}-permeable channel. *Nature (London)* **355**, 356–358.

Lytton, J., Westlin, M., and Hanley, M. R. (1991). Thapsigargin inhibits the sarcoplasmic or endoplasmic reticulum Ca^{2+}–ATPase family of calcium pumps. *J. Biol. Chem.* **266**, 17,067–17,071.

Lytton, J., Westlin, M., Burk, S. E., Shull, G. E., and MacLennan, D. H. (1992). Functional comparisons between isoforms of the sarcoplasmic or endoplasmic reticulum family of calcium pumps. *J. Biol. Chem.* **267**, 14,483–14,489.

Maruyama, K., and MacLennan, D. H. (1988). Mutation of aspartic acid-351, lysine-352, and lysine-515 alters the Ca^{2+} transport activity of the Ca^{2+}–ATPase expressed in COS-1 cells. *Proc. Natl. Acad. Sci. U.S.A.* **85**, 3314–3318.

Mason, M. J., Garcia-Rodriguez, C., and Grinstein, S. (1991). Coupling between intracellular Ca^{2+} stores and the Ca^{2+} permeability of the plasma membrane. *J. Biol. Chem.* **266**, 20,856–20,862.

McLennan, D. H., Toyofuku, T., and Lytton, J. (1992). Structure–function relationships in sarco-plasmic or endoplasmic reticulum (SERCA) type Ca^{2+} pumps. *In* "Ion-Motive ATPases: Structure, Function, and Regulation" (A. Scarpa, E. Carafoli, and S. Papa, eds.), pp. 1–10. New York: New York Academy of Sciences.

Menniti, F. S., Bird, G. St. J., Takemura, H., Thastrup, O., Potter, B. V. L., and Putney, J. W., Jr. (1991). Mobilization of calcium by inositol trisphosphates from permeabilized rat parotid acinar cells. *J. Biol. Chem.* **266,** 13,646–13,653.

Merritt, J. E., Armstrong, W. P., Benham, C. D., Hallam, T. J., Jacob, R., Jaxa-Chamiec, A., Leigh, B. K., McCarthy, S. A., Moores, K. E., and Rink, T. J. (1990). SK&F 96365, a novel inhibitor of receptor-mediated calcium entry. *Biochem. J.* **271,** 515–522.

Meyer, T., and Stryer, L. (1990). Transient calcium release induced by successive increments of inositol 1,4,5-trisphosphate. *Proc. Natl. Acad. Sci. U.S.A.* **87,** 3841–3845.

Missiaen, L., Wuytack, F., Raeymaekers, L., DeSmedt, H., Droogmans, G., Declerck, I., and Casteels, R. (1991). Ca^{2+} extrusion across plasma membrane and Ca^{2+} uptake by intracellular stores. *Pharmacol. Ther.* **50,** 191–232.

Missiaen, L., De Smedt, H., Droogmans, G., and Casteels, R. (1992). 2,5-Di-(*tert*-butyl)-1,4,-benzohydroquinone and cyclopiazonic acid decrease the Ca^{2+} permeability of endoplasmic reticulum. *Eur. J. Pharmacol.* **227,** 391–394.

Neher, E. (1992). Controls on calcium influx. *Nature (London)* **355,** 298–299.

Norup, E., Smitt, U. W., and Christensen, S. B. (1986). The potencies of thapsigargin and analogues as activators of rat peritoneal mast cells. *Planta Medica 251–255.*

Ochsner, M., Creba, J., Walker, J., Bentley, P., and Muakkassah-Kelly, S. F. (1990). Nafenopin, a hypolipidemic and non-genotoxic hepatocarcinogen increases intracellular calcium and transiently decreases intracellular pH in hepatocytes without generation of inositol phosphates. Biochem. Pharmacol. 40, 2247–2257.

Ohuchi, K., Sugawara, T., Watanabe, M., Hirasawa, N., Tsurrufuji, S., Fujiki, H., Christensen, S. B., and Sugimura, T. (1988). Analysis of the stimulative effect of thapsigargin, a non-TPA-type tumor promoter, on arachidonic acid metabolism in rat peritoneal macrophages. *Br. J. Pharmacol.* **94,** 917–923.

Polverino, A. J., Hughes, B. P., and Barritt, G. J. (1991). Inhibition of Ca^{2+} inflow cases an abrupt cessation of growth-factor-induced repetitive free Ca^{2+} transients in single NIH-3T3 cells. *Biochem. J.* **278,** 849–855.

Preston, S. F., and Berlin, R. D. (1992). An intracellular calcium store regulates protein synthesis in HeLa cells, but it is not the hormone-sensitive store. *Cell Calcium* **13,** 303–312.

Putney, J. W., Jr. (1986). A model for receptor-regulated calcium entry. *Cell Calcium* **7,** 1–12.

Rossier, M. F., and Putney, J. W., Jr. (1991). The identity of the calcium-storing, inositol 1,4,5-trisphosphate-sensitive organelle in non-muscle cells: Calciosome, endoplasmic reticulum, . . . or both? *Trends Neurosci.* **14,** 310–314.

Sagara, Y., Wade, J. B., and Inesi, G. (1992). A conformational mechanism for formation of a dead-end complex by the sarcoplasmic reticulum ATPase with thapsigargin. *J. Biol. Chem.* **267,** 1286–1292.

Sasaki, T., Inui, M., Kimura, Y., Kuzuya, T., and Tada, M. (1992). Molecular mechanism of regulation of Ca^{2+} pump ATPase by phospholamban in cardiac sarcoplasmic reticulum. *J. Biol. Chem.* **267,** 1674–1679.

Scharff, O., Foder, B., Thastrup, O., Hoffman, B., Moller, J., Ryder, L. P., Jacobsen, K. P., Langhoff, E., Dickmeiss, E., Christensen, S. B., Skinhoj, P., and Svejgaard, A. (1988). Effect of thapsigargin on cytoplasmic Ca^{2+} and proliferation of human lymphocytes in relation to AIDS. *Biochim. Biophys. Acta* **972,** 257–264.

Schonthal, A., Sugarman, J., Heller-Brown, J., Hanley, M. R., and Feramisco, J. P. (1991). Regulation of c-*fos* and c-*jun* protooncogene expression by the Ca^{2+}-ATPase inhibitor thapsigargin *Proc. Natl. Acad. Sci. U.S.A.* **88,** 7096–7100.

Seidler, N. W., Jona, I., Vegh, M., and Martonosi, A. (1989). Cyclopiazonic acid is a specific inhibitor of the Ca^{2+}–ATPase of sarcoplasmic reticulum. *J. Biol. Chem.* **264**, 17,816–17,823.

Silverstein, F. S., and Nelson, C. (1992). The microsomal calcium ATPase inhibitor thapsigargin is a neurotoxin in perinatal rodent brain. *Neurosci. Lett.* **145**, 157–160.

Sumbilla, C., Lu, L., Sagara, Y., and Inesi, G. (1993). Specific Ca^{2+} activation and thapsigargin inhibition of Ca^{2+} and Na^+,K^+ ATPase chimeras. *Biophys. J. Abstr.* **64**, A335.

Suzuki, C. K., Bonifacino, J. S., Lin, A. Y., Davis, M. M., and Klausner, R. D. (1991). Regulating the retention of T-cell receptor alpha chain variants within the endoplasmic reticulum: Ca^{2+}-dependent association with BiP. *J. Cell Biol.* **114**, 189–205.

Tada, M., and Kadoma, M. (1989). Regulation of the Ca^{2+} pump ATPase by cAMP-dependent phosphorylation of phospholamban. *BioEssays* **10**, 157–163.

Takemura, H., Hughes, A. R., Thastrup, O., and Putney, J. W., Jr. (1989). Activation of calcium entry by the tumor promoter thapsigargin in parotid acinar cells. *J. Biol. Chem.* **264**, 12,266–12,271.

Tao, J., and Haynes, D. H. (1992). Actions of thapsigargin on the Ca^{2+}-handling systems of the human platelet. *J. Biol. Chem.* **267**, 24,972–24,992.

Thastrup, O., Foder, B., and Scharff, O. (1987). The calcium mobilizing and tumor promoting agent, thapsigargin, elevates the platelet cytoplasmic free calcium concentration to a higher steady state level. A possible mechanism of action for the tumor promotion. *Biochem. Biophys. Res. Commun.* **142**, 654–660.

Thastrup, O., Cullen, P. J., Drobak, B. K., Hanley, M. R., and Dawson, A. P. (1990). Thapsigargin, a tumor promoter, discharges intracellular Ca^{2+} stores by specific inhibition of the endoplasmic reticulum Ca^{2+}–ATPase. *Proc. Natl. Acad. Sci. U.S.A.* **87**, 2466–2470.

Thomas, A. P., and Delaville, F. (1991). The use of fluorescent indicators for measurements of cytosolic-free calcium concentration in cell populations and single cells. *In* ''Cellular Calcium: A Practical Approach'' (J. G. McCormack and P. H. Cobbold, eds.), pp. 1–54. Oxford: IRL Press.

Toyofuku, T., Kurzydlowski, K., Lytton, J., and MacLennan, D. H. (1992). The nucleotide binding/hinge domain plays a crucial role in determining isoform-specific Ca^{2+} dependence of organellar Ca^{2+}–ATPases. *J. Biol. Chem.* **267**, 14,490–14,496.

Toyofuku, T. Kurzydlowski, K., Tada, M., and McLennan, D. H. (1993). Identification of regions in the Ca^{2+}–ATPase of sarcoplasmic reticulum that affect functional association with phospholamban. *J. Biol. Chem.* **268**, 2809–2815.

Tsien, R. W., and Tsien, R. Y. (1990). Calcium channels, stores, and oscillations. *Annu. Rev. Cell Biol.* **6**, 715–760.

Verboomen, H., Wuytack, F., De Smedt, H., Himpens, B., and Casteels, R. (1992). Functional difference between SERCA 2a and SERCA 2b Ca^{2+} pumps and their modulation by phospholamban. *Biochem. J.* **286**, 591–596.

Verma, A., Hirsch, D. J., Hanley, M. R., Thastrup, O., Christensen, S. B., and Snyder, S. H. (1990). Inositol trisphosphate and thapsigargin discriminate endoplasmic reticulum stores of calcium in rat brain. *Biochem. Biophys. Res. Commun.* **172**, 811–816.

Wictome, M., Michelangeli, F., Lee, A. G., and East, M. J. (1992). The inhibitors thapsigargin and 2,5-di(*tert*-butyl)-1,4-benzohydroquinone favour the E2 forms of the Ca^{2+},Mg^{2+}–ATPase. *FEBS Lett.* **304**, 109–113.

Xuan, Y. T., Wang, O. L., and Whorton, A. R. (1992). Thapsigargin stimulates Ca^{2+} entry in vascular smooth muscle cells: Nicardipine-sensitive and -insensitive pathways. *Am. J. Physiol.* **262(5)**, C1258–C1265.

Zacchetti, D., Clementi, E., Fasolato, C., Lorenzon, P., Zottini, M., Grohovaz, F., Fumagalli, G., Pozzan, T., and Meldolesi, J. (1991). Intracellular Ca^{2+} pools in PC12 cells. *J. Biol. Chem.* **266**, 20,152–20,158.

PART II

Microelectrode Techniques for Measuring $[Ca^{2+}]_i$ and Ca^{2+} Fluxes

CHAPTER 4

How to Make and Use Calcium-Specific Mini- and Microelectrodes

Stéphane Baudet,★ Leis Hove-Madsen,† and Donald M. Bers‡

★ Laboratoire de Cardiologie Expérimentale
URA CNRS 1340
Hôpital Laënnec BP1005
44035 Nantes, France
† Cardiologie Cellulaire et Moléculaire
Faculté de Pharmacie
INSERM 92-11
F-92296 Chatenay-Malabry, France
‡ Department of Physiology
Stritch School of Medicine
Loyola University at Chicago
Maywood, Illinois 60153

I. Introduction

Detection and measurement of intracellular calcium concentration ($[Ca^{2+}]_i$) have relied on various methods, the popularity of which depends on their ease of use and applicability to different cell types. Historically, Ca^{2+}-selective electrodes have been used concomitantly with absorption indicators such as arsenazo III, but interest in these indicators has been eclipsed by the introduction of fluorescent probes such as fura-2 and indo-1 (Grynkiewicz *et al.*, 1985). In this chapter, we emphasize the utility of Ca^{2+}-selective electrodes and show that their use is complementary to use of fluorescent indicators; indeed, each method has advantages and disadvantages. We first describe the construction and use of Ca^{2+}-selective minielectrodes, based on the Ca^{2+} ligand ETH 129 (Schefer *et al.*, 1986). We have used these electrodes in several applications. The second part of the chapter is dedicated to ETH 129-based Ca^{2+}-selective microelectrodes (MEs), which we have been using to determine $[Ca^{2+}]_i$ in cardiac cells. Since numerous reviews and books have been dedicated to the theoretical aspects of ion-selective ME principles and technology, this chapter is not intended for investigators who have no experience with MEs. For more basic reference to electrode technology and electrophysiology, we suggest monographs by Thomas (1982), Purves (1981), and Ammann (1986).

II. General Considerations

A. Implications of Main Characteristics of Ca^{2+}-Selective Electrodes

The advantage of the Ca^{2+}-selective electrodes is the wide dynamic range of their response (e.g., from pCa 9 to 1) compared, for example, with the ranges of fluorescent and metallochromic Ca^{2+} indicators (Fig. 1). The response of these electrodes is based on a semi-empirical equation (Nicolski–Eisenman equation) derived from the Nernst equation:

$$E_x = E_0 + RT/Z_xF \ln(a_x + K_{xy}^{pot}a_y^{z_x/z_y}) \tag{1}$$

where E_x is the ion-selective electrode potential, E_0 is a constant, R is the molar gas constant, T is the temperature in °K, Z is the valence of the ion, F is the Faraday constant, a_x is the activity of the ion that is measured (activity, a, is related to concentration, C, by the relationship $a = \gamma C$, where γ is the activity coefficient), and K_{xy}^{pot} is the selectivity coefficient. This expression is strictly valid for activities only, but if the activity coefficients do not change, the equation can be used with free concentrations also. This adaptation is often for convenience, since solutions and chemical equilibria are described more often in these concentrations term. So, if x is Ca^{2+} and y is Na^+ (the most common interfering cation for the Ca^{2+}-selective ligand), the relationship becomes, for 30°C and changing to $\log_{10}$:

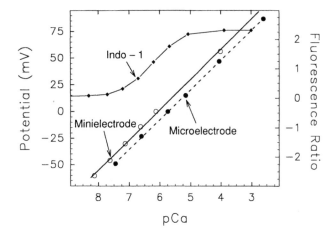

Fig. 1 Dynamic range of Ca^{2+}-selective electrodes and indo-1. The electrode potential of Ca^{2+}-selective mini- and microelectrodes is shown with the fluorescence ratio (400/470) for indo-1. Measurements were performed in a KCl buffer containing 140 mM KCl, 10 mM HEPES, 10 mM NaCl, and 1 mM EGTA. Notice that indo-1 is suitable for measurements between pCa 7.5 and 5, whereas the dynamic range for the Ca^{2+} electrode is wider, ranging from pCa 9 to 1 for minielectrodes and 7.5 to 1 for microelectrodes.

$$E_{Ca} = E_0 + 30 \log([Ca^{2+}] + K^{pot}_{NaCa}[Na^+]^2) \qquad (2)$$

Thus, for the case of no interfering ion and extracellular $Ca^{2+} = [Ca^{2+}]_{ref}$, the potential difference (ΔE) between two solutions of different $[Ca^{2+}]$ reduces to the Nernst equation:

$$\Delta E = 30 \log([Ca^{2+}]/[Ca^{2+}]_{ref}) \qquad (3)$$

The response of an ion-selective electrode to changes in free $[Ca^{2+}]$ is much slower than that of the fluorescent, bioluminescent, and metallochromic Ca^{2+} indicators. Thus, Ca^{2+} electrodes are ideal for measurements of slow changes over wide ranges of $[Ca^{2+}]$, but not as good for very rapid changes of free $[Ca^{2+}]$ (although the Ca^{2+} electrodes can respond in the millisecond range at higher $[Ca^{2+}]$; Bers, 1983).

Finally, as long as the Ca^{2+} ligand ETH 1001 (Ammann, 1986) was the only one available, determinations of $[Ca^{2+}]$ at less than 1 μM were imprecise because the ligand had a very poor sensitivity below this $[Ca^{2+}]$ value. That is, the slope of the calibration curve tended to flatten out. About the time that indo-1 and fura-2 were synthesized (Grynkiewicz *et al.*, 1985), a ligand of improved selectivity (ETH 129) was introduced (Schefer *et al.*, 1986), making measurement of low Ca^{2+} levels more realistic since the electrode response was Nernstian to pCa 8–9 (depending on the ionic background). However, the availability and popularity of the fluorescent indicators have limited further use and characterization of ETH 129.

B. Ca²⁺-Selective Electrodes and Fluorescent Indicators

We will not present an extensive review of the advantages and disadvantages of Ca^{2+}-selective electrodes over fluorescent indicators, but will show how, in the laboratory, we have combined the advantages of both approaches. For example, one problem with indicators is the conversion of the fluorescent signal to actual $[Ca^{2+}]$. In fact, *in vitro* calibration curves have been shown not to be applicable to *in vivo* conditions because indicators have been shown to bind to intracellular constituents, modifying their excitation/emission spectrum and decreasing the apparent affinity of Ca^{2+} for the probe (Konishi *et al.,* 1988; Blatter and Wier, 1992; Hove-Madsen and Bers, 1992). *In vivo* calibration curves are even more difficult to obtain and are highly dependent on the cell type under study. Ca^{2+}-selective electrodes are still among the most straightforward techniques used to measure and quantify Ca^{2+} because (1) their behavior is not altered appreciably by the intracellular milieu (ions, proteins); (2) they can be included easily in an electrophysiological setup and do not require the extensive and expensive apparatus used for fluorescence measurements; (3) their linear (Nernstian) response simplifies the conversion of the voltage signal to $[Ca^{2+}]_i$ (4) their behavior allows determination of wide ranges of pCa (see previous discussion); and (5) the Ca^{2+} ligand itself does not change (or buffer) $[Ca]_i$.

Ca^{2+}-selective electrodes have been used to prepare calibration solutions for Ca^{2+} determinations and measurements of dissociation constants for Ca^{2+}-binding compounds such as ethylene glycol bis(β-amino ethyl ether)-N,N,N',N'-tetraacetic acid (EGTA), indo-1, 1,2-bis(*o*-aminophenoxy)ethane-N,N,N',N'-tetraacetic acid (BAPTA), and oxalate under different experimental conditions (Bers, 1982; Harrison and Bers, 1987,1989; Hove-Madsen and Bers, 1992,1993a). We also have used the Ca^{2+} electrodes to measure cellular Ca^{2+} buffering and changes in the free $[Ca^{2+}]$ in cell suspensions (Hove-Madsen and Bers, 1993a,b). Ca^{2+} electrodes are easy to use to measure free $[Ca^{2+}]$ in experimental solutions. Indeed, for individuals studying Ca^{2+}-dependent processes, no practical reason exists for Ca^{2+} electrodes not to be used as routinely as pH electrodes.

III. Minielectrodes

A. Preparation

Ca^{2+}-selective minielectrodes can be prepared by dipping polyethylene (PE) tubes (typically ~5 cm) in a membrane solution (see composition in a subsequent section). We have tried other types of tubing but polyvinyl chloride (PVC) tubing appears to absorb ETH 129 from the Ca^{2+}-selective membrane, resulting in a faster loss of sensitivity than with PE tubing. On the other hand, materials such as Teflon tubing absorb little ETH 129, but the PVC membrane does not adhere well to the tubing. As a result, the electrodes are damaged more easily, although they may have a longer lifetime if handled with care.

The dimensions of the tubing vary from a diameter less than 1 mm to ~3 mm. Electrodes prepared with the membrane solution described subsequently result in Ca^{2+}-selective membranes that are a few hundred μm thick. With diameters larger than 5 mm, the Ca^{2+}-selective membrane bursts more easily during handling, but this problem may be overcome by inserting a ceramic plug into the tubing before dipping it into the membrane solution, as described by Orchard *et al.* (1991). For general purposes, we have used inner electrode diameters of 1.67 mm (PE 240; Clay Adams, Parsippany, NJ).

After dipping the PE tubing in the membrane solution, the Ca^{2+}-selective membrane is allowed to dry overnight. Then the electrode is filled with an appropriate filling solution which should correspond to experimental conditions (see subsequent section). After filling the electrode, allow it to equilibrate for at least 3 days in a glass vial containing the filling solution (but see subsequent discussion).

B. Preparation and Use of the Ca^{2+}-Selective Ligand

The Ca^{2+}-selective membrane can be prepared as described by Schefer *et al.* (1986; Table I). ETH 129 is dissolved in *N*-phenyl-octyl ether (NPOE) under vigorous stirring in a small glass vial (Solution 1). At the same time, PVC is dissolved in tetrahydrofuran (THF); when completely dissolved, potassium tetrakis chlorophenyl borate (TCPB) is added (Solution 2). When the components of solutions 1 and 2 are dissolved completely, the two solutions are mixed; the membrane solution is ready to use, or it can be stored in a glass vial that is closed with a Teflon screw cap and protected from light. If THF evaporates from the membrane solution during storage, a small amount of THF can be added to achieve the desired viscosity of the membrane solution.

Table I
Preparation of the PVC-Based
Ca^{2+}-Selective Ligand
for Minielectrodes[a]

Component	Amount
Solution 1	
ETH 129	25 mg
N-Phenyl-octyl ether (NPOE)	451.5 μl
Solution 2	
Polyvinyl chloride (PVC)	250 mg
Potassium tetrakis chlorophenyl borate (TCPB)	12.9 mg
Tetrahydrofuran (THF)	~5 ml

[a] All solutions are obtained from Fluka Chemical Corporation (Ronkonkoma, New York).

The filling solution used for the minielectrode depends on the experimental solutions. Generally, the ionic composition should mimic the environment in which measurements of Ca^{2+} are planned; the $[Ca^{2+}]$ of the filling solution also should be in the range of the measured values. With measurements of low $[Ca^{2+}]$, Ca^{2+} in the filling solution can be buffered with EGTA to the desired free $[Ca^{2+}]$. However, we have not obtained good results with filling solutions with a pCa higher than 7.5. We typically use a Ca^{2+}-EGTA buffer of 1 μM as a filling solution.

C. Electrode Characteristics

The resistance of the minielectrodes is 1–2 MΩ, which normally makes it possible to use a standard pH/ion meter to monitor the electrode potential. We have used either a commercial pH/ion meter (Orion pH/ion analyzer, Boston, MA) or a preamplifier (A311J; Analog Devices, Norwood, MA). Using commercial pH/ion meters normally necessitates an adapter cable that connects the minielectrode to the meter input. We have used a chart recorder or an electronic data acquisition device for continuous monitoring of the electrode potential.

The lifetime of the minielectrode depends on the $[Ca^{2+}]$ the electrode is used to measure, and on the composition of the experimental solutions. For measurements in solutions without protein or interfering ions, the detection limit for the electrodes increases slowly with time; the electrodes will have a detection limit in the subnanomolar range and a Nernstian response down to 10 nM for at least 1 month after preparation of the electrodes. Figure 2 shows the change in the response of a Ca^{2+} electrode with time. The electrode response was still "super-Nernstian" at low $[Ca^{2+}]$ (i.e., >29 mV per 10-fold change in $[Ca^{2+}]$) 2 days after filling, but normalized within 7 days of filling. Note that we still obtained a Nernstian response down to pCa 8 for 2 months after filling the electrode. The response time at low $[Ca^{2+}]$, however, slowed with time; measurements below 30 nM are only practical with fairly fresh electrodes. We generally fill electrodes once a week to obtain the best results. However, if the Ca^{2+} electrodes are used to measure micromolar or higher $[Ca^{2+}]$ in protein-free solutions, the same electrode can be used for longer periods (up to several months).

In the presence of cellular proteins, a small offset in the electrode response is seen at the first exposure to protein. Then, no further alteration of the electrode response occurs, but the response time of the Ca^{2+} electrode is increased after exposure to protein. Measurements of free $[Ca^{2+}]$ below 10 nM are more difficult in the presence of protein concentrations higher than 10 mg/ml.

The response time of the Ca^{2+} electrodes can be of critical importance in some applications. Figure 3 compares the response times of a Ca^{2+} electrode and indo-1 fluorescence to a decrease in the $[Ca^{2+}]$ in a suspension of permeabilized myocytes (3 mg/ml), in which cellular Ca^{2+} uptake processes have been blocked with thapsigargin and ruthenium red. We examined the response to a decrease in $[Ca^{2+}]$, since this test may be more stringent than an increase in $[Ca^{2+}]$. Note

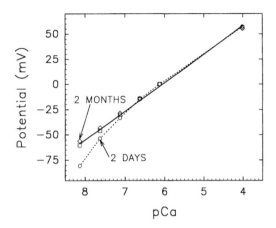

Fig. 2 Electrode potential of a Ca^{2+}-selective minielectrode, 2, 7, and 60 days after filling. Measurements were performed in a KCl buffer as in Fig. 1. Notice the "super-Nernstian" response of electrodes 2 days after filling (circles). Electrode response was linear down to a free $[Ca^{2+}]$ of less than 10 nM, 7 days after filling (squares); The slope of the regression line was -28.4 mV/pCa. Electrode response was linear to pCa 7.5, 2 months after filling (diamonds), but response time was slowed at high pCa.

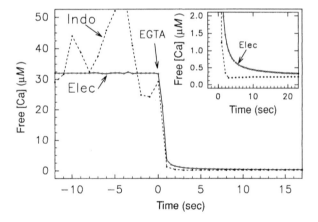

Fig. 3 Comparison of response time of a Ca^{2+}-selective minielectrode and indo-1. Free $[Ca^{2+}]$ was measured in a suspension of permeabilized rabbit ventricular myocytes. Ca^{2+} uptake into sarcoplasmic reticulum and mitochondria was inhibited with thapsigargin and ruthenium red, respectively. The initial free $[Ca^{2+}]$ was 32 μM, which is near saturation for indo-1, resulting in a very noisy trace because a small change in fluorescence ratio corresponds to a large change in $[Ca^{2+}]$ at this level (see Fig. 1). At the arrow, 2 mM EGTA was added to the cell suspension to lower free $[Ca^{2+}]$. Both the electrode and the indo-1 signal were more than 90% complete in 1 sec. *Inset* The response of indo-1 and the Ca^{2+} electrode at low $[Ca^{2+}]$. Notice that the indo-1 signal was 100% complete in 2 sec (and actually undershot slightly) whereas the final completion of the electrode response was slower.

that, when 2 mM EGTA was added to lower the free [Ca^{2+}], the electrode response was 94% complete in 1 sec whereas the indo-1 signal was 97% complete in 1 sec. A slow final phase, lasting several seconds, is apparent in the electrode signal only (see amplified inset).

In Fig. 3, the response time was examined under experimental conditions in which spatial inhomogeneities in the myocyte suspension are minimized by buffering Ca^{2+} with indo-1 and oxalate. However, under some experimental conditions, an apparently slower electrode response may result from inhomogeneities rather than from a slower electrode response per se. This effect is illustrated in Fig. 4, for which free [Ca^{2+}] was monitored with a Ca^{2+} electrode and indo-1, simultaneously, in a myocyte suspension in the absence and presence of oxalate. In Fig. 4A, Ca^{2+} addition to the cells causes a rapid increase in [Ca^{2+}], which subsequently is sequestered by the sarcoplasmic reticulum (SR). In the absence of 10 mM oxalate (Fig. 4A), the electrode response appears to be slower than the corresponding indo-1 signal. However, when oxalate is added subsequently to the cell suspension (Fig. 4B), the measured change in free [Ca^{2+}] after a Ca^{2+} addition is similar for indo-1 and the Ca^{2+} electrode. Note that oxalate not only buffers the free [Ca^{2+}], but also

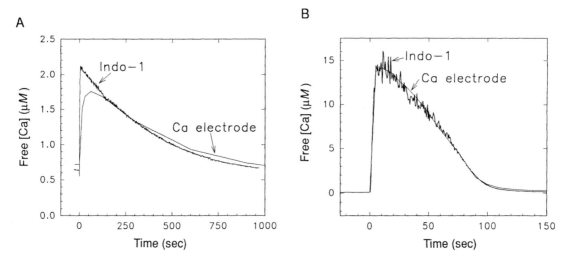

Fig. 4 Effect of inhomogeneities in [Ca^{2+}] on a Ca^{2+}-selective minielectrode and indo-1 response in permeabilized rabbit ventricular myocytes. (A) Simultaneous measurements of Ca^{2+} uptake in digitonin-permeabilized myocytes with both a Ca^{2+}-selective minielectrode and indo-1. Ca^{2+} uptake in mitochondria was inhibited with ruthenium red. Ca^{2+} was added at time 0 and was accumulated largely by the SR. Under these conditions, the response of the Ca^{2+} electrode was slower than that of indo-1. (B) Ca^{2+} uptake enhanced by oxalate, which precipitates intra-SR Ca^{2+} and thereby prevents build-up of a Ca^{2+} gradient. Notice that Ca^{2+} uptake is much faster in the presence of oxalate, with no apparent difference between electrode and indo-1 response. (Reproduced from Hove-Madsen and Bers, 1993a, with permission.)

increases the Ca^{2+} uptake rate in the SR and, thereby, the removal of Ca^{2+} from the cell suspension. Thus, despite inducing a faster rate of change in free $[Ca^{2+}]$, oxalate eliminates the difference between the Ca^{2+} electrode and the indo-1 signal by eliminating spatial inhomogeneities in free $[Ca^{2+}]$ in the myocyte suspension. Indeed, indo-1 is expected to be less sensitive to spatial inhomogeneities as it diffuses into the permeabilized cells and binds to cellular proteins (Hove-Madsen and Bers, 1992,1993a). In contrast, the Ca^{2+} electrode can measure only the Ca^{2+} outside the permeabilized cells; inhomogeneities during uptake or release of Ca^{2+} from the cells are therefore likely to occur, resulting in erroneous measurements with the Ca^{2+} electrode.

D. Storage

After PE tubes have been dipped in an ETH 129 membrane solution and been allowed to dry overnight, the dry electrodes can be stored in a closed glass vial for long periods of time. We have filled minielectrodes that had been stored for 3 yr; the electrodes made with PE tubing still had a resistance of 1–2 MΩ with a linear response down to less than 10 nM Ca^{2+} after filling. Electrodes made with PVC tubing had higher resistance ($\sim$50 MΩ) but were also functional, although slower and less sensitive. Storage of the electrodes in plastic vials results in "Ca^{2+}-selective plastic containers," since the ETH 129 slowly diffuses into the container. Once the Ca^{2+} electrodes are filled with the filling solution, however, the response time increases and the electrodes gradually lose sensitivity.

E. Application

Minielectrodes can be used for a number of purposes. The most straightforward application is the preparation of solutions in which Ca^{2+} is buffered with chelators such as EGTA, ethylene diamine tetraacetic acid (EDTA), or BAPTA, as described by Bers (1982). We have developed a spreadsheet that allows calculation of the actual pCa of these solutions, based on the Nernstian response of the minielectrodes (see Chapter 1). Further, the spreadsheet allows determinations of the K_d and the purity of the Ca^{2+} chelator used to prepare the solution. Thus, we have used the minielectrodes to determine the K_d for EGTA, BAPTA, and oxalate in buffer solutions (Bers, 1982; Harrison and Bers, 1987, 1989; Hove-Madsen and Bers, 1993a). A more complete program to calculate the amount of Ca^{2+} and Ca^{2+} buffer needed to prepare the solutions was developed by C. W. Patton (Hopkins Marine Station, Stanford University; cf. Chapter 1).

We also have used the minielectrodes to characterize the binding of Ca^{2+} to indo-1 *in vitro* and in cell suspensions, to calibrate the indo-1 signal when used in cell suspensions (Hove-Madsen and Bers, 1992). In agreement with previous studies of fura-2 (Konishi *et al.*, 1988), we found that indo-1 binds extensively to cellular proteins and causes a ~4-fold increase in the K_d for Ca^{2+}–indo-1 in permeabilized myocytes.

We have used the minielectrodes to titrate the passive Ca^{2+} binding sites in permeabilized myocytes when the cellular Ca^{2+} uptake and release processes were inhibited. Using the same titration method, we have measured total Ca^{2+} uptake in the SR in permeabilized myocytes by inhibiting Ca^{2+} uptake in the mitochondria and release of Ca^{2+} through the SR Ca^{2+} release channels (Hove-Madsen and Bers, 1993a).

Finally, we have used the minielectrodes in conjunction with indo-1 for on-line measurements of the Ca^{2+} uptake rate in the SR in permeabilized ventricular myocytes, and we have determined the inhibition of Ca^{2+} uptake by thapsigargin to measure the number of SR Ca^{2+} pump sites (Hove-Madsen and Bers, 1993b). In experiments measuring Ca^{2+} uptake rates, caution should be taken when using minielectrodes because of the possibilities of inhomogeneities in $[Ca^{2+}]$ in cell suspensions and slowing of the electrode response.

IV. Microelectrodes

A. Preparation of Glass Tubing

Initially we used filamented, thick-walled glass to make the electrodes because we thought it might facilitate ME filling with the conducting electrolyte. Although this idea was valid, the presence of the inside filament may contribute to a larger tip diameter, which may increase cell damage during impalements. Moreover, the presence of the filament may create an electrical shunt that would hamper the ME performance (Ammann, 1986). Therefore, we subsequently used nonfilamented capillaries, which also worked well (150-μm outer diameter, 15 cm long; Clark Electromedical Instruments, Reading, UK; World Precision Instruments (WPI), Sarasota, FL). The glass is cut in the middle with a diamond pen or glass scorer. Care must be taken to put the micropipettes on a clean surface during the cutting procedure and to hold them by the ends to minimize deposition of dirt. Both ends of the glass tubes (now 7.5 cm long) are fire-polished lightly; a whole batch can be prepared and kept in a small glass beaker, preferably in a dust-proof container.

Cleaning the micropipettes prior to pulling has been a matter of debate but, since 99.8% of the glass after pulling is newly exposed (Deyhimi and Coles, 1982), we do not find such procedures to be necessary.

B. Pulling and Silanization

Classically, two organizational strategies are used to prepare MEs: the "on-demand" and the "batch" methods. The first one consists of preparing (i.e., pulling and silanizing) MEs on the experimental day, which has the obvious advantage of having freshly prepared electrodes and being able to manufacture as many as desired each day. Protocols for this purpose have been described already (reviewed by Ammann, 1986). The other method, which we have chosen, is probably the most common and consists of preparing a batch of MEs that can be kept for several days in a dry and dust-proof container.

MEs are pulled on a programmable horizontal stage puller (Model P80 PC; Sutter Instruments, Novato, California). By trial and error, and according to the type of biological preparations studied, a satisfactory shape of ME can be found. However, several aspects of ME pulling must be taken into account when designing its shape. Since ion-selective MEs have an intrinsically high resistance (i.e., >50–100 GΩ), the signal-to-noise ratio must be minimized, in one of two ways. The first, and most obvious, is increasing the tip diameter, within certain limits that are dictated by the size of the cell type. In our case, we impale cardiac cells in whole muscle. Cell length is typically between 50 and 150 μm and tip diameters less than 0.5 μm are needed. However, a trade-off exists in increasing the tip diameter and the detection limit of the electrode (see subsequent discussion), that is, increasing tip diameter improves electrode response (in terms of detection limit and speed of response) but is more likely to damage the cell during impalement. Another simple way to decrease the resistance is to decrease the length of the ME shank. This procedure further reduces the capacitative artifacts that are encountered when the level of physiological solution fluctuates in the experimental bath (Vaughan-Jones and Kaila, 1986), and helps electrolyte filling as well (see subsequent discussion). In contracting muscular preparations, the shank also should possess some flexibility to avoid dislodging the ME during a contraction. Again, by trial and error, an adequate shape that fulfills all these requirements can be found. The shape of MEs was designed to reduce their resistance by making them strongly tapered, having a shank length of approximately 150 μm and diameter of 20 μm (10 μm above the tip). Under light microscopic observation, the tip diameter was estimated to be ~0.5 μm.

Microelectrodes are dehydrated, tip up in an aluminum block, at 200°C for 12 hr. Our experience, in accordance with results from Vaughan-Jones and Wu (1990), has been that better silanization and a longer lifetime of the MEs were achieved this way. The silanization protocol consists of spritzing 300 μl of N,N-dimethyltrimethylsilylamine (Fluka, Ronkonkoma, New York) onto the aluminum block and rapidly placing a glass lid on top of the dish. Care must be taken not to inhale vapor from the silane vial or during its introduction into the baking dish. In our laboratory, once opened, the silane vial can be kept for at least 2 months without losing its properties. The silanization procedure lasts 90 min. Then the lid is removed and the MEs are baked for another 60 min, which drives off the excess silane vapor. The aluminum block and the MEs then are placed into an air-tight plastic container, also containing desiccant. We advise not keeping the MEs more than 1 week, because repetitive openings of the container and insufficient seal quality will cause the electrodes to lose their hydrophobicity progressively.

C. Electrolyte Filling

The filling or conducting electrolyte is introduced into the ME from the back (back-filling) but, because of the hydrophobicity of the glass wall (due to the silanization), the tip is not filled immediately. The strategy to fill the ME completely depends on the absence or presence of the ligand at its very tip.

When the ME is filled with the electrolyte first, traditionally, the solution can be advanced in the ME shank by twisting an animal whisker as far as possible, with local heating using a microforge. Our experience has shown that cat whiskers seem to work best, but rabbit whiskers are also suitable. However, this method is a little tricky and has some disadvantages: it tends to create air bubbles or to add debris, decreasing the number of usable electrodes. This disadvantage becomes particularly critical when the "batch" method for ME manufacture is chosen. Another classical method to fill the tip is to tap the ME gently, but problems with air bubbles still can be encountered.

We prefer an alternative method which we find works consistently. Right after filling, *without tapping the ME*, the back of the ME is connected, via flexible tubing, to a 50-ml syringe whose plunger previously has been pulled to 5 ml. Positive pressure is then applied and, as assessed by microscopic observation, the electrolyte creeps along the ME wall and fills the tip. The biggest air bubbles are removed by gently tapping the ME, held tip down. Although we initially applied this procedure to filamented MEs, a similar success rate (more than 90%) has been obtained with nonfilamented MEs. This observation was surprising because ease of filling has been considered as an index of poor silanization. If this were the case, the electrodes would not fill with the ligand, a phenomenon we have not observed. Although we have no explanation for this observation, the long dehydration time (by an unknown mechanism) may facilitate the filling. Note also that the shank of our ME is rather short, which obviously facilitates filling.

Whatever the method used, if the electrolyte column is interrupted by air bubbles, gentle heating of the tip of the ME under microscopic control with a tungsten-platinum wire, according to the device described by Thomas (1982), can remedy the problem. Once filled with electrolyte to the tip, the hydrophobic ligand can be introduced (see subsequent discussion).

When the electrolyte is added after the Ca^{2+} ligand, additional problems exist. In fact, caution must be taken to avoid the presence of air at the electrolyte–ligand interface. If a whisker is used, great care must be taken not to disrupt the column of ligand accidentally, an action that can lead to mixing of oil and water, creating unstable ion-selective MEs. If a heating filament is used, care must be taken not to heat the ligand because the local high temperature is likely to damage the ligand properties.

We have used a filling solution that has an ionic composition mimicking the intracellular medium: 10 mM Na^+, 140 mM, K^+, 10 mM HEPES, 10 mM EGTA, pH 7.1 (at 30°C) and pCa 7. This solution is, in fact, identical to the calibrating solution of the same pCa (see subsequent text and Orchard *et al.*, 1991, for additional comments).

In our experience, it is best to minimize the time between electrolyte and ligand filling. We prefer to fill the ME with the ligand as soon as the ion-selective ME is filled with the electrolyte (although we managed to draw the ligand into the tip 2–3 hr after electrolyte filling). If MEs are left overnight with the filling solution, drawing the ligand into the tip can be very difficult; this observation could be explained by a progressive glass hydration, causing it to lose its ability to retain the ligand.

D. Preparation and Use of the Ca^{2+}-Selective Ligand

Subsequent to the work by Tsien and Rink (1981), investigators have determined that, for repetitive and long-lasting Ca^{2+} measurements with Ca^{2+}-selective MEs, dissolving the Ca^{2+} ligand in a "cocktail" containing PVC is useful. Although the cocktail available from Fluka [Cocktail II, containing 94% (w/w) NPOE, 5% (w/w) ETH 129, and 1% (w/w) sodium TPB] is satisfactory (Amman *et al.*, 1987), the small volume provided (0.1 ml) does not facilitate handling. Therefore, we prefer to make up our own cocktail, in a larger NPOE volume with the same proportions. The composition for 500 μl NPOE is given in Table II. Because of the small volumes and the required stirring, working with flat-bottomed, small-volume glass vials and miniature stirring bars is preferable. ETH 129 and TPB are dissolved with vigorous stirring in NPOE in a 2-ml glass vial (Solution 1). Solution 1 can be kept at room temperature for several months, covered with a Teflon screw cap and protected from light. When the final cocktail (Solution 2) is prepared, PVC is dissolved in THF with stirring; 0.2 ml Solution 1 is added, stirred, and finally sonicated. THF is allowed to evaporate

Table II
Preparation of the PVC-Based
Ca^{2+}-Selective Ligand
for Microelectrodes

Component[a]	Amount
Solution 1	
ETH 129	27.5 mg
NPOE	500 μl
Sodium tetraphenylborate (TCPB)	5.53 mg
Solution 2	
Solution 1	0.2 ml
PVC	36 mg
THF	~0.4 ml

[a] For abbreviations, see Table I.

partially to approximately half the initial volume; the cocktail finally is poured into a 0.5-ml conical vial (Clark Electromedical Instruments).

Before dipping the ME tip, THF is allowed to evaporate until the mixture has the consistency of a thick syrup. Experience is the only criterion that helps determine the adequate consistency of the ligand. In fact, if not enough THF has evaporated, the ETH 129 sensor is too dilute and evaporation in the ME may cause retraction of the gel, yielding poor responses. On the other hand, in some instances we have managed to fill the electrodes even if the ligand appeared to be solidified (as a rule, the thickest mixture that will fill the tip is best).

Because of the small diameter of the tip and the viscosity of the ligand, negative pressure must be applied to the back of the electrode, which is achieved using a 50-ml syringe connected, via a 3-way stopcock, to flexible Teflon tubing connected to the back of the electrode with soft tubing. Vacuum then is applied by pulling the plunger out and blocking it with a rod or collar placed along the plunger (precautions must be taken to check the air-tightness of the elements involved regularly). Observation of the ME and measurements of the ligand column height are performed under microscopic observation and control using a microforge (e.g., as described by Thomas, 1982, pp. 27–28). In brief, the microscope body is laid on its back so the stage (removed) would be vertical and the eyepieces oriented upward. A long working distance objective (40×) is used. The ME and the ligand vial are held independently by two micromanipulators.

With sharp filamented electrodes, at least 30 min negative pressure is required to fill the tip. However, in nonfilamented MEs, columns of at least 50 μm can be obtained in 10 min, allowing several MEs to be prepared in a short time period. Vacuum is released slowly before lifting the electrode from the ligand. As shown by Vaughan-Jones and Kaila (1986), a column height of less than 300 μm is preferable because it decreases electrode sensitivity to changes in temperature and in the level of the bath in the experimental chamber. Our experience is that, depending on the tip diameter, column heights between 50 and 250 μm yield acceptable electrode responses.

Once the electrode is filled with both the ligand and the electrolyte, we prefer not to let the ME equilibrate in Tyrode or high pCa solutions, because this favors the deposition of dirt on the tip of the ME and might contribute to clogging. Rather, we place the MEs tip up in a drilled plastic plate, protected by an upside-down glass beaker. Before calibration, the column height is rechecked because THF in the column continues to evaporate, leading to shrinkage of the PVC gel.

As THF evaporates, the stock cocktail solution becomes even thicker. Periodically, enough THF must be added to decrease the viscosity of the mixture. This process is hastened by mixing with a glass rod and then on a vortex mixer (maximal setting). Sonication can be used also, but does not give better results.

E. Calibration

1. Calibrating Bath and Solution Perfusion

Calibrating ion-selective MEs in the experimental chamber in which measurements are made or having the calibrating bath as close as possible in design and proximity to the experimental chamber is preferred. Our calibration chamber is a "flow-through" type (volume, 0.1 ml), immediately adjacent to the experimental chamber. In contrast with many other design, the experimental chamber is viewed from the front, not from above.

2. Calibration Procedure

The bath electrode is either an Ag wire (chlorided by dipping it in bleach for 15–20 min) or an agar bridge. Ideally, a conventional electrode (3 M KCl-filled) also should be immersed in the bath, and the differential voltage (ion-selective ME minus conventional) should be read. We use a commercial amplifier (FD-223; WPI) or a home built amplifier using varactor bridge preamplifiers (AD311J; Analog Devices), as described by Thomas (1982).

We have adopted the following method to select suitable MEs quickly. The ME is mounted in its holder and advanced into the calibrating bath, allowing the trace to stabilize. If the device used to measure the signal has a resistance measurement feature, this parameter is worth measuring. In fact, our experience has been that MEs with resistances ranging between 100 and 250 GΩ were suitable for our experiments, in terms of linearity and detection limit of the calibration curve (Fig. 5). Moreover, Fig. 5 shows that, *within this range,* the higher the resistance, the lower the detection limit for a given batch of MEs. At resistances higher than 300 GΩ, the detection limit decreased sharply, probably because of the small tip diameter (Ammann, 1986). Finally, although low resistance MEs also tended to have low detection limits, they were not suitable for our experiments because of the large tip diameter that could damage the cell membrane seriously during impalement. In contrast, we have found the height of the ligand column not to be a valuable predictor of ME performance (although this height was always kept at 50–250 μm). In our experience, a ME can be used a few minutes after ligand filling.

After equilibration in the control physiological solution (in our case, a HEPES-based Tyrode, containing 2 mM Ca^{2+}), ion-selective ME potential is adjusted to 0 mV. Our 2 mM Ca^{2+} Tyrode gives a voltage reading corresponding to an intracellular calibrating solution of pCa 2.6. Then, flow is switched to a solution of high pCa (between 7.5 and 9). At 30°C (our experimental temperature), the theoretical slope of the relationship between voltage and pCa is −30 mV/pCa so, between pCa 2.6 and pCa 8, the theoretical voltage should be −162 mV. However, since the electrode detection limit tends to decrease at high pCa, we consider that readings more negative than −140 mV are accept-

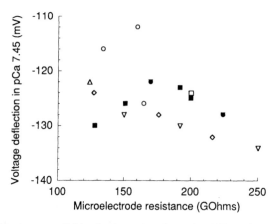

Fig. 5 Relationship between Ca^{2+}-selective microelectrode (ME) resistance and voltage deflection when the calibrating solutions are switched from normal Tyrode (pCa 2.6) to pCa 7.45. The theoretical deflection should be -145 mV, considering a slope of -30 mV/pCa at 30°C. The MEs used in this example were pulled with the same settings, but came from two different batches. Each symbol corresponds to one microelectrode in a given experimental day. Clearly, within the range presented (100–250 GΩ), the higher the resistance, the closer the deflection to the theoretical value. MEs with resistances > 300 GOhms yielded a deflection < -110 mV.

able. A Ca^{2+}-selective ME meeting these criteria then may be calibrated over a wider range of $[Ca^{2+}]$. Since PVC matrix-based Ca^{2+}-selective MEs do not exhibit hysteresis, solutions of increasing Ca^{2+} concentrations are measured. If any drift (difference between the voltage in control solution before and after calibration) has occured, it is usually linear with time. After the calibration is completed, the ME is moved into the experimental chamber and equilibrated until stable. Conventional 3 M KCl-filled MEs are pulled from the same glass and with the same characteristics as the Ca^{2+} electrodes, but are not silanized.

F. Results

Figure 6 is an illustration of what we consider an acceptable record, in terms of impalement quality. First, the conventional ME is impaled (lower trace) and the membrane potential stabilizes at -82 mV. Shortly after, the Ca^{2+}-selective ME is impaled (upper trace), producing a deflection that stabilizes within several minutes (depending on the rate of membrane sealing around the tip of the electrode) at -139 mV. Since the Ca^{2+}-selective ME records both the signal generated by $[Ca^{2+}]_i$ *and* the membrane potential, the latter must be subtracted to obtain the electrochemical potential of Ca^{2+} ions (differential record, middle trace) which, in this example, has a value of -57 mV. When compared with the *in vitro* calibration curve corresponding to this ME (slope, -24 mV/pCa), pCa is

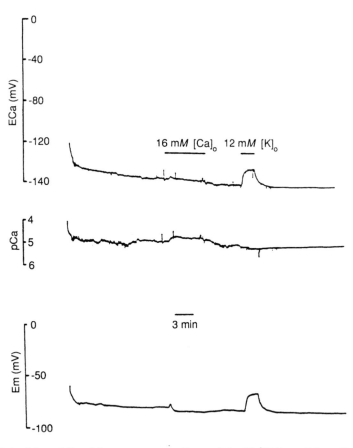

Fig. 6 Tests of the validity of the measurement of intracellular $[Ca^{2+}]_i$ in a rabbit papillary muscle. The upper trace is the Ca^{2+}-selective microelectrode (ME) signal (E_{Ca}); the lower trace, the conventional ME voltage (E_m); and the middle trace, the differential signal ($E_{diff} = E_{Ca} - E_m$), calibrated in pCa values as obtained from the calibration curve (slope = -24mV/pCa). In this example, MEs were impaled ~200 μm apart. After equilibration (~10 min after impalement), pCa was ~5. Since this value was too high, damage caused by the Ca^{2+}-selective ME was suspected, a hypothesis that was tested first by increasing extracellular Ca^{2+}($[Ca^{2+}]_o$) to 16 mM. Initially, an increase in pCa was seen that may signal a leakage of Ca^{2+} inside the cell, but with time, the trace tended back toward its pretest value, meaning the ME was not damaging the cell significantly. Afterward, the possibility that the Ca^{2+}-selective ME sensed a different potential than the conventional ME was tested by depolarizing the preparation, by increasing extracellular potassium to 12 mM. This maneuver did not alter pCa (the deflection on the E_{Ca} and E_m traces were of similar amplitude), showing that both MEs sensed the same membrane potential. At the end of these tests, the steady-state value of pCa was higher than initially (5.5), probably because increasing $[Ca^{2+}]_o$ is known to "stabilize" the membrane and may have helped sealing it around the ME tip.

~5, that is, [Ca]$_i$ = 10 μM, which is ~50 times higher than the expected value (~0.2 μM).

This type of observation has been consistent throughout our project, that is, a good calibration curve (almost Nernstian between pCa 7 and 8) and higher than expected diastolic [Ca^{2+}] values (pCa = 5.34 $\pm$ 0.08; mean $\pm$ SD; n = 53 determinations). Several possibilities can account for these observations. A first concern (usual with MEs) is that the Ca^{2+}-selective ME damages the sarcolemma, producing a leak of extracellular Ca^{2+} into the cell. To test this possibility, extracellular Ca^{2+} is raised to 16 mM. If a leak is responsible for the high [Ca^{2+}]$_i$, the Ca^{2+}-selective ME trace should display a fast rising deflection. This test produced a deflection of +4mV at the end of the differential signal (not shown), which usually is considered a good criterion for sarcolemmal integrity (Marban *et al.*, 1980).

Another explanation is that the Ca^{2+}-selective ME records a lower membrane potential than the conventional ME. One way to test this point is to depolarize the preparation: if both MEs sense the same potential, the differential signal should not change. Figure 6 shows that, when external potassium is increased from 6 to 12 mM, E$_m$ depolarizes by 18 mV and E$_{Ca}$ by 16 mV, yielding a differential signal of -2 mV. This result indicates that the Ca^{2+}-selective ME and the conventional ME sense the same membrane potential.

One possible explanation for such high [Ca^{2+}]$_i$ values is that ETH 129 reacts with intracellular constituents. Although results from minielectrodes in cardiac cell suspensions make this possibility seem unlikely (Hove-Madsen and Bers, 1992), we nevertheless have tested the effects of 10 mM taurine, 2 mM carnitine, and 10 mM caffeine, the latter as a representative of imidazole compounds that are known to be present in high concentration in cardiac muscle (O'Dowd *et al.*, 1988). These compounds were diluted in a pCa 7 (assumed diastolic [Ca^{2+}]) calibrating solution. The electrodes first were equilibrated in control pCa 7 and each of them was tested, with a return to control pCa 7. The effect of these molecules on the Ca^{2+} signal was not sufficient (less than ±5 mV) to explain the large difference from the expected value (at least +30 mV in our experiments).

Intracellular proteins are known to affect Ca^{2+} electrodes, although at high concentrations (Hove-Madsen and Bers, 1992). Although preincubation of freshly prepared Ca^{2+}-selective MEs in solutions containing 20 mg/ml bovine serum albumin (BSA) tended to decrease their detection limit, the magnitude of the effect was again insufficient to explain the difference that we recorded. Moreover, if proteins significantly interfered with the electrodes (by adhering to the glass and/or the membrane), this effect could be expected to be long lasting and therefore present even after withdrawal of the ME from the cell. Consequently, the postimpalement calibration curve should show some loss of sensitivity at high pCa. Although we did observe this phenomenon, the magnitude was too small to explain our results. Therefore, we were left with no explanation for the surprisingly high [Ca^{2+}]$_i$ estimated from these experiments.

A final test performed used ETH 1001-based MEs prepared in the same manner as ETH 129 MEs. Although the detection limit for ETH 1001 is poorer than that for ETH 129 at low [Ca^{2+}] (Schefer *et al.*, 1986), we have been able to record Ca^{2+} signals with ETH 1001 that, when calibrated, gave resting Ca^{2+} concentrations closer to those reported with this ligand (pCa = 6.34 ± 0.15; mean ± SD; $n = 10$ determinations).

Finally, ETH 1001 also has been used in muscle cells by two groups. One report (from the J. Lopez group, in skeletal muscle cells; Allen *et al.*, 1992) shows that the MEs are suitable for measuring changes in resting [Ca^{2+}]$_i$, although they were prepared differently than ours. These investigators used the cocktail provided by Fluka (no PVC matrix) and the Ca^{2+}-selective ME was *back-filled* with the ligand. This same group has not used ETH 129 in any cardiac preparation (J. Lopez, personal communication, February 1992). In the other report, by Rodrigo and Chapman (1990), ETH 129 was used in patch-clamp-type MEs in isolated myocytes; the reported values of resting [Ca^{2+}]$_i$ are in line with previous measurements in multicellular preparations. J. A. S. McGuigan (personal communication, December 1991), in Bern, also has attempted to measure Ca^{2+} in multicellular preparations, but has encountered problems with calibration and measurements that led him to abandon the use of ETH 129.

G. Conclusions

We have managed to design and make Ca^{2+}-selective MEs, based on the ligand ETH 129 in a PVC matrix, that give excellent responses during *in vitro* calibration (before and after impalement). However, we are concerned about an undetermined artifact producing higher than expected [Ca^{2+}]$_i$ in cardiac muscle. Our results with the Ca^{2+} ligand ETH 1001 seem to show that this artifact may be caused by an effect of the ETH 129 ligand itself, rather than by nonspecific damage of the cell membrane during ME impalement.

V. Troubleshooting

We present a list of the most commonly encountered problems during manufacture and calibration of Ca^{2+}-selective MEs. For more specific problems, see Amman (1986).

The ME cannot be filled with the ligand.

1. The ligand may be too thick because most of the THF has evaporated. Redilute PVC by adding small amounts of THF and stirring the mixture to homogeneity.

2. The tip diameter is too small. One simple rule is, when the shape of the ME is designed, prepare batches of electrodes with different shapes and test them (ligand filling and calibration), in parallel, on the same day. Relying

on the ME resistance may give misleading results because this parameter also is affected by the geometry of the shank.

3. The silanization of the ME is insufficient. Several explanations are possible; in order of decreasing likelihood, these are: insufficient time of silanization and/or insufficient sealing of the beaker during exposure to silane vapors, old silane, rehydration of the glass (storage problem), insufficient dehydration.

The ME gives bad calibration curves, that is sub-Nernstian slopes or low detection limits. Although these problems are somewhat different (Amman, 1986), simple possibilities can be checked.

1. Make sure that the calibrating solutions are adequate. Check them with commercial macroelectrodes or the ETH 129-based minielectrodes, as described in previously published methods (e.g., Bers, 1982).

2. The ligand may be too old. This explanation is common. Perhaps the ligand is old because the vial containing the ligand has been opened and exposed to light, leading to degradation of the ETH 129. Make sure to limit exposure to light.

3. The ETH 129 is too dilute. Ammann (1986) showed that the ligand should have a minimal concentration to be effective. Too dilute a ligand may occur by not allowing THF to evaporate sufficiently before ligand filling (the mixture is then very liquid). In this case, allowing all the THF to evaporate, and re-adding small amounts (few tens of μl) of THF and stirring the mixture until a syrup-like solution is obtained is better than having too dilute a solution. Also, if the cocktail is not mixed to homogeneity (by vortexing or sonication) after adding THF to a gelled cocktail, the ligand may be too dilute in some areas.

4. During preparation of the ligand, either an insufficient amount of ETH 129 has been weighed or the proportion of NPOE to ETH 129 is too high.

Acknowledgments

This work was supported by a grant from the National Institutes of Health (HL30077). S. Baudet was recipient of postdoctoral fellowship from the American Heart Association, California Affiliate (91-47), and L. Hove-Madsen was supported by a grant from the University of Aarhus (Denmark) and the Danish Research Academy.

References

Allen, P. D., Lopez, J. R., Sanchez, V., Ryan, J. F., and Sreter, F. A. (1992). EU 4093 decreases intracellular [Ca^{2+}] in skeletal muscles fibers from control and malignant hyperthermia-susceptible swine. *Anesthesiology* **76**, 132–138.

Ammann, D. (1986). "Ion-Selective Microelectrodes. Principles, Design and Application." New York: Springer-Verlag.

Ammann, D., Bührer, T., Schefer, U., Müller, M., and Simon, W. (1987). Intracellular neutral carrier-based Ca^{2+} microelectrode with subnanomolar detection limit. *Pflügers Arch.* **409**, 223–228.

Bers, D. M. (1982). A simple method for the accurate determination of free $[Ca^{2+}]$ in Ca–EGTA solutions. *Am. J. Physiol.* **242**, C404–C408.

Bers, D. M. (1983). Early transient depletion of extracellular Ca^{2+} during individual cardiac muscle contractions. *Am. J. Physiol.* **244**, H462–H468.

Blatter, L., and Wier, W. G. (1992). Intracellular diffusion, binding, and compartmentalization of the fluorescent calcium indicators indo-1 and fura-2. *Biophys. J.* **58**, 1491–1499.

Deyhimi, F., and Coles, J. A. (1982). Rapid silylation of a glass surface: Choice of reagent and effect of experimental parameters on hydrophobicity. *Helv. Chim. Acta* **65**, 1752–1759.

Grynkiewicz, G., Poenie, M., and Tsien, R. Y. (1985). A new generation of Ca^{2+} indicators with greatly improved fluorescence properties. *J. Biol. Chem.* **260**, 3440–3450.

Harrison, S. M., and Bers, D. M. (1987). The effect of temperature and ionic strength on the apparent Ca-affinity of EGTA and the analogous Ca-chelators BAPTA and di-bromo BAPTA. *Biochim. Biophys. Acta* **925**, 133–143.

Harrison, S. M., and Bers, D. M. (1989). Correction for absolute stability constants of EGTA for temperature and ionic strength. *Am. J. Physiol.* **256**, C1250–C1256.

Hove-Madsen, L., and Bers, D. M. (1992). Indo-1 binding to protein in permeabilized ventricular myocytes alters its spectral and Ca binding properties. *Biophys. J.* **63**, 89–97.

Hove-Madsen, L., and Bers, D. M. (1993a). Passive Ca^{2+} buffering and SR Ca^{2+} uptake in permeabilized rabbit ventricular myocytes. *Am. J. Physiol.* **264**, C677–C686.

Hove-Madsen, L., and Bers, D. M. (1993b). SR Ca^{2+} uptake and thapsigargin sensitivity in permeabilized rabbit and rat ventricular myocytes. *Circ. Res.* **73**, 820–828.

Konishi, M., Olson, A., Hollingsworth, S., and Baylor, S. M. (1988). Myoplasmic binding of Fura-2 investigated by steady-state fluorescence and absorbance measurements. *Biophys. J.* **54**, 1089–1104.

Marban, E., Rink, T. J., Tsien, R. W., and Tsien, R. Y. (1980). Free calcium in heart muscle at rest and during contraction measured with Ca^{2+}-sensitive microelectrodes. *Nature (London)* **286**, 845–850.

O'Dowd, D. J., Robins, D. J., and Miller, D. J. (1988). Detection, characterization and estimation of carnosine and other histidyl derivatives in cardiac and skeletal muscle. *Biochim. Biophys. Acta* **967**, 241–249.

Orchard, C. H., Boyett, M. R., Fry, C. H., and Hunter, M. (1991). The use of electrodes to study cellular Ca^{2+} metabolism. *In* "Cellular Calcium. A Practical Approach" (J. G. McCormack and P. H. Cobbold, eds.), pp. 83–113. New York: Oxford University Press.

Purves, R. D. (1981). "Microelectrode Methods for Intracellular Recording and Ionophoresis," Biological Techniques Series, Vol. 6. New York: Academic Press.

Rodrigo, G. C., and Chapman, R. A. (1990). A novel resin-filled ion-sensitive micro-electrode suitable for intracellular measurements in isolated cardiac myocytes. *Pflügers Arch.* **416**, 196–200.

Schefer, U., Ammann, D., Pretsch, E., Oesch, U., and Simon, W. (1986). Neutral carrier based Ca^{2+}-selective electrode with detection limit in the subnanomolar range. *Anal. Chem.* **58**, 2282–2285.

Thomas, R. C. (1982). "Ion-Selective Microelectrodes: How to Make and Use Them," Biological Techniques Series. New York: Academic Press.

Tsien, R. Y., and Rink, T. J. (1981). Ca^{2+}-selective electrodes: A novel PVC-gelled neutral carrier mixture compared with other currently available sensors. *J. Neurosci. Meth.* **4**, 73–86.

Vaughan-Jones, R. D., and Kaila, K. (1986). The sensitivity of liquid sensor, ion-selective microelectrodes to changes in temperature and solution level. *Pflügers Arch.* **406**, 641–644.

Vaughan-Jones, R. D., and Wu, M. L. (1990). pH dependence of intrinsic H^+ buffering power in the sheep cardiac Purkinje fibre. *J. Physiol. (London)* **425**, 429–448.

CHAPTER 5

The Vibrating Ca^{2+} Electrode: A New Technique for Detecting Plasma Membrane Regions of Ca^{2+} Influx and Efflux

Peter J. S. Smith, Richard H. Sanger, and Lionel F. Jaffe★

NIH National Vibrating Probe Facility and
★ Calcium Imaging Laboratory
Marine Biological Laboratory
Woods Hole, Massachusetts 02543

I. Introduction

Biological systems generate significant steady-state transmembrane currents, driven by differences in the distribution of ion channels and pumps in different regions. The ion movements underlying these currents can derive from numerous biological functions such as transport or the establishment of polarity and growth. The latter frequently involves the asymmetric movement of calcium ions across the plasma membrane. The extracellular vibrating calcium-selective probe is uniquely suited to the study of such currents.

The fundamental problem in measuring weak spatial distributions of voltage generated by steady current flow is the noise and instability found in the electrode. This problem is solved by both the vibrating voltage-sensitive and ion-selective probes, by moving a single self-referencing electrode between two points that are microns apart. Whereas the voltage-sensitive probe measures net current flow, the ion-selective system measures a transmembrane flux attributable to a single ion species. The similarities and differences between the two techniques are considered in this chapter. The sensor of the ion system consists of a glass microelectrode with ionophore in the tip, located close to the cell membrane. With signal averaging, vibration of this self-referencing system produces a sensitivity on the order of microvolts, which equates to a measure of picomolar fluxes. The advantages of a vibrating ion-selective probe over conventional techniques for the study of steady-state currents are numerous. First, such a probe is orders of magnitude more sensitive and measures fluxes due to non-conductive mechanisms. Second, the electrodes are noninvasive, located within microns of the cell surface, and have a spatial resolution determined by the electrode tip (~ 4 μm; Fig. 1). Further, these extracellular probes allow the study of naturally occurring current patterns not only from several regions of a growing cell (or tissue) but also over extended periods of time, from hours to days. Finally, the probe technique is not restricted to calcium but can incorporate a broad range of ionophores that allow the measurement of a number of ionic species.

For comparison, considering the resolution of more familiar techniques in the study of relatively steady transmembrane ionic currents is useful. For example, with the patch-clamp technique, contrasting the activity from different regions of a cell membrane gives a unique insight into the physiological characteristics of the channels within the area examined. However, this technique has limitations. Patch recordings frequently involve driving changes in the natural membrane potential; the technique is inevitably invasive and, in practice, one could not hope to compare more than a few regions in any one cell. In addition, clamping techniques can only be applied to cells where GΩ seals can be achieved. Another well-known approach to the study of ionic asymmetries in cells is imaging intracellular patterns of specific free ions. Again, this approach can yield extraordinary insights into free ion patterns but has obvious limitations in assessing transmembrane ion flux, the most serious one of which is the inability to discriminate between internal and external ion sources, a problem the calcium probe could resolve usefully. Another disadvantage is that free ion imaging appears to be applicable only to ions present at very low intracellular concentrations (notably calcium and protons). The probes can, in contrast, measure fluxes of several ions—most notably calcium, protons, potassium,

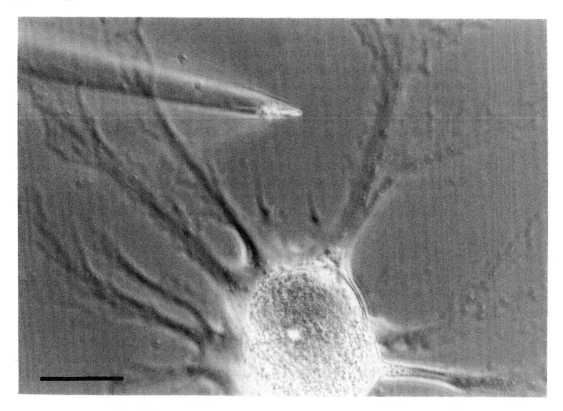

Fig. 1 An illustration of the relative proportion of the ion-selective vibrating probe to a single cell. Here an electrode has been photographed in the background position during a study of the calcium flux from an isolated *Aplysia* bag cell neuron (see Fig. 8). As discussed in Section V, the location of the electrode is under computer control and can be brought toward the cell surface with movements of submicron accuracy. Vibration can be either perpendicular or parallel to the surface of the culture dish. Bar, 30 μm.

chloride, and potentially others—although clearly they cannot resolve changes in intracellular ion distribution.

II. Principles of Vibrating Electrodes

Steady transcellular currents generate very minute voltages in conductive extracellular fluids. The aim of using vibrating electrodes is to map these voltage fields, then deducing the underlying current. One could attempt to measure the field around a cell by placing two glass electrodes, filled with a liquid electrolyte, at a known distance from the cell and a known distance apart. Simply using the continuous medium version of Ohm's Law:

$$E = J\rho \qquad (1)$$

one could calculate the current from the potential difference recorded between the two electrodes. The voltage gradient, E, would be in volts/cm, the current density, J, in amps/cm^2, and the resistivity of the medium, ρ, in ohm $\cdot$ cm.

This idea is simple in theory, but impossible in practice. First, the Johnson noise (thermal noise) in a liquid-filled electrode would make the small voltage difference sought visible only with the most tedious averaging procedure; second, the two electrodes would drift independently of each other, making any such analysis impossible. Therefore, an alternative approach is necessary. To overcome these problems, a method for determining transcellular currents from extracellular voltage gradients and Ohm's Law had to be developed. A radically new method for measuring voltage differences down to nanovolts—a signal far below the noise of conventional electrodes—was developed: the voltage-sensitive vibrating electrode (Jaffe and Nuccitelli, 1974).

A. Original Vibrating Voltage–Sensitive Probe

Neurobiologists are familiar with extracellular recordings of fast action potentials using metal-filled microelectrodes. This tool works because enough capacitance exists at the interface of metal and water to reduce impedance and, hence, the noise generated by this interface, allowing the measurement of high frequency events such as action potentials. In theory, we could measure steady voltage differences by vibrating a bare metal electrode between the points in the voltage field to be compared, at a frequency so high that the electrode capacitance would reduce the interface impedance to negligible levels. This approach would have the additional advantage of being a self-referencing system, avoiding the gross drift in steady voltage at the surface of all real electrodes. However, the frequency needed is on the order of 10 kHz, introducing unacceptable mechanical disturbance. This problem can be overcome by plating the metal electrode with a high capacitance ($\approx$10 nF) platinum black ball. The resulting reduction in interface impedance reduces the unavoidable thermal noise according to the Johnson noise formula. Indeed, in practice, vibration at only a few hundred Hz of an electrode tipped by a 10-μm platinum black ball is found to exhibit noise that is explained fully by the access resistance to the probe. In other words, this simple maneuver reduces interface noise to negligible levels. The probe is vibrated over twice the diameter of the ball, giving a spatial resolution of approximately 20–30 μm. (For further details on the two-dimensional voltage-sensitive vibrating probe, the reader is referred to Scheffey, 1988).

Voltage fields now have been recorded from numerous preparations, providing information on current flow associated with growth, development, and repair (see Nuccitelli, 1986; Borgens *et al.*, 1989), sites of focused ion transport, (Scheffey *et al.*, 1983), paracellular ion movements under voltage clamp conditions (Nagel *et al.*, 1993), brain homeostasis (Smith and Shipley, 1990), and synaptic channel densities (Betz *et al.*, 1984). (For a review of voltage-sensitive probe studies, see Nuccitelli, 1990.) Unfortunately, the voltage-sensitive probe measures only net current flow, and provides no information on the directionality of specific ion movements or on the ions making up the net current. Accessing this information requires fairly exhaustive and relatively indirect studies with ion substitution or pharmacology. The ion-selective probe takes great strides toward making this analysis much simpler and more direct.

B. Ion-Selective Vibrating Probe

The ion-selective probes, first described by Jaffe and Levy (1987) and in a more developed form by Kühtreiber and Jaffe (1990), are quite different from the voltage probe but rely on related principles.

Ordinarily, ions move through the extracellular medium by simple diffusion governed by Fick's Law, so

$$J = -D(dc/dx) \tag{2}$$

where J is the ion flux (μmol $\cdot$ cm^{-2} $\cdot$ sec^{-1}), D is the diffusion constant (which, for calcium, is 8×10^{-6} cm^2 $\cdot$ sec^{-1}), and dc/dx is the concentration gradient [the change in concentration, dc (μmol $\cdot$ cm^{-3}), over a known distance, dx (cm)]. We wish to calculate the ion flux and therefore need to know the value dc/dx.

Again, as for the voltage probe, any attempt to measure dc with two static ion-selective electrodes would be impossible because of noise and independent drift. Again the solution is to vibrate an ion-selective electrode, creating a self-referencing system the output of which can be signal-averaged extensively (see Section V). This system not only has the advantage of selectivity over the voltage-sensitive probe but also has greater spatial resolution, the latter being determined by the diameter of a glass electrode tip of approximately 4 μm.

Although the operating principles behind the voltage and ion-selective vibrating probes are somewhat similar, the details are quite different. The voltage probe does not depend on the movement of any specific ion in the external milieu but only on the voltage field created by the net movement of ions; mixing has no effect on the resultant fields. For the ion-selective probe, however, avoiding the result of mixing is critical. Here we measure the specific gradient of a single ion. Vibration frequencies must, therefore, be low; the electrode must remain at each extreme of displacement long enough for the gradient to reestablish itself. This frequency is 0.3–0.5 Hz for calcium ions (see subsequent discussion). The ion gradient then can be calculated using the measured ΔV (change in voltage), known distances, and background ion concentrations in conjunction with the Nernstian properties of the ionophore electrode. The latter is measured easily by calibrating the electrode, without vibration, in solutions of known calcium concentration. A three-point calibration is best, with values bracketing the expected background in the experimental medium.

Using the technique of electrode construction described in Section III, we generally find the values to be close to Nernstian, being 28 mV over a 10-fold change in concentration, as expected for the calcium-sensitive ionophore used (Fig. 2). [Voltage fields would not be expected to contribute significantly to the output from these electrodes (Kühtreiber and Jaffe, 1990)]. Based on these characteristics, we can calculate the meaning of ΔV by the simple equation

$$\Delta V = S \log(C_2/C_1) \tag{3}$$

ΔV is the voltage difference (mV) between the extremes of the vibration [the analysis setup in DVIS3 (see subsequent text) expresses an outward flux from the source as a negative voltage change], S is the Nernstian slope of the electrode in mV, and C_1 and C_2 are the calcium concentrations at the extremes of the vibration distance. (Concentrations are expressed throughout these calculations as $\mu mol \cdot cm^{-3}$.) This equation is equivalent to

$$\Delta V = \frac{S \ln(C_2/C_1)}{2.3} \tag{4}$$

Further,

$$Ln(C_2/C_1) = \ln(C_1 + \Delta C)/C_1 = \ln(1 + \Delta C/C_1) \tag{5}$$

When values of $\Delta C/C_1$ are small then, according to Kühtreiber and Jaffe (1990),

$$\ln(C_2/C_1) \approx \Delta C/C_{average} \tag{6}$$

We can now substitute Eq. 6 into Eq. 4, so

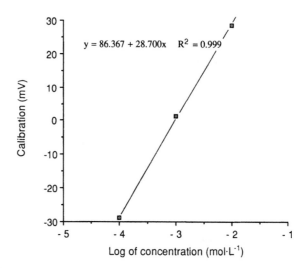

Fig. 2 This figure illustrates a three-point Nernst calibration for a glass microelectrode with a short column of ionophore in the tip ($\approx 15 \ \mu m$). Note that the slope of the regression line approximates the expected value of 28 mV. This line can be used to calibrate the characteristics of a calcium source when a static electrode is placed at known distances from the same source.

$$\Delta V = \frac{S\,(\Delta C/C_{average})}{2.3} \tag{7}$$

When we are dealing with small signals, as is often the case with a calcium flux, using the background concentration of calcium [C_B (see Eq. (12))] instead of $C_{average}$ is permissible. We can simplify Eq. 7 to

$$\Delta V \approx \frac{S\,(\Delta C/C_B)}{2.3} \tag{8}$$

This equation can be rearranged to the form

$$\Delta C \approx \frac{2.3(\Delta V C_B)}{S} \tag{9}$$

ΔV is the voltage output from the vibrating probe in millivolts, C_B is the background concentration of calcium (μmol $\cdot$ cm^{-3}) as calculated from the static mV reading at various distances from a point source (see Fig. 3), and ΔC is the concentration difference between the extremes of vibration, also measured in μmol $\cdot$ cm^{-3}. The latter value now can be calculated and substituted into the Fick Equation (*dc;* Eq. 1) to calculate the flux:

$$\text{Flux } (J) = -D(\Delta C/\Delta r) \tag{10}$$

Flux will be in μmol $\cdot$ cm^{-2} $\cdot$ sec^{-1} (although in reality this parameter is more

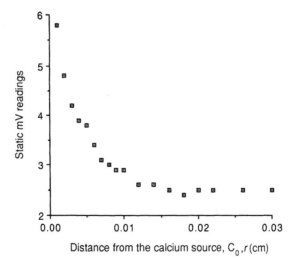

Fig. 3 To calculate the characteristics of a calcium gradient, static mV readings are taken at known distances from that source (C_0). This figure illustrates a practical example which will be used in Eq. 12 and Fig. 5. The behavior of the static mV readings also allows the estimation of C_B, the background concentration of calcium.

likely to be expressed in pmol values), ΔC is as defined earlier, and Δr is the vibrational amplitude in cm.

III. Construction of Ion-Selective Probes

A. Electrode

Micropipettes are pulled from 1.5-mm diameter borosilicate glass without a filament (TW150-4; World Precision Instruments (WPI), New Haven, Connecticut). The electrode puller must be capable of a reproducible two-step pull. At our facility we use a BB-CH puller (Mecanex, Geneva, Switzerland). A model P-87 Flaming–Brown puller (Sutter Instruments, Novato, California) is also suitable, and perhaps better (see Kochian *et al.*, 1992). The shape for which to aim is a stubby electrode, not dissimilar to a patch electrode, with a short shank (≈ 1 mm), a tip of approximately 2–3 mm, and a final tip diameter of between 2 and 4 μm (Fig. 4). A slightly conical tip appears to convey greater mechanical stability to the cocktail column. This shape makes it possible to front-fill the

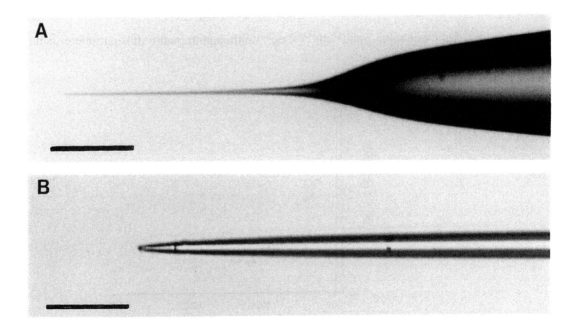

Fig. 4 Ion-selective electrodes are produced by a two-step pull, generating a stubby electrode (A) into which a controlled length of ionophore can be drawn by front-filling (B). (A) The general shape of the electrode is apparent. (B) A higher magnification of the tip. The short ionophore column is visible in the tip of the electrode. (Bars, (A) 300 μm; (B) 30 μm.

electrodes with 10- to 50-μm columns of ionophore (Fig. 4B). Once a batch of electrodes has been pulled to the correct dimensions, the electrodes are silanized. Since the ionophore cocktail is hydrophobic, this procedure helps stabilize the short column in the tip of the electrode.

B. Silanization

Numerous different silanization procedures and silanes are available. The procedure we use has, from experience, given us the best stability but we have not explored the different silanes available extensively. The type we use is N,N-dimethyltrimethylsilylamine (41716; Fluka Chemical, Ronkonkoma, New York). Petrarch Systems (Bristol, Pennsylvania) commercially supplies a wide range of silanes.

Silanization chemically changes the surface of the glass so hydroxyl groups on the glass surface condense with unstable silanol generated from the silane (see Ammann, 1986). Successful silanization depends critically on drying the glass beforehand. To achieve this drying, we place our pulled electrodes (40 at a time) in a stainless steel electrode rack, lying nearly horizontal. The electrodes are dried at 180°C for at least 1 hr. Overnight drying gives better results. After drying, the rack is placed in a presilanized glass chamber (approximate volume of 1 liter) and injected with 50 μl silane. After 15 min, the chamber cover is removed and the oven door left open for 30 sec to allow the silane vapor to escape. The rack is then recovered and the electrodes are dried for at least 1 hr. Again the longer the drying time, the better. Electrodes can be stored, covered, at either 180°C or in a chamber over desiccant. Silanization is best done in a fume hood.

We had problems silanizing the electrodes when the oven walls were shedding dust into the chamber. Some attention to the cleanness of the glass is, therefore, recommended. Usually if this is a problem, particulate matter can be seen when front-filling with the ionophore. We have not found it necessary to acid-clean the glass.

C. Ionophores and Their Insertion

Prior to inserting the ionophore by front-filling, the electrolyte must be introduced into the micropipette. We have explored several types of electrolyte to improve the stability of the ionophore in the tip. Currently, we use a solution of 100 mM $CaCl_2$ in a 0.5% agar gel. The solution is filtered through a 0.2-μm filter disk to remove particulate matter. An agar solution confers significantly better stability on the ionophore column during vibration than a simple aqueous $CaCl_2$ electrolyte. By back-filling through a polystyrene syringe, pulled out over a cool flame, only the minimal electrolyte column length is placed in the micropipette. When back-filling, the electrolyte does not flow to the end of the electrode but an air space remains that should be forced out just before loading the electrode with the ionophore cocktail (see subsequent discussion).

When the ion-selective technique was developed, only one commercial calcium ionophore was easily available. This ionophore was purchased from Fluka and is now used routinely both in our facility and by Kochian *et al.* (1992). This ionophore (FLUKA Calcium Ionophore I, Cocktail A) contains a calcium-selective neutral carrier (ETH 1001) in a mix of 89% 2-nitrophenyl octyl ether and 1% sodium tetraphenylborate. To date we have not made a systematic survey of the other ionophores now available from such companies as Sigma (St. Louis, Missouri), WPI, and CalBiochem (La Jolla, California).

Commercially available ionophores are not ideal. The vibrating probe ionophore requires less specificity and more conductance than those used for other purposes, for example, intracellular electrodes. The principle reason for this requirement is that competing ions such as sodium, which might generate a voltage with a less selective ionophore, are present at such high concentrations that the fractional difference in competing ion concentrations is relatively insignificant. A higher conductance, and therefore lower resistance, ionophore would suffer less from thermal noise. Noise from this source generates a clear limitation for the sensitivity and speed of the technique.

Front-filling the micropipette is done under an upright microscope with a 40× objective. A silanized pipette, with the final taper broken back to an aperture of 50–100 μm, is used as an ionophore reservoir. Pressure is applied to this pipette so the ionophore bulges out and the microscope can be focused on its leading edge. The probe is inserted into a standard microelectrode holder half-cell (WPI MEH2SW15; WPI, Sarasota, Florida). This holder comes fitted with a silver wire, the end of which can be plated electrolytically with silver chloride. (To plate, immerse a length of the silver wire in a 1 N NaCl solution for approximately 1 min with the electrode at 9 V positive to a silver return through an IKΩ resistor). This holder type has a side arm through which pressure can be adjusted. Electrode and holder are mounted to a micromanipulator; then the tip of the electrode is brought to the same focal plane as the ionophore reservoir. Pressure is now applied through the side arm so the electrolyte runs to the tip of the electrode. As soon as the tip fills, it is pushed into the ionophore and negative pressure is applied. Ionophore should enter the tip easily in a controlled manner, allowing the accurate regulation of the column length ($\approx$20 μm). The finished electrode now can be installed in the preamplifier with a 3 M KCl in 1% agar return.

IV. Calibration and Sensitivity

An important step in the development of the ion-selective electrode was the demonstration that the technique could measure an artificially generated gradient reliably. The steps behind this test have been described by Jaffe and Levy (1987) and by Kühtreiber and Jaffe (1990). The measurement was accomplished by filling a blunt micropipette (tip diameter, approximately 10 μm) with a solution of 100 mM CaCl$_2$ in a 0.5% agar solution. The agar blocks bulk water flow. This pipette was placed in a petri dish containing the same concentration of MgCl$_2$ with approximately 1 mM CaCl$_2$ (C$_B$). The presence of MgCl$_2$ at the same concentration as the source provides counter diffusion and minimizes osmotic water flow. After 0.5 hr, a steady state was established between the calcium concentration deep in the pipette (C$_S$) and the large sink in the dish. Convective disturbances in the gradient could be minimized by placing the source close to the bottom of the dish ($\approx$100 μm from the bottom).

Once steady state is established, note that the physical characteristics of the source pipette will reduce the concentration at the opening (C$_O$) greatly relative to the concentration in the body of the source (C$_S$). However, this concentration is measured easily using the Nernst equation. Once the slope of the electrode has been established, by a three-point calibration, a series of static millivolt readings can be taken at the opening of the source and at known distances from this source (Fig. 3). Typically, C$_O$ is 0.5–3% of C$_S$. Using these millivolt readings and the distances from the source, the absolute concentrations can be calculated from the equation of the regression line for the log concentration of the three-point Nernst calibration against the mV values:

$$\log C = a + m \cdot \text{static V} \tag{11}$$

The characteristics of the gradient now can be defined by regressing the concentrations (in μmol $\cdot$ cm^{-3}) against the inverse of the distance away from the source (r, in cm) according to the equation

$$C = C_B + K/r \tag{12}$$

C$_B$ is the background concentration of calcium (μmol $\cdot$ cm^{-3}). This equation facilitates calculation of the empirical constant K (μmol $\cdot$ cm^{-2}), which defines the diffusion properties of the source. In selecting the data for the calculation of K, two data sets must be excluded. First, the regression should stop when the static mV readings plateau at background. Second, the first few points away from C$_O$ should be excluded. Data here are complicated by two factors. An inevitable error will exist in estimating the value r so close to the source. As the electrode moves away, this error becomes proportionately less significant. Additionally, when the source electrode is close to the bottom of the dish that surface will act as a reflective surface, generating complex gradient behaviors modeled more easily as two point sources.

Equation 12 can be differentiated to yield the concentration at any distance from the source.

$$dC/dr = -K/r^2 \tag{13}$$

We now can substitute Eqs. 12 and 13 into Eq. 8, so

$$\Delta V = \frac{S[(-K\Delta r)/(C_B r^2 + Kr)]}{2.3} \tag{14}$$

All units are as defined earlier, and Δr is the vibrational amplitude (cm). Using this equation, one now can check the accuracy of the measured voltage difference (ΔV_{meas}) between the extremes of vibration and the theoretical value for the gradient at the same position (ΔV_{theo}). As illustrated in Fig. 5, the measured signal faithfully tracks the theoretical. At intermediate values of ΔV, the theoretical and measured values are nearly the same but deviate from the ideal as weaker signals more typical of biological signals are approached. All electrodes examined to date follow the theoretical behavior, although duration from the theoretical can be between 0.5 and 1.1 of the ideal, even at equivalent distances from C_0. This variability should be borne in mind when determining the possible range of fluxes measured from a biological system but the accuracy attained is still impressive given the assumptions made in the calculation and the inevitable experimental error that must have been introduced. ΔV can be substituted into Eq. 3 to give a measured flux (Fig. 5).

With the knowledge that the technique can measure small voltage changes associated with transmembrane calcium flux, a systematic study of the electrode characteristics and how they will influence the noise and sensitivity of the technique is possible. Several factors should be considered, such as the ionophore column length, the electrolyte, and the exact method of data acquisition. The last aspect will be dealt with in a subsequent section.

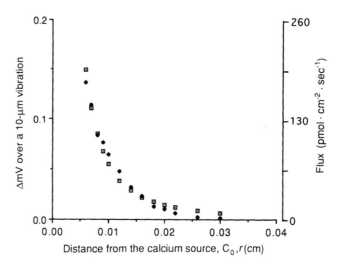

Fig. 5 Data shown in Figs. 2 and 3 are brought together in this figure to calculate the theoretical gradient of calcium (□) coming from a known source (C_0) at different distances (r in cm) and calculated by Eq. 14. These values can be compared directly with the measured ΔmV readings obtained with a vibrating calcium-selective probe (◆). These measured ΔmV readings can be converted to flux values by incorporation into Eqs. 2 and 3. Here, a 50% efficiency is assumed for the electrode.

Column length and the resulting resistance originally were proposed to be the limiting factors on noise reduction with this technique (Jaffe and Levy, 1987). One can measure the resistance of the electrodes with the aid of the formula

$$R_e = R_p(V_o/V_i - 1) \tag{15}$$

where, R_e is the electrode resistance, R_p is a known resistance (e.g., 1 GΩ) switched in parallel with the electrode, and V_o/V_i is the resultant voltage drop. A 1-GΩ resistance is used, since this value approximates the expected electrode resistance and minimizes the current that flows through the ionophore during the calibration. Current flowing through the electrode is destabilizing with these short ionophore lengths. Normally, current flow through the probe is kept to a minimum by the high input impedance preamplifier (see subsequent discussion). Electrode resistances range from 500 MΩ to 3 GΩ.

Although the column length is clearly relevant (for example a 100-μm column of sensor can generate noise of 1.43 ± 0.5 μV dropping to 0.54 ± 0.2 at 50 μm), we find that the reduction in the electrolyte column is also important. This factor influences the detection of noise through unwanted capacitance along the electrode length. Reduction of this parameter greatly cuts down this effect and further reduces the possibility of small breaks in the length of silver chloride wire generating additional noise. More improvements may be possible but are, as yet, untried. We could insulate the electrode further toward the tip using a layer of Sylgard (Dow Corning, Midland, MI) and could coat the electrode with metal to extend the driven shield from the preamplifier to the saline interface. We expect that these simple improvements would further reduce noise levels, increase the sensitivity, and thus speed the rate of data acquisition.

V. Data Acquisition

A. Probe Positioning and Vibration

The original design published by Kühtreiber and Jaffe (1990) positioned the probe using Newport linear motors; the vibration resulted from three piezo-electric microstages (PZS-100; Burleigh Instruments, Fishers, New York) stacked in an orthogonal array and holding the preamplifier. This system had several drawbacks: obviously the size of the array located so close to the microscope condenser, the weight on the manipulator, and, with respect to the linear motors, the accuracy of positioning and lack of computer control. Our latest design relies on an orthogonal array of stepper motors that control both vibrational angle and length (in x, y, and z planes), but also allow computer-controlled movement from 3–4 cm down to a submicron accuracy (0.4 μm). Both probe and preamplifier are vibrated together, preventing capacitative changes in the linking cable and mechanically generated noise. Movement is driven by a set of Newport 310 series translator stages fitted with size 23 stepper motors. The motors are driven by DVIS3 software (National Vibrating Probe Facility, Woods Hole, Massachusetts) through the computer printer port.

The motion control system interfaces easily with a 386 PC-AT running the DVIS3 ion probe software package, permitting keyboard control of vibration parameters, as well as coarse and fine positioning. We foresee this system facilitating automatic scans of complex biological surfaces. Presently, one problem with scanning the asymmetric surfaces of cells or tissue lies in the vibration angle. Ideally, the probe should be vibrated only at right angles to the long axis of the electrode, minimizing what we presume to be a pressure-dependent offset on the signal output and induced instabilities on the ionophore. However, cells seldom oblige the investigator by presenting their most interesting features perpendicular to the line of attack. The solution may be to rotate the specimen relative to the probe rather than vice versa. We hope to develop this rotational feature in the near future.

The signal that can be measured within a gradient depends on both the background concentration (C_B) and the amplitude of vibration Δr. C_B is obviously dependent on the biological preparation. We have been surprised by how tolerant even animal cells can be of a lower saline calcium level, provided another divalent cation is substituted. As a rule, magnesium is suitable cultured. *Aplysia* bag cells (hormone-secreting neurons from the abdominal ganglion) will regenerate well in artificial seawater when the calcium is reduced from 11 mM to <0.1 mM (P. J. S. Smith, unpublished observations). Isolated pancreatic β cells also survive at calcium levels as low as 0.1 mM (C. A. Leech, personal communication). With respect to the optimal value for Δr, Kühtreiber and Jaffe (1990) assert that the value of ΔV increases linearly with Δr up to an amplitude of ≈ 30 μm. We routinely use amplitudes between 10 and 30 μm. A trade-off exists, however, between vibrational amplitude and spatial resolution.

In addition to the amplitude, the other important parameter of vibration is the frequency. For this parameter, the logic is quite simple. If the electrode vibrates too rapidly, the ion gradient will not re-establish itself for the period of measurement; if too slowly, the problems inherent to stationary electrodes will become more prominent and slow drift will obscure any gradient ΔV. Our current electronics also incorporate both high pass and low pass filters (see Section V,B) which, in concert with the other attributes, optimize the signal input at 0.3–0.5 Hz.

B. Data Collection and Processing

The head stage for the ion-selective vibrating probe contains an operational amplifier with an input impedance of 10^{15} Ω (Fig. 6; AD 515; Analog Devices) and a unity gain. As a voltage follower, this op-amp is configured as a voltage follower draws minimal current through the ionophore. The preamplifier is housed in a metal cylinder, the latter arranged as a driven shield to minimize the effects of unwanted capacitances in the head stage as well as to shield the amplifier from external signals. The front end of the amplifier is fitted with a 2-mm female connector allowing the use of standard Ag/AgCl half cells (Fig. 7). Preamplifier, probe, and the preparation all are contained in a box-like Faraday cage, mounted on an air table. This arrangement minimizes electrical and mechanical interference as well as any effect of air turbulence. The ion system is more sensitive in this regard than are conventional electrophysiological setups. The cells and probes are viewed remotely via a video camera mounted on the side-port of a Zeiss IM35, as in Fig. 1.

Signals from the amplifier, representing the voltages measured at each extreme of vibration, are fed to a 386 PC-AT by the analog-to-digital (A–D) board (DT 2800 series; Data Translations). The DT 2800 has a dynamic range of ± 10 V and a resolution of 16 bits, enabling the computer to process voltage changes of 350 μV. Since we want to resolve small μV differences, an additional amplifier is placed between the head stage and the computer with a gain of 1000-fold. Low and high pass filters remove signals of frequencies faster than approximately 30 Hz as well as the large static voltages associated with the Nernstian behavior of the electrode. We plan to improve this system by incorporating an offset–null method to minimize artifacts from large, fast voltage transients such as static discharge.

The software sets the sampling rate of the A–D board at 1000 Hz. Our current software is a modified version (DVIS3; R. Sanger) of the original written by W. Kühtreiber then at the NVFP. The collected data points are divided up and averaged to yield 10 mean values for each probe position. For example, if we vibrate the probe at 0.5 Hz (i.e., 1 sec per probe position), every 100 data points entering the A–D board are averaged. Of these averages, 10 are processed for each position of the probe. Each of the 10 values is compared with the overall mean from the previous position and 10 separate ΔVs are calculated and fed into a running average. The process is then repeated. The running average can contain over 100 data points with each new entry causing the oldest value to be discarded. The experimenter controls the size of the running buffer since its optimal setting depends largely on signal strength and background noise. The results are displayed during collection and written to disk. A blanking function allows the selective omission of data collected during electrode movement and the period of gradient re-establishment. Data collection takes approximately 0.5–1 min with typical settings for the averages, blanking, and vibration frequency.

VI. Data Analysis and Examples

Once the data are written to disk, they can be imported into several commercially available spreadsheet and analysis packages. Software programs such as *Quattro Pro*, *DaDisp*, *Plot It*, and *Grapher* are all appropriate. Most are more

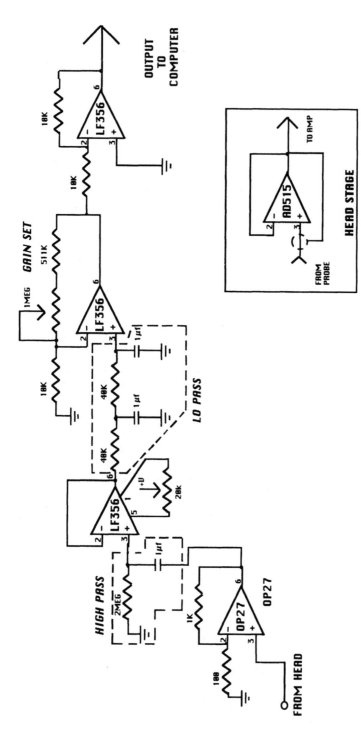

Fig. 6 The electronic amplifiers required for the ion-selective vibrating probes involve a unity gain preamplifier, incorporating an FET operational amplifier, and a 1000× amplifier located before the analog-to-digital board and the computer analysis. The circuit also incorporates high- and low-pass filters to remove large voltage changes attributable to the Nernstian characteristics of the electrode and unwanted signals with frequencies above 30 Hz.

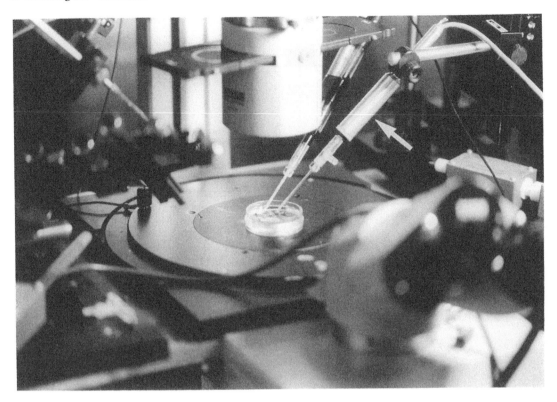

Fig. 7 The experimental setup used for the ion-selective vibrating electrode is mounted over a Zeiss IM35. Both the electrode and the headstage preamplifier (arrow) are vibrated together by stepper motors under computer control. Images are viewed remotely by video through the microscope side-port. The whole assembly is enclosed in a box-like Faraday cage.

powerful than required. In effect, the only manipulation required is to average the data values from the different positions with a deviation and to convert them to ion flux values. These values can be plotted for the different experimental conditions or cell positions. Figures 8 and 9 present results from recent studies at the National Vibrating Probe Facility.

Figure 8 illustrates the steady-state calcium efflux measured from the soma of a cultured *Aplysia* bag cell (neuron). The pattern shown is characteristic of these cells between 24 and 48 hr of *in vitro* growth (P. J. S. Smith, unpublished observations). Effluxes of this magnitude generally can be measured over all surface areas examined to date. The arrow indicates the point at which a new electrode is used; the drop in flux is likely to be a reflection of a different electrode efficiency. As mentioned earlier, this problem is unavoidable with the technique, when experimental electrode efficiencies cannot be characterized individually. In this figure, as well as in Fig. 9, a 50% efficiency was used in the

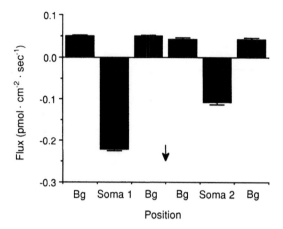

Fig. 8 Effluxes of calcium can be recorded routinely from the soma of cultured *Aplysia* bag cells 24–48 hr after isolation. These cells are neurons from the abdominal ganglion. The arrow represents the point at which data were collected with a new electrode. The drop in the flux may illustrate a difference in the electrode efficiency as discussed in Section IV. Bg represents values collected at background and should be compared with the two sets of soma data from the same position on the cell membrane (P. J. S. Smith, unpublished results).

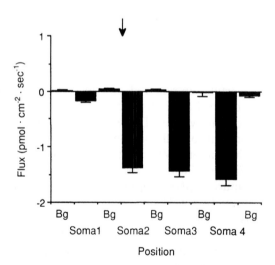

Fig. 9 Calcium effluxes can be recorded across the plasma membrane of rat pancreatic β cells in primary culture. These effluxes can be increased greatly after a 30-sec perfusion with the peptide GLP1(7–37) (arrow) (C. A. Leech, unpublished results).

conversion of ΔV to a flux value. Preliminary studies tentatively indicate that calcium enters into the growth cone; this is the subject of an ongoing research project. We also are exploring the possibility of combining probe work with direct electrical stimulation and clamping of neurons.

Bag cells make excellent preparations partly because of their large size (50–70 μm in culture). The probes can, however, follow a steady efflux from much smaller cells, such as the rat pancreatic β cell (10 μm; Fig. 9). These cells were dissociated in primary culture and exhibited a small steady efflux that could be enhanced greatly by a 30-sec superperfusion with glucagon-like peptide 1(7–37) (GLP1; arrow, Fig. 9; C. A. Leech, unpublished results). This peptide acts in concert with glucose to stimulate calcium-dependent action potentials and the subsequent elevation of intracellular calcium (Holz *et al.*, 1993). A calcium rise occurs in seconds; presumably the efflux recorded by the probe represents the cell recovering its normal intracellular calcium level by pumping the ion across the plasma membrane. This situation is a good example of how the probe technique could be combined usefully with both intracellular recording and calcium imaging.

The calcium probe also has been applied to plant tissue and cells. These samples frequently have the advantage of growing in very dilute media with low calcium concentrations. Kochian *et al.* (1992) provide an excellent illustration of the use of the calcium probe in an intact system, in this case the root hairs of maize. These authors also studied the potassium and proton fluxes using the same technique. E. S. Pierson *et al.* (unpublished results) have examined the inward calcium flux at the growing tip of a lily (*Lilium longiflorum*) pollen grain. Clear influxes related to the forward extension of the tip are recorded and can be eliminated by the intracellular injection of the calcium shuttle buffer 4,4-difluoro-1,2-bis(*o*-aminophenoxy)ethane-*N*,*N*,*N'*,*N'*-tetraacetic acid (BAPTA) (100 m*M*). This treatment also arrests cell growth but not cytoplasmic streaming.

Acknowledgments

The authors thank C. A. Leech (Massachusetts General Hospital) and E. S. Pierson (University of Siena, Italy) for making available unpublished material and L. Kochian (Cornell University), H. Fishman (University of Texas) and A. Cubitt (University of California at San Diego) for helpful discussion during the preparation of the manuscript.

The National Vibrating Probe Facility is funded by the National Institutes of Health Division of Biomedical Research Resources (Grant No. P41RR01395). Part of its role is to collaborate with outside researchers in applying the probe technology to their systems. Application for research time should be made to P. J. S. Smith at the NVPF, Marine Biological Laboratory, Woods Hole, Massachusetts 02543.

References

Ammann, D. (1986). "Ion-Selective Microelectrodes: Principles, Design, and Application." New York: Springer-Verlag.

Betz, W. J., Caldwell, J. H., and Kinnamon, S. C. (1984). Increased sodium conductance in the synaptic region of rat skeletal muscle fibers. *J. Physiol.* (*London*) **352**, 189–202.

Borgens, R. B., Robinson, K. R., Vanable, J. W., Jr., McGinnis, M. E., and McCaig, C. D. (1989). "Electric Fields in Vertebrate Repair: Natural and Applied Voltages in Vertebrate Regeneration and Healing." New York: Liss.

Holz, G. G., IV, Kühtreiber, W. M., and Habener, J. F. (1993). Pancreatic beta-cells are rendered glucose-competent by the insulinotropic hormone glucagon-like peptide-1(7–37). *Nature* (*London*) **361**, 362–365.

Jaffe, L. F., and Levy, S. (1987). Calcium gradients measured with a vibrating calcium-selective electrode. *Proc. 9th Ann. Conf. IEEE* **9**, 779–781.

Jaffe, L. F., and Nuccitelli, R. (1974). An ultrasensitive vibrating probe for measuring steady extracellular currents. *J. Cell Biol.* **63**, 614–628.

Kochian, L. V., Shaff, J. E., Kuhtreiber, W. M., Jaffe, L. F., and Lucas, W. J. (1992). Use of an extracellular, ion-selective, vibrating microelectrode system for the quantification of K^+, H^+, and Ca^{2+} fluxes in maize roots and maize suspension cells. *Planta* **188**, 601–610.

Kühtreiber, W. M., and Jaffe, L. F. (1990). Detection of extracellular calcium gradients with a calcium-specific vibrating electrode. *J. Cell Biol.* **110**, 1565–1573.

Nagel, W., Shipley, A., and Smith, P. J. S. (1993). Vibrating probe analysis of voltage-activated Cl^- current across mitochondrial-rich cells of toad skin. *FASEB J.* **7(3)**, A351(2035).

Nuccitelli, R. (1986). "Ionic Currents in Development." New York: Liss.

Nuccitelli, R. (1990). Vibrating probe technique for studies of ion transport. *In* "Noninvasive Techniques in Cell Biology," (J. K. Foskett and S. Grinstein, eds.) pp. 273–310. New York: Wiley-Liss.

Scheffey, C. (1988). Two approaches to construction of vibrating probes for electrical current measurement in solution. *Rev. Sci. Instr.* **59**, 787–792.

Scheffey, C., Foskett, J. K., and Machen, T. E. (1983). Localization of ionic pathways in the teleost opercular membrane by extracellular recording with a vibrating probe. *J. Membrane Biol.* **75**, 193–203.

Smith, P. J. S., and Shipley, A. (1990). Regional variation in current flow across an insect blood–brain barrier. *J. Exp. Biol.* **154**, 371–382.

CHAPTER 6

Application of Patch Clamp Methods to the Study of Calcium Currents and Calcium Channels

Colin A. Leech and George G. Holz, IV

Laboratory of Molecular Endocrinology
Massachusetts General Hospital
Harvard Medical School
Boston, Massachusetts 02114

I. Introduction

This chapter will not provide a detailed description of patch clamp methodology, since this information can be found in many other sources. The most comprehensive description of these techniques is likely to be *Single Channel Recording* (Neher and Sakmann, 1983). Other useful reference works include volumes edited by Standen *et al.* (1987), Conn (1991), Kettenmann and Gratyn (1992), and Bean (1992). We describe how these techniques can be applied to the study of Ca^{2+} currents as well as some of the advantages and potential pitfalls of the various recording configurations.

The first indication of the presence of Ca^{2+} channels came from studies on crustacean muscle by Fatt and Katz (1953). Our awareness of different types of Ca^{2+} channels has expanded rapidly; a useful subdivision based on some properties of voltage-dependent Ca^{2+} channels was proposed by Nowycky *et al.* (1985). These authors proposed three groups of channels—T-, L-, and N-types—based on biophysical and pharmacological properties of Ca^{2+} channels in chick dorsal root ganglion neurons. Subsequent studies on central nervous system neurons revealed at least one additional class, P-type, found at high concentrations in cerebellar Purkinje neurons (Llinas *et al.*, 1989). Biophysical studies also have suggested the presence of voltage-independent metabolically regulated Ca^{2+} channels of the G-type (Rojas *et al.*, 1990; Cena *et al.*, 1991).

As summarized in Fig. 1, parallel biochemical studies have revealed the presence of multiple subunits in Ca^{2+} channel structures (for review, see Catterall, 1988). The application of molecular biological techniques has revealed an increasing number of subunit isotypes (for review, see Tsien *et al.*, 1991; Miller, 1992). These types of study have been particularly successful in the investigation of intracellular Ca^{2+} channels such as the inositol trisphosphate (IP_3) receptor family and the ryanodine receptor family, which have a tetrameric structure that has been reviewed by Berridge (1993). This tetrameric structure contrasts with data on voltage-sensitive Ca^{2+} channels, which appear to be composed of five different subunits (for review, see Catterall, 1988; Fig. 1). Despite this pentameric structure, α_1 subunits alone appear to form functional voltage-dependent Ca^{2+} channels (Mikami *et al.*, 1989; Perez-Reyes *et al.*,

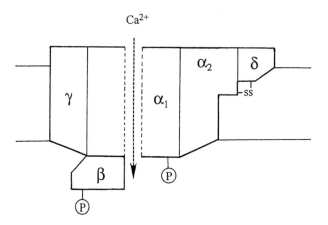

Fig. 1 Subunit structure of a voltage-dependent Ca^{2+} channel. The α_1 subunit, which forms the pore, has a molecular mass of 175 kDa; $\alpha2$ is 143 kDa and is disulfide linked to δ (27 kDa); the β subunit is 54, Da and γ is 30 kDa. (Reproduced with permission from Catterall, 1988.)

1989). In addition to the heterogeneous structure of Ca^{2+} channels, individual channels are able to exhibit multiple modes of activity (e.g., Hess *et al.*, 1984; Delcour *et al.*, 1993). This diversity of channel structure and behavior requires investigators to be cautious in the interpretation of data from Ca^{2+} channels and highlights the need to record for reasonable periods of time to distinguish between the presence of more than one type of channel and possible shifts in mode of activity.

II. Pipettes and Solutions

Numerous considerations must be borne in mind when fabricating pipettes. Most cells seem to have a preference for particular glass types, which will, as often as not, be the deciding factor in making a choice. However, being aware that different glass types have been reported to influence channel behavior, although not directly for Ca^{2+} channels is important. Cota and Armstrong (1988) reported that, in rat pituitary cells, the use of certain types of soft glass with low (0.5 mM) ethylene glycol bis(β-amino ethylether)-N,N,N^1,N^1-tetraacetic acid (EGTA) buffer concentrations could induce fast inactivation of potassium channels. These effects were prevented with high (20 mM) EGTA concentrations or by using hard glass pipettes. The inactivation was suggested to result from channel block by di- or multivalent ions leeching from the glass. Zuazaga and Steinacker (1990) also reported variations in channel behavior attributable to differences in glass type, as observed in studies of acetylcholine-gated channels in patches from *Xenopus* myocytes. In this case, hard glass pipettes caused the preferential dropout of 40 pS channels compared with 60 pS channels. These investigators also showed that longer openings of the smaller channel dropped out first, distorting open time distributions.

Patch clamp techniques have been applied to a large number of cells from different sources, although the type of glass used to fabricate pipettes often is omitted from methods sections of papers. However, capillaries usually can be bought in small quantities, at least for the more common type of glass, and the process of trial and error can resolve the choice between hard- and soft-type glass reasonably quickly. Detailed discussion of the properties of different types of glass can be found in Corey and Stevens (1983). The majority of, if not all, electrode pullers can be adapted to pull patch pipettes. Pipettes conventionally are pulled in two stages to produce a more steeply tapered electrode shank, although a single-stage pull can be used. The heating coil temperature is usually the main determinant of the pipette tip size, the next factor that must be considered. Choice of tip size may be restricted by the ability to form and maintain a stable seal. Whole-cell recording generally employs a lower resistance pipette than is used for single channel experiments. The lower the resistance of the pipette, the lower the access resistance to the cell whereas, for single channel records, higher resistance electrodes which, as a rule of thumb, are likely to give a smaller patch of membrane may reduce the possibility of multiple channels being present in the patch. However, no simple direct relationship exists between tip size, pipette resistance, and patch area (Sakmann and Neher, 1983). Some characteristics of membrane patches sealed into pipettes and the influence of transmembrane pressure gradients are discussed by Sokabe and Sachs (1990).

Single Ca^{2+} channels often have a relatively small conductance, especially under physiological ion gradients, as well as brief openings. These properties require low noise recording at reasonably high bandwidths, conditions that can be achieved more easily by coating the shank of the electrode with one of several materials. The most commonly used material is Sylgard 184 (Dow-Corning), a silicone polymer. This compound can be premixed, partially cured to a thick stringy consistency, and then aliquoted and stored in a freezer for several weeks. The Sylgard is painted onto the pipette, close to the tip and up the shank to a point above the solution level in the bath, and cured by heating in a coil or by oven baking. The design of a jig suitable for this procedure is given by Corey and Stevens (1983). The pipette tip usually must be fire-polished after Sylgard coating to remove any material from the tip, a process that may not be necessary with uncoated pipettes. A modest coat of Sylgard can reduce recording noise substantially. Some laboratories have noted a qualitative improvement in ease of seal formation but a reduction in seal stability using Sylgard coated pipettes; whether this effect is the result of incomplete curing or reflects some property of fully cured material is unclear.

Ca^{2+} currents can be enhanced by elevation of extracellular $[Ca^{2+}]$. For some types of Ca^{2+} channel substitution of other ions, commonly barium, also can enhance currents, although not all Ca^{2+} channels have a greater permeability to Ba^{2+} than to Ca^{2+}. Most experiments are performed at elevated concentrations of permeant ion, commonly raised to levels between 10 times the physiological concentration and isotonic solutions. Ca^{2+} currents can be enhanced further by designing salines to block other ionic currents, especially K^+ currents in cells that contain voltage- and Ca^{2+}-activated K^+ channels. These solutions must be tailored to the cell under investigation to suit the complement of channels in that cell type, which can be very variable.

III. Patch Configurations

The different patch configurations (Hamill *et al.*, 1981; see Fig. 2) have been described in many reviews and will be familiar to most readers. Descriptions of how to monitor seal formation and obtain these preparations are provided in the handbooks supplied with commercial patch amplifiers and will not be described here, since these details are hardware dependent.

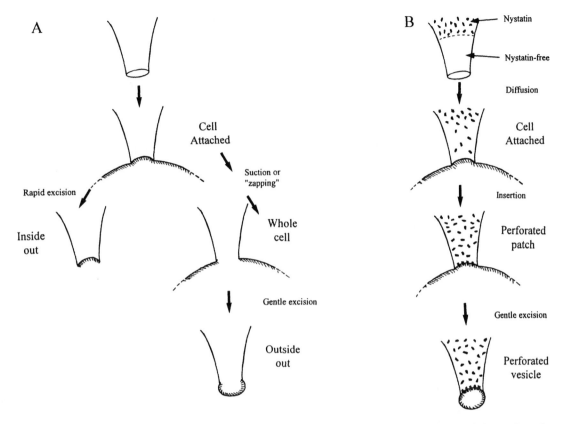

Fig. 2 Different patch configurations. (A) When the recording pipette is sealed onto the cell, a cell-attached patch is achieved. Rapid excision results in an inside-out patch (*left*). Suction results in a whole-cell patch (*right*) where gentle excision results in an outside-out patch. (B) Use of nystatin in a procedure similar to the one shown in A results in formation of a perforated patch or, with gentle excision, a perforated vesicle.

A. Cell–Attached Recording

The first patch configuration to be obtained is the cell-attached mode, in which the recording pipette is sealed onto the cell. This mode allows control of the patch membrane potential relative to the resting potential of the cell; the exact value of the resting potential is unknown unless measured or estimated independently. A common technique is to bathe the cell in isotonic KCl to set the membrane potential to zero artifically, a procedure that also can eliminate artifacts caused by action currents in excitable cells. The membrane potential also can be estimated if the cell contains a type of channel with defined properties. For example, if a channel is known to activate at -40 mV and is seen to be activated by a step of 20 mV, the cell resting potential is -60 mV. Despite this limitation, the cell-attached patch is useful because the cell is not disrupted and its normal intracellular environment is maintained. Calcium channels are known to be modulated by intracellular second messenger systems and these mechanisms are not disturbed. The effects of agents that act through second messenger systems also can be studied by application of these agents to the cell membrane outside the patch or by inclusion of such agents in the pipette filling solution.

Successful recordings in the cell-attached mode (and in the two whole-cell modes described in the next section) require a stable physical relationship between the pipette tip and the cell. Given the use of a suitable manipulator and vibration isolation, one common source of electrode drift appears to be the joint between the pipette and the electrode holder. Patch pipette holders have a rubber gasket to seal the shank of the pipette into the holder, which allows the application of pressure to the inside of the pipette. These gaskets appear to age and their regular replacement aids stability.

B. Whole Cell and Perforated Patch

Whole-cell and perforated-patch configurations permit the same type of macroscopic current recording from a cell but are distinctly different. The whole-cell method involves the physical disruption of the patch membrane from the cell-attached mode, which can be achieved by suction or by electronic "zapping"—application of a brief large-amplitude voltage pulse to the pipette. The choice of method sometimes is dictated by the cell type; some cells do not respond reliably to suction as a means of gaining electrical access, that is, the seal breaks more often than the membrane. Once the patch has been disrupted, the membrane potential of the cell can be measured directly (in current clamp) or can be controlled (for suitable cells, under voltage clamp). The cell is dialyzed with the pipette solution in whole-cell recording, that is, the cell contents are reasonably well known (however, see subsequent discussion) and various compounds can be introduced to the cell interior. The disadvantage is that normal second messenger systems also can be dialyzed out, and some types of Ca^{2+} current are very susceptible to wash-out. The pipette tip appears to form the major barrier to diffusion and equilibration of the pipette and intracellular solutions (see, for example, Mathias *et al.*, 1990). Under conditions of active transport across the cell membrane, by Na^+/K^+ pumps or Na^+/Ca^{2+} exchangers, the intracellular ion concentrations may not be equal to pipette concentrations, especially in regions close to the plasma membrane.

The perforated-patch method was introduced by Horn and Marty (1988). This method involves briefly dipping the pipette tip in normal pipette solution and then back-filling with solution to which, most commonly, nystatin has been added. Other antibiotics such as amphotericin B also have been used (Rae *et al.*, 1991). The duration of the tip dipping will depend on the geometry of the pipette; we use 5–10 sec for soft glass pipettes with a resistance of $\sim$4 MΩ. The nystatin solution we use is prepared by dissolving 6 mg nystatin in 100 μl dimethylsul-foxide (DMSO) by vortexing, then holding for 30 sec in a bath sonicator. From this stock solution, 20 μl is added to 5 ml filtered pipette solution, vortexed, and sonicated for 30 sec. This final solution should be used within 2–3 hr. Once the pipette has been back-filled with the nystatin solution, a seal must be obtained on the cell as quickly as possible. Seal formation becomes more difficult, if not impossible, once the nystatin has diffused to the pipette tip. Some laboratories recommend using positive pressure as the pipette is passed through the air–saline interface; others do not. We tend not to use positive pressure, especially when using nystatin, and routinely do not have difficulty forming seals. Once the seal has formed and the nystatin diffuses to the pipette tip, it begins to insert in the patch membrane. This process can be observed as a slow decrease in the access resistance and increase of the cell capacitance. For a given pipette, the series resistance will be higher than under standard whole-cell conditions, so particular care must be taken to compensate for this difference during experi-ments to investigate voltage-dependent Ca^{2+} channels. The time required to obtain good electrical access can vary from 1–2 min to 30 min. These delays can put people off from using this technique, but for the study of Ca^{2+} channels the advantages are numerous. Nystatin pores are permeant to monovalent ions, but not to divalents; hence normal intracellular Ca^{2+} buffering is maintained. Sec-ond messenger systems also are not washed out; nystatin pores are impermeant to compounds with a molecular weight of more than $\sim$200. These factors allow the study of Ca^{2+} currents that are lost rapidly during standard whole-cell recording. For example, Kaczmarek (1986) showed that phorbol esters that activate protein kinase C induced activity of a second type of Ca^{2+} channel in cell-attached patches from *Aplysia* bag cells. This activation was lost if the phorbol ester was applied after dialysis of the cell in the whole-cell mode.

C. Outside–Out and Perforated Vesicle

Outside-out and perforated vesicle methods allow the recording of single channel activity from a small area of membrane with the external face of the membrane exposed to the bath solution. Both methods allow the bath applica-tion of experimental solutions to the external face of the patch and investigation of the effects of physiological modulators, or pharmacological agents, on chan-nel function. The outside-out patch is obtained from a cell initially under whole-cell recording conditions and the perforated vesicle (Levitan and Kramer, 1990) from a cell under perforated-patch conditions; each has advantages and disad-vantages similar to those of the initial configurations. The amount, and content, of the cytoplasm held within the perforated vesicle will be quite variable, and the presence or absence of wash-out of Ca^{2+} channels will be similarly variable. Internal perfusion of patch electrodes is used rarely, if ever, with outside-out patches since these tend to be quite fragile (bath perfusion of these patches also requires extra care). The nystatin-permeated membrane of the perforated vesicle also would limit the range of compounds that could be introduced to the cytoplasm of the patch.

D. Inside–Out Patch

The inside-out patch configuration permits the recording of single channels and bath perfusion of the cytoplasmic face of the membrane. Potential problems include wash-out of channel activity. Patch excision also has been shown to affect channel kinetics, at least for ligand-gated channels (Trautmann and Siegelbaum, 1983; Covarrubias and Steinbach, 1990). Perfusion of the cytoplasmic face of the membrane allows the investigation of how various agents can reverse run-down or influence inactivation of the channels. Inside-out patches can be remarkably stable and perfusing the pipette is possible (Cull-Candy *et al.*, 1980), although few laboratories routinely use this technique. The remarkable stability (at least for some preparations) is demonstrated also by Sokabe and Sachs (1990), who showed that inside-out patches, which often show an Ω-profile inside the pipette, could be everted from the pipette tip under pressure.

E. Loose Patch

The loose-patch method is applied to tissues for which the tight seals between pipette and membrane required for standard cell-attached recording are not possible, for example, to avoid the extensive enzyme treatment sometimes required to "clean" the membrane. Commercial amplifiers are available for standard loose-patch recording; these generally have analog compensation for the lack of seal resistance (Stuhmer *et al.*, 1983). A more advanced form of loose-patch was described by Almers *et al.*, (1984); this technique uses concentric electrodes to eliminate the problem of potential gradients across the rim of the recording pipette. These concentric electrodes are, however, difficult to produce and the technique is used infrequently.

The primary use of the loose-patch method is mapping currents along large cells, the same pipette being used to take many recordings. The method is noninvasive to the cell and, hence, has similar advantages to cell-attached recording in terms of maintenance of normal intracellular mechanisms. The pipettes used generally have a larger tip diameter than for tight seal configurations, and records are typically from several hundred channels. As with cell-attached recording, the membrane potential is not known unless measured or estimated independently.

IV. Reconstitution of Purified Channel Proteins

Ca^{2+} channels are abundant in locations not amenable to direct patch clamp recording. To access these channels, membrane vesicle preparations can be made or channel proteins can be purified biochemically. Much work using reconstitution methods has been directed at investigation of intracellular organelles involved in Ca^{2+} release, for example, the sarcoplasmic reticulum (SR), and at the study of the transverse tubular system (T tubules) of vertebrate skeletal muscle, which contains high concentrations of dihydropyridine-sensitive Ca^{2+} channels.

Two configurations commonly are used to record from reconstituted channels: the planar lipid bilayer and microelectrode-based recording from giant liposomes or bilayers formed directly onto the pipette by tip dipping. The bilayer method is well known and has been used for single channel recording of a number of types of channel. Single Ca^{2+} channel records from lipid bilayers have been reported by several groups (e.g. Smith *et al.*, 1986) from a number of preparations. Giant liposome preparations also have been described extensively (e.g., Tank and Miller, 1983) but have not been applied as extensively as bilayers to the study of single channel Ca^{2+} currents. Liposome suspensions with Ca^{2+} channels incorporated have, however, been widely used for flux studies.

The advantage of lipid bilayers for the study of Ca^{2+} channels lies in the ability to change solutions on either side of the membrane easily; this ability can be helpful since the insertion of channels into the membrane may not always be in the same orientation. Under these circumstances, the ability to manipulate the ion gradient across the membrane by changing either side can be advantageous. Although perfusing the interior of patch pipettes is possible when recording from a patch excised from a liposome, we have noticed that these patches tend to be more fragile than native membranes. We also find that liposomes seem to lift quite readily when attempting to excise patches. One additional advantage of the bilayer seems to be that the membrane can be monitored until the Ca^{2+} channels become incorporated, whereas the channel is either present or not in traditional microelectrode patches. Attempts to balance the lipid : protein ratio of liposomes to optimize the probability of hitting a channel over getting multiple channels can be frustrating. In practice, the bilayer has remained the method of choice for recording single channel currents through Ca^{2+} channels.

Whichever method is used, another choice also must be made. The Ca^{2+} channel can be in a native membrane vesicle or can be purified and reconstituted into artificial vesicles. The use of native membrane vesicles requires less manipulation and possibly reduces the risk of losing channel activity, but the vesicle may contain other types of channel. The use of purified components allows investigation of the action of combinations of individual subunits. However, the risk of denaturation or the absence of some component required for the normal interaction of subunits remains, although this problem does not seem to be major.

Two main methods are available for generating giant liposomes suitable for patch clamp recording. The first is the freeze–thaw method (Kasahara and Hinkle, 1977; Tank and Miller, 1983) and the second is dehydration/rehydration (Criado and Keller, 1987). Which method works (best) depends on the mix of lipids used in the liposomes. For example, we find that phosphatidylserine-based liposomes reliably fuse with dehydration/rehydration, but not with freeze–thaw.

With the increasing use of molecular biology to isolate genes and express channel subunits, either alone or in combination, the use of reconstitution may become less important at least for Ca^{2+} channels found in the plasma membrane, since conventional patch technology can be applied readily to many cells used for expression studies. Reconstitution studies will remain useful for direct investigation of electrophysiological properties of intracellular Ca^{2+} channels which, when expressed, may be directed for insertion into intracellular membranes.

V. Expression Cloning of Channel Subunits

The use of *Xenopus* oocytes for studies directed at the cDNA cloning and functional expression of Ca^{2+} channel subunits has been reviewed by Snutch (1988). Detailed methodology regarding preparation of Ca^{2+} channel subunit mRNA (or cRNA) for purposes of microinjection is found in Snutch and Mandel (1992). Methodology detailing the construction of mammalian cell lines (CHO cells, HEK293 cells, mouse L cells) that stably express L- and N-type Ca^{2+} channel subunit cDNAs is found in the studies by Bosse *et al.* (1992), Williams *et al.* (1992), and Lacerda *et al.* (1991). The construction of a Ca^{2+} channel subunit cDNA expression vector for transfection of primary cell cultures of adrenal chromaffin cells is described by Ma *et al.* (1992). In general, these approaches have allowed coexpression studies using two electrode voltage clamps which, for example, have demonstrated that the β subunits derived from skeletal muscle accelerate not only the activation, but also the inactivation of Ca^{2+} channels formed by the α_1 subunits (Varardi *et al.*, 1991; Perez-Reyes *et al.*, 1992). In addition, more detailed macroscopic current analysis of the L-type α_1 subunit has revealed a voltage-sensor sequence (S4 region) and an activation domain (Tanabe *et al.*, 1991). Although the application of patch clamp methodology to the analysis of Ca^{2+} channels derived by recombinant DNA technology is still in its infancy, clearly only such an approach at the single channel level will allow the identification of functional domains in the channel that influence transitions between different forms of modal gating behavior.

VI. Pancreatic β Cell: A Model System for Analysis of Ca^{2+} Signaling

To illustrate the usefulness of an integrated approach to the study of intracellular Ca^{2+} signaling, we have selected the pancreatic β cell, a specialized endocrine cell that secretes the hormone insulin in a Ca^{2+}-dependent fashion (reviewed by Ashcroft and Rorsman, 1989; Rajan *et al.*, 1990; Boyd, 1992).

β Cells are noteworthy because insulin secretion is a tightly regulated process that is subject to stimulatory and inhibitory modulation by a remarkably large number of nutrients (glucose, amino acids), hormones (glucagon-like peptides, arginine vasopressin, gastric inhibitory peptide), and neurotransmitters (norepinephrine, somatostatin, galanin). The combined application of patch clamp (e.g., Holz *et al.*, 1993) and Ca^{2+}-imaging analysis (e.g., Grapengiesser *et al.*, 1991; Valdeolmillos *et al.*, 1992) to the study of signaling pathways that mediate these modulatory influences is one approach currently being used by investigators seeking to define the subcellular basis for stimulus–secretion coupling in this system.

The primary physiological stimulus that induces insulin secretion from pancreatic β cells is the rise in blood glucose concentration that results after ingestion of a meal (reviewed by Holz and Habener, 1992). Glucose stimulates insulin secretion via a sequence of events that requires its metabolism by aerobic glycolysis to generate ATP (Ashcroft *et al.*, 1984) and cyclic ADP–ribose (Takasawa *et al.*, 1993), two metabolites that act via a signaling cascade to depolarize β cells and to increase the concentration of cytosolic Ca^{2+} (see Fig. 3). Since vesicular insulin secretion generally is recognized to result from Ca^{2+}-dependent exocytosis, considerable attention has focused on exactly how the glucose-induced rise in intracellular Ca^{2+} is achieved. In this regard, the patch clamp technique has proven a very useful tool for analysis of Ca^{2+} channels that mediate the glucose-induced entry of Ca^{2+} across the plasma membrane. Perforated-patch studies (see Sala *et al.*, 1991, for a review of this technique as it applies to β cells) have revealed the existence of voltage-gated Ca^{2+} channels that open in response to glucose-induced depolarization (Falke *et al.*, 1989); these channels correspond to the dihydropyridine-sensitive (L-type) channels, as demonstrated by cell-attached patch recordings of unitary Ca^{2+} currents (Smith *et al.*, 1989). Rat β cells also contain a second type of voltage-dependent Ca^{2+} channel that is not susceptible to wash-out (Hiriart and Matteson, 1988). Interestingly, the β cell also expresses low conductance voltage-independent Ca^{2+} channels that open in response to glucose and have been classified as G channels by Rojas and co-workers (Rojas *et al.*, 1990).

Patch clamp analysis also has been applied to the study of second messenger pathways that mediate the infuence of hormones and neurotransmitters on β-cell Ca^{2+} channels. Arginine vasopressin, a stimulator of insulin secretion and an activator of protein kinase C, was reported to facilitate L-type Ca^{2+} currents in a transformed β cell line (Thorn and Peterson, 1991). Conversely, inhibitors of insulin secretion such as norepinephrine, somatostatin, and galanin reportedly inhibit β-cell L-type Ca^{2+} currents via a signaling pathway that involves a pertussis toxin-sensitive GTP-binding protein (Keahey *et al.*, 1989; Homaidan *et al.*, 1991; Hsu *et al.*, 1991; Schmidt *et al.*, 1991). Significantly, the facilitatory and inhibitory modulation of L-type Ca^{2+} currents by these transmitters was observed under conditions of whole-cell dialysis and recording, as expected if these responses are mediated by membrane-delimited signaling pathways that

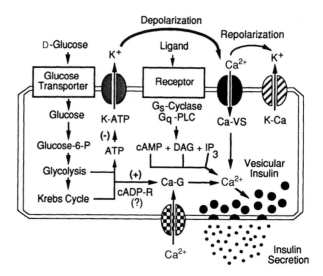

Fig. 3 Signaling systems that regulate intracellular Ca^{2+} and insulin secretion from pancreatic β cells. Illustrated are essential features of the glucose signaling system (*left*); the hormonally regulated 3',5'-cyclic adenosine monophosphate (cAMP), inositol trisphosphate (IP_3), and diacylglycerol (DAG) signaling systems (*middle*); and the cyclic ADP–ribose signaling system (*bottom*) that act in concert to trigger insulin secretion. The initial uptake of glucose is facilitated by the type-2 glucose transporter, whereas the conversion of glucose to glucose 6-phosphate is catalyzed by glucokinase. Stimulation of aerobic glycolysis generates multiple signals, one of which is an increased ratio of intracellular ATP relative to ADP. Binding of ATP to ATP-sensitive potassium channels (K–ATP) induces closure of the channels and membrane depolarization, which is necessary for the opening of L-type voltage-sensitive Ca^{2+} channels (Ca–VS). The glucose-induced rise in Ca^{2+} also results from the action of glucose-derived metabolites to open G-type voltage-independent Ca^{2+} channels (Ca–G). In this figure, one such metabolite is proposed to be cyclic ADP–ribose (cADP–R). Entry of Ca^{2+} across the plasma membrane triggers vesicular insulin secretion by Ca^{2+}-dependent exocytosis. Repolarization results from the action of cytosolic Ca^{2+} to activate Ca^{2+}-dependent potassium channels (K–Ca) and to inhibit voltage-sensitive Ca^{2+} channels. Also illustrated are hormonally regulated second messenger signaling systems that act through GTP-binding protein (G_s,G_q)-coupled receptors to stimulate production of cAMP (a messenger generated by adenylyl cyclase) and IP_3 or DAG (messengers generated by phospholipase C, PLC). These messengers increase cytosolic Ca^{2+} by facilitating the opening of Ca–VS and by mobilizing Ca^{2+} from intracellular Ca^{2+} stores.

do not involve cytosolic second messengers. This situation contrasts with that which has been described for the facilitatory action of glucose on the L-type Ca^{2+} current. As mentioned earlier, the L-type Ca^{2+} channels in β cells are known to open in response to glucose-induced cellular depolarization. In addition, a second more direct facilitatory action of glucose has been described that involves modulation of the channels through influences on the kinetics of channel gating (Smith *et al.*, 1989). Under conditions of whole-cell dialysis and recording, both responses to glucose are abolished ("wash-out"), as expected given that the β-cell glucose signaling system requires the functional integrity of numerous cytosolic enzymes, cofactors, and substrates. Only under conditions of perforated-patch recording are these actions of glucose observed, a finding that underscores the importance of carefully considering which patch clamp configuration to employ when designing an experimental study.

An important methodological advance that is generally relevant to the study of stimulus–secretion coupling is the attempt by investigators (e.g., Rorsman *et al.*, 1992) to combine the fura-2 Ca^{2+}-imaging technique (Grynkiewicz *et al.*, 1985) with perforated-patch clamp analysis. The advantage of this approach is illustrated clearly by studies conducted in our laboratory using perforated-patch and fura-2 recordings. As illustrated in Fig. 4, fura-2 recordings obtained from β cells exposed to a near-threshold stimulatory concentration of glucose (7.5 mM) revealed large periodic oscillations in the concentration of intracellular Ca^{2+}.

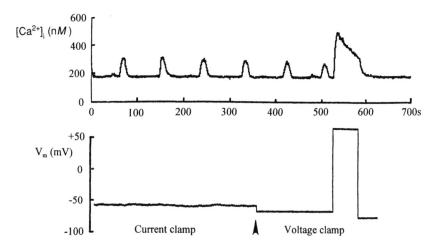

Fig. 4 Simultaneous records of intracellular calcium and membrane potential from a rat pancreatic β cell. The cell was loaded with 1 μM fura-2/AM at 37°C for 30 min. The cell was excited at 350 nm and 380 nm and the emitted light was collected at 510 nm using an Ionoptix imaging system. Intracellular $[Ca^{2+}]$ was calculated from

$$[Ca^{2+}]_i = K_d\beta(R - R_{min})/(R_{max} - R)$$

where K_d = 224 nM, the dissociation constant of fura-2 for Ca^{2+}; β = ratio of free/bound fura-2 fluorescences at 380 nm; R = ratio of 350/380 nm fluorescences measured; R_{min} = ratio of 350/380 fluorescences in Ca^{2+}-free solution; and R_{max} = ratio of 350/380 fluorescences in saturating $[Ca^{2+}]$. (*Top*) Intracellular $[Ca^{2+}]$. Oscillations of $[Ca^{2+}]$ can be seen with the cell bathed in saline containing 7.5 mM glucose, with no other form of stimulation. (*Bottom*) The record of membrane potential obtained with the perforated patch technique. Initially the cell was under current clamp with a resting potential near −60 mV. At the point marked by the arrowhead, the cell was switched to voltage clamp and held at −70 mV. The oscillations of intracellular $[Ca^{2+}]$ can be seen to occur without membrane depolarization and continue with the cell held at −70 mV. After two elevations of intracellular $[Ca^{2+}]$, a depolarizing voltage step was applied to activate voltage dependent Ca^{2+} channels and the cell responded with a large increase of intracellular $[Ca^{2+}]$.

Simultaneous perforated-patch recordings revealed that the oscillations were independent of detectable changes in membrane potential, and also were observed when the membrane potential was voltage clamped and held at $-70\,\mathrm{mV}$. The oscillations are, therefore, unlikely to result from influx of Ca^{2+} through voltage-gated (L-type) Ca^{2+} channels. Instead, the oscillations may result from the mobilization of intracellular Ca^{2+} stores and/or the activity of voltage-independent Ca^{2+} channels (G channels). Using this combined patch clamp and imaging approach, we can begin to assess the relative contribution of voltage-dependent and -independent processes that govern intracellular Ca^{2+} homeostasis.

One final example illustrates the impressive flexibility of patch clamp recording for analysis of Ca^{2+}-dependent processes. The perforated-patch configuration was reported to be suitable for measurement of changes in cellular capacitance that are attributable to exocytotic fusion of secretory vesicles with the plasma membrane (Gillis and Misler, 1992). Although this capacitance tracking methodology has yet to be exploited fully, it should allow an assessment of the temporal pattern of vesicle fusion in response to chemical stimuli (glucose, hormones, neurotransmitters) or voltage clamp stimuli (step depolarizations). Certainly the full potential of this technique will be realized in future studies that combine Ca^{2+}-imaging analysis with perforated-patch recordings of Ca^{2+} currents and cellular capacitance.

References

Almers, W., Roberts, W. M., and Ruff, R. L. (1984). Voltage clamp of rat and human skeletal muscle. Measurements with an improved loose-patch technique. *J. Physiol.* **347,** 751–768.

Ashcroft, F. M., and Rorsman, F. (1989). Electrophysiology of the pancreatic β-cell. *Prog. Biophys. Mol. Biol.* **54,** 87–143.

Ashcroft, F. M., Harrison, D. E., and Ashcroft, S. J. H. (1984). Glucose induces closure of single potassium channels in isolated rat pancreatic β-cells. *Nature (London)* **312,** 446–448.

Bean, B. P. (1992). Whole-cell recording of calcium channel currents. *Meth. Enzymol.* **207,** 181–193.

Berridge, M. J. (1993). Inositol triphosphate and calcium signalling. *Nature (London)* **361,** 315–325.

Bosse, E., Bottlender, R., Kleppisch, T., Hescheler, J., Welling, A., Hofman, F., and Flockerzi, V. (1992). Stable and functional expression of the calcium channel α_1 from smooth muscle in somatic cell lines. *EMBO J.* **11(6),** 2033–2038.

Boyd, A. E., III. (1992). The role of ion channels in insulin secretion. *J. Cell. Biochem.* **48,** 234–241.

Catterall, W. A. (1988). Structure and function of voltage-sensitive ion-channels. *Science* **242,** 50–61.

Cena, V., Brocklehurst, K. W., Pollard, H. B., and Rojas, E. (1991). Pertussis toxin stimulation of catecholamine release from adrenal medullary chromaffin cells: Mechanism may be by direct activation of L-type and G-type calcium channels. *J. Membrane Biol.* **122,** 23–31.

Conn, P. M. (ed.) (1991). "Electrophysiology and Microinjection," Methods in Neurosciences, Vol. 4. San Diego: Academic Press.

Corey, D. P., and Stevens, C. F (1983). Science and technology of patch-recording electrodes. *In* "Single Channel Recording" (E. Neher and B. Sakmann, eds.), pp. 53–68. New York: Plenum Press.

Cota, G., and Armstrong, C. M. (1988). Potassium channel "inactivation" induced by soft-glass patch pipettes. *Biophys. J.* **53**, 107–109.

Covarrubias, M., and Steinbach, J. H. (1990). Excision of membrane patches reduces the mean open time of nicotinic acetylcholine receptors. *Pflügers Arch.* **416**, 385–392.

Criado, M., and Keller, B. U. (1987). A membrane fusion strategy for single channel recordings of membranes usually non-accessible to patch-clamp electrodes. *FEBS Lett.* **224**, 172–176.

Cull-Candy, S. G., Miledi, R., and Parker, I. (1980). Single glutamate-activated channels recorded from locust muscle fibres with perfused patch-clamp electrodes. *J. Physiol.* **321**, 195–210.

Delcour, A. H., Lipscombe, D., and Tsien, R. W. (1993). Multiple modes of N-type calcium channel activity distinguished by differences in gating kinetics. *J. Neurosci.* **13**, 181–194.

Falke, L. C., Gillis, K. D., Pressel, D. M., and Misler, S. (1989). "Perforated patch" recording allows long-term monitoring of metabolite-induced electrical activity and voltage-dependent Ca currents in pancreatic β-cells. *FEBS Lett.* **251**, 167–172.

Fatt, P., and Katz, B. (1953). The electrical properties of crustacean muscle fibres. *J. Physiol.* **120**, 171–204.

Gillis, W. D., and Misler, S. (1992). Single cell assay of exocytosis from pancreatic β-cells. *Pflügers Arch.* **420**, 121–123.

Grapengiesser, E., Gylfe, E., and Hellman, B. (1991). Cyclic AMP as a determinant for glucose-induction of fast Ca^{2+} oscillations in isolated pancreatic β-cells. *J. Biol. Chem.* **266**, 12207–12210.

Grynkiewicz, G., Poenie, M., and Tsien, R. (1985). A new generation of Ca^{2+} indicators with greatly improved fluorescence properties. *J. Biol. Chem.* **260**, 3440—3450.

Hamill, O. P., Marty, A., Neher, E., Sakmann, B., and Sigworth, F. J. (1981). Improved patch-clamp techniques for high resolution current recording in cells and cell-free membrane patches. *Pflügers Arch.* **391**, 85–100.

Hess, P., Lansman, J. B., and Tsien, R. W. (1984). Different modes of Ca-channel gating behaviour favoured by dihydropyridine Ca-agonists and antagonists. *Nature (London)* **311**, 538–544.

Hiriart, M., and Matteson, D. R. (1988). Na channels and two types of Ca^{2+} channels in rat pancreatic β-cells identified with the reverse hemolytic plaque assay. *J. Gen. Physiol.* **91**, 617–639.

Holz, G. G., and Habener, J. F. (1992). Signal transduction crosstalk in the endocrine system: Pancreatic β-cells and the glucose competence concept. *Trends Biochem. Sci.* **7**, 388–393.

Holz, G. G., Kuhtreiber, W. M., and Habener, J. F. (1993). Pancreatic β-cells are rendered glucose-competent by the insulinotropic hormone glucagon-like peptide-1(7-37). *Nature (London)* **361**, 362–365.

Homaidan, F. R., Sharp, G. W., and Nowak, L. M. (1991). Galanin inhibits a dihydropyridine-sensitive Ca^{2+} current in the RINm5f cell line. *Proc. Natl. Acad. Sci. U.S.A.* **88**, 8744–8788.

Horn, R., and Marty, A. (1988). Muscarinic activation of ionic currents measured by a new whole-cell recording method. *J. Gen. Physiol.* **92**, 145–159.

Hsu, W. H., Xiang, H. D., Rajan, A. S., Kunze, D. L., and Boyd, A. E. (1991). Somatostatin inhibits insulin secretion by a G-protein-mediated decrease in Ca entry through voltage-dependent Ca channels in the beta-cell. *J. Biol. Chem.* **266**, 837–843.

Kaczmarek, L. K. (1986). Phorbol esters, protein phosphorylation and the regulation of ion-channels. *J. Exp. Biol.* **124**, 375–392.

Kasahara, M., and Hinkle, P. C. (1977). Reconstitution and purification of the D-glucose transporter from human erythrocytes. *J. Biol. Chem.* **252**, 7384–7390.

Keahey, H., Boyd, A. E., and Kunze, D. L. (1989). Catecholamine modulation of calcium currents in clonal pancreatic β-cells. *Am. J. Physiol.* **257**, C1171–C1176.

Kettenmann, H., and Gratyn, R. (eds.) (1992). "Practical Electrophysiological Methods. A Guide for *In Vitro* Studies in Vertebrate Neurobiology." New York: Wiley-Liss.

Lacerda, A-E., Kim, H-S., Ruth, P., Perez-Reyes, E., Flockerzi, V., Hofmann, F., Birnbaumer, L., and Brown, A-M. (1991). Normalization of current kinetics by interaction between the α_1 and β subunits of skeletal muscle dihydropyridine-sensitive Ca^{2+} channel. *Nature (London)* **352**, 527–530.

Levitan, E. S., and Kramer, R. H. (1990). Neuropeptide modulation of single calcium and potassium channels detected with a new patch clamp configuration. *Nature (London)* **348,** 546–547.

Llinas, R., Sugimori, M., Lin, J. W., and Cherksey, B. (1989). Blocking and isolation of a calcium channel from neurones in mammals and cephalopods utilizing a toxin fraction (FTX) from funnel-web spider poison. *Proc. Natl. Acad. Sci. U.S.A.* **86,** 1689–1693.

Ma, W. J., Holz, R. W., and Uhler, M. D. (1992). Expression of a cDNA for a neuronal calcium channel alpha 1 subunit enhances secretion from adrenal chromaffin cells. *J. Biol. Chem.* **267,** 22728–22732.

Mathias, R. T., Cohen, I. S., and Oliva, C. (1990). Limitations of the whole cell patch-clamp technique in the control of intracellular concentrations. *Biophys. J.* **58,** 759–770.

Mikami, A., Imoto, K., Tanabe, T., Niidome, T., Mori, Y., Takeshima, H., Narumiya, S., and Numa, S. (1989). Primary structure and functional expression of the cardiac dihydropyridine-sensitive calcium channel. *Nature (London)* **340,** 230–233.

Miller, R. J. (1992). Voltage-sensitive Ca^{2+} channels. *J. Biol. Chem.* **267,** 1403–1406.

Neher, E., and Sakmann, B. (1983). "Single Channel Recording." New York: Plenum Press.

Nowycky, M. C., Fox, A. P., and Tsien, R. W. (1985). Three types of neuronal calcium channel with different calcium agonist sensitivity. *Nature (London)* **316,** 440–443.

Perez-Reyes, E., Kim, H. S., Lacerda, A. E., Horne, W., Wei, X., Rampe, D., Campbell, K. P., Brown, A. M., and Birnbaumer, L. (1989). Induction of calcium currents by the expression of the α_1-subunit of the dihydropyridine receptor from skeletal muscle. *Nature (London)* **340,** 233–236.

Perez-Reyes, E., Castellano, A., Kim, H. S., Bertrand, P., Baggstrom, E., Lacerda, A. E., Wei, X. Y., and Birnbaumer, L. (1992). Cloning and expression of a cardiac/brain beta subunit of the L-type calcium channel. *J. Biol. Chem.* **267,** 1792–1797.

Rae, J., Cooper, K., Gates, P., and Watsky, M. (1991). Low access resistance perforated patch recordings using amphotericin B. *J. Neurosci. Meth.* **37,** 15–26.

Rajan, A. S., Aguilar-Bryan, L., Nelson, D. A., Yaney, G. C., Hsu, W. H., Kunze, D. L., and Boyd, A. E. (1990). Ion channels and insulin secretion. *Diabetes Care* **13,** 340–363.

Rojas, E., Hidalgo, J., Carroll, P. B., Li, M. X., and Atwater, I. (1990). A new class of calcium channel activated by glucose in human pancreatic β-cells. *FEBS Lett.* **261,** 265–270.

Rorsman, P., Ammala, C., Berggren, P.-O., Bokvist, K., and Larsson, O. (1992). Cytoplasmic calcium transients due to single action potentials and voltage-clamp depolarizations in mouse pancreatic β-cells. *EMBO J.* **11,** 2877–2884.

Sakmann, B., and Neher, E. (1983). Geometric parameters of pipettes and membrane patches. *In* "Single Channel Recording," (E. Neher and B. Sakmann, eds.), pp. 37–51. New York: Plenum Press.

Sala, S., Parsey, R. V., Cohen, A. S., and Matteson, D. R. (1991). Analysis and use of the perforated patch technique for recording ionic currents in pancreatic β-cells. *J. Membrane Biol.* **122,** 177–187.

Schmidt, A., Hescheler, J., Offermanns, S., Spicher, K., Hinsch, K. D., Klinz, F. J., Codina, J., Birnbaumer, L., Gausepohl, H., Frank, R., Schultz, G., and Rosenthal, W. (1991). Involvement of pertussis toxin-sensitive G-proteins in the hormonal inhibition of dihydropyridine-sensitive Ca^{2+} currents in an insulin-secreting cell line (RINm5F). *J. Biol. Chem.* **266,** 18025–18033.

Smith, J. S., Coronado, R., and Meissnes, G. (1986). Single channel measurements of the calcium release channel from skeletal muscle sarcoplasmic reticulum. Activation by Ca^{2+} and ATP and modulation by Mg^{2+}. *J. Gen. Physiol.* **88,** 573–588.

Smith, P. A., Rorsman, P., and Ashcroft, F. M. (1989). Modulation of dihydropyridine-sensitive Ca^{2+} channels by glucose metabolism in mouse pancreatic β-cells. *Nature (London)* **342,** 550–553.

Snutch, T. P. (1988). The use of Xenopus oocytes to probe synaptic communication. *Trends Neurosci.* **11,** 250–256.

Snutch, T. P., and Mandel, G. (1992). Tissue RNA as a source of ion channels and receptors. *Meth. Enzymol.* **207,** 297–309.

Sokabe, M., and Sachs, F. (1990). The structure and dynamics of patch-clamped membranes: A study using differential interference contrast light microscopy. *J. Cell Biol.* **111,** 599–606.

Standen, N. B., Gray, P. T. A., and Whitaker, M. J. (eds.) (1987). "Microelectrode Techniques. The Plymouth Workshop Handbook." Cambridge, The Company of Biologists.

Stuhmer, W., Roberts, W. M., and Almers, W. (1983). The loose-patch technique. *In* "Single Channel Recording" (E. Neher and B. Sakmann, eds.), pp. 123–132. New York: Plenum Press.

Takasawa, S., Nata, K., Yonekura, H., and Okamoto, H. (1993). Cyclic ADP-ribose in insulin secretion from pancreatic β-cells. *Science* **259,** 370–373.

Tanabe, T., Adams, B. A., Numa, S., and Beam, K. G. (1991). Repeat 1 of the dihydropyridine receptor is critical in determining calcium channel activation kinetics. *Nature (London)* **352,** 800–803.

Tank, D. W., and Miller, C. (1983). Patch-clamped liposomes. Recording reconstituted ion-channels. *In* "Single Channel Recording" (E. Neher and B. Sakmann, eds.), pp. 91–105. New York: Plenum Press.

Thorn, P., and Peterson, O. H. (1991). Activation of voltage-sensitive Ca^{2+} currents by vasopressin in an insulin-secreting cell line. *J. Membrane Biol.* **124,** 63–71.

Trautmann, A., and Siegelbaum, S. A. (1983). The influence of membrane patch isolation on single acetylcholine-channel current in rat myotubes. *In* "Single Channel Recording" (E. Neher and B. Sakmann, eds.), pp. 473–480. New York: Plenum Press.

Tsien, R. W., Ellinor, P. T., and Horne, W. A. (1991). Molecular diversity of voltage-dependent Ca^{2+}-channels. *Trends Pharmacol. Sci.* **12,** 349–354.

Valdeolmillos, M., Nadal, A., Contreras, D., and Soria, B. (1992). The relationship between glucose-induced K-ATP channel closure and the rise in Ca^{2+} in single mouse pancreatic β-cells. *J. Physiol. (London)* **455,** 173–186.

Varardi, G., Lory, P., Schultz, D., Varadi, M., and Schwartz, A. (1991). Acceleration of activation and inactivation by the beta subunit of the skeletal muscle calcium channel. *Nature (London)* **352,** 159–162.

Williams, M. E., Brust, P. F., Feldman, D. H., Patthi, S., Simerson, S., Maroufi, A., McCue, A. F., Velicelebi, G., Ellis, S. B., and Harpold, M. M. (1992). Structure and functional expression of an omega-conotoxin-sensitive human N-type calcium channel. *Science* **257,** 389–395.

Zuazaga, C., and Steinacker, A. (1990). Patch-clamp recording of ion channels: Interfering effects of patch pipette glass. *News Physiol. Sci.* **5,** 155–158.

PART III

Fluorescence Techniques for Imaging $[Ca^{2+}]_i$

CHAPTER 7

Practical Aspects of Measuring $[Ca^{2+}]$ with Fluorescent Indicators

Joseph P. Y. Kao

Medical Biotechnology Center and
Department of Physiology
University of Maryland School of Medicine
Baltimore, Maryland 21201

I. Introduction

In the application of any measurement technique, a body of practical knowledge is possessed by all experienced practitioners. Although important for making successful measurements, such lore, which sometimes seems arcane, often is not described explicitly or explained in journal publications. In this respect, measuring free $[Ca^{2+}]$ with fluorescent indicators is no exception. The purpose of this chapter is to gather in one place some of the most common and useful practical information relevant to the use of fluorescent Ca^{2+} indicators. Such a collection of information is hoped to alleviate the frustration of individuals who are novices at using fluorescent indicators.

II. Fluorescent Ca^{2+} Indicators

The commonly available fluorescent indicators for Ca^{2+} fall into two operational classes: single-wavelength intensity-modulating dyes and dual-wavelength ratiometric dyes (Table I), which are referred to as single-wavelength (SW) indicators and ratiometric indicators, respectively. For SW indicators, changes in $[Ca^{2+}]$ bring about changes in the intensity of their fluorescence excitation and emission spectra[1] whereas the spectral maxima remain essentially unchanged. Chemical structures of some of the indicators listed in Table I are shown in Fig. 1. Excitation spectra of fluo-3 (Minta *et al.*, 1989), a SW indicator, at saturating and "zero" $[Ca^{2+}]$ are shown in Fig. 2. Ratiometric indicators exhibit not only intensity changes with changing $[Ca^{2+}]$ but the Ca^{2+}-free and Ca^{2+}-bound forms of the indicator actually have distinct spectra, the maxima in which are located at different wavelengths (the spectra show wavelength shifts). The two ratiometric indicators most commonly used are fura-2 and indo-1 (Grynkiewicz *et al.*, 1985). For fura-2, significant shifts are observed in the excitation spectra (Fig. 3) but not in the emission spectra. Indo-1 shows a significant shift primarily in its emission spectra. For SW indicators, because intensity monitored at a single wavelength is the only experimental measurement that is related to $[Ca^{2+}]$, intensity changes arising from factors unrelated to changes in $[Ca^{2+}]$ (e.g., changes in cell thickness, loss of indicator from a cell by leakage) can confound interpretation of the intensity data. In contrast, because the Ca^{2+}-free and Ca^{2+}-bound forms of ratiometric indicators are characterized by spectral peaks at different wavelengths, intensity measurements can be made at two different wavelengths and a ratio can be obtained (Grynkiewicz *et al.*, 1985). Obtaining a ratio minimizes the effect of many artifacts that are unrelated to changes in $[Ca^{2+}]$. The two commonly used ratiometric indicators, fura-2 and indo-1, require excitation in the UV range whereas most of the common SW dyes use visible excitation light. Although the ratiometric dyes can be calibrated more reliably (Section V), sometimes

---

[1] An excitation spectrum is taken by monitoring fluorescence emission intensity at a fixed wavelength while excitation light is scanned through a wavelength range over which the sample can absorb light. The emission intensity is plotted as a function of excitation wavelength. To collect an emission spectrum, excitation light at a fixed wavelength is delivered to the sample while the emission intensity is monitored over a wavelength range. Here, the emission intensity is plotted as a function of emission wavelength.

Table I
Properties of Common Fluorescent Ca^{2+} Indicators[a]

Indicator type	K_d (nM)	Absorption maxima (nm) Ca^{2+}-free	Absorption maxima (nm) Ca^{2+}-bound	Emission maxima (nm) Ca^{2+}-free	Emission maxima (nm) Ca^{2+}-bound
Single-wavelength intensity-modulating					
Monomeric					
Quin2	115[b]	352	332	492	498
Fluo-3	400	503	506	526	526
Calcium Green-1™	189	506	506	534	533
Calcium Green-2™	574	506	506	531	531
Calcium Green-5N™	3.3[c]	506	506	531	531
Rhod-2	1.0[c]	556	553	576	576
Calcium Orange™	328	554	555	575	576
Calcium Crimson™	205	588	588	611	611
Dextran-conjugated					
Calcium Green-1™ dextran[d]	~250–350	506	509	531	532
Dual-wavelength ratiometric					
Monomeric					
Fura-2	224[b]	362	335	512	505
Fura Red™	133	472	436	645	640
Indo-1	250[b]	349	331	485	410
Dextran-conjugated					
Fura(-2) dextran[d]	~340–440	~366	~341	~501	~495
Indo(-1) dextran[d]	~250–360	~356–361	~341	~465	~408

[a] Data from Tsien (1980), Grynkiewicz et al. (1985), Minta et al. (1989), and Haugland (1992).

[b] Effective K_d in the presence of 1 mM Mg^{2+}.

[c] μM.

[d] The K_d and absorption and emission maxima of dextran-conjugated indicators can vary from lot to lot and is dependent on the molecular weight of the dextran used as well.

avoiding using UV light for excitation may be necessary (e.g., UV can excite significant autofluorescence in some biological preparations and can photolyze photosensitive caged compounds). Clearly, in practice, instrumentation for using ratiometric indicators is more complex than that for SW indicators.

Quin2 (Tsien, 1980; Tsien et al., 1982) is the archetypal tetracarboxylate indicator listed in Table I. Its properties and applications as a SW indicator have been reviewed in detail (Tsien and Pozzan, 1989). However, quin2 has been superseded by the new generations of SW and ratiometric indicators. Of the SW dyes listed in Table I, fluo-3 and the Calcium Green™ dyes incorporate fluorescein chromophores and are, therefore, excited at wavelengths typical for fluoresceins. Fluo-3 and Calcium Green-2 exhibit the largest intensity changes in their transition from Ca^{2+}-free to Ca^{2+}-bound forms (40- to 100-fold; Minta et al., 1989; Haugland, 1992). This change can be an advantage because, compared with other SW indicators, similar changes in [Ca^{2+}] result in larger changes in

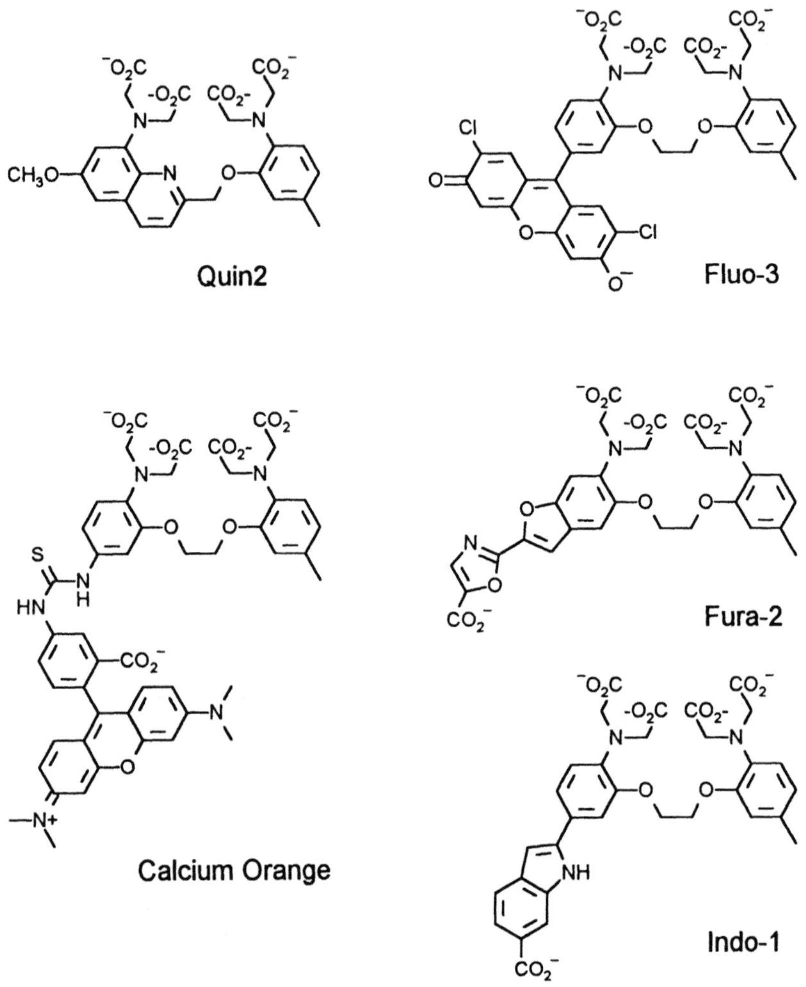

Fig. 1 Structures of selected fluorescent Ca^{2+} indicators. All indicators are represented in their polycarboxylate Ca^{2+}-sensitive forms.

brightness for fluo-3 and Calcium Green-2. Because fluorescence quantum efficiency[2] can range only from 0 to 1, the large intensity difference between Ca^{2+}-bound and Ca^{2+}-free forms implies that the Ca^{2+}-free forms of the two indicators must be only weakly fluorescent. This fact sometimes can be slightly annoying because cells with relatively low resting $[Ca^{2+}]_i$ (cytosolic free Ca^{2+}

---

[2] Fluorescence quantum efficiency (Φ_F) is the fraction of total light absorbed that is emitted as fluorescence.

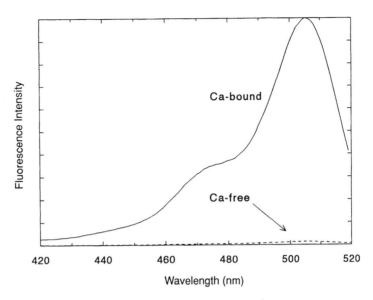

Fig. 2 Excitation spectra of fluo-3 ($\lambda_{emission}$ = 525 nm). The Ca^{2+}-free form of fluo-3 is more than 40 times less bright than the Ca^{2+}-bound form.

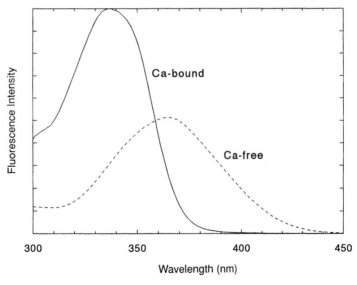

Fig. 3 Excitation spectra of Ca^{2+}-bound and Ca^{2+}-free forms of fura-2 ($\lambda_{emission}$ = 505 nm).

concentration) would have most of the indicator in the Ca^{2+}-free form and therefore would be very dim.

Rhod-2, Calcium Orange, and Calcium Crimson are indicators that incorporate rhodamine-type chromophores and therefore are excited at much longer wavelengths than are fluo-3 and Calcium Green dyes. When the acetoxymethyl (AM) ester is used to load cells, rhod-2 seems to load well into mitochondria; up to 80% of the intracellular dye is located in that organelle (B. Bacskai, personal communication).

A choice of ratiometric indicator can be made on practical grounds. Typically, fura-2 is excited alternately at two different wavelengths whereas the emission is collected at a single fixed wavelength. Therefore the pair of intensity measurements, whether in imaging or in single-cell microfluorometry, must be collected sequentially. Indo-1, on the other hand, usually is excited at a fixed wavelength whereas emission is monitored simultaneously at two different wavelengths, that is, emission from the Ca^{2+}-bound and Ca^{2+}-free forms of the indicator can be collected simultaneously. Therefore, indo-1 potentially can give better temporal resolution. However, in conventional imaging, indo-1 can be more difficult to use because the two emission images, usually collected through slightly different optical paths, can be difficult to keep in spatial registration. At present, fura-2 is the most widely used Ca^{2+} indicator, both in conventional imaging and in single-cell measurements. Indo-1 has, however, been used with success in confocal imaging applications by Tsien and colleagues (Wong *et al.,* 1992). Fura Red™ is touted as a ratiometric indicator the excitation and emission wavelengths of which are both in the visible range. This indicator suffers from having very low fluorescence quantum efficiency (~0.013 in the Ca^{2+}-free form; J. P. Y. Kao, unpublished results[3]). Fura Red differs from the other ratiometric indicators because its fluorescence intensity is diminished by Ca^{2+} binding. The relatively low quantum efficiency implies that higher indicator concentrations and/or higher excitation intensities are required.

The dextran-conjugated dyes are biopolymers with pendant indicator molecules. Those compounds listed in Table I are available with dextran molecular weights of 10,000 and 70,000 (Molecular Probes, Eugene, Oregon). Being membrane impermeant, dextran conjugates must be loaded into cells by an invasive technique such as microinjection. Whereas the monomeric indicators can leak out of cells at a steady rate (Section III,C), dextran-conjugated indicators tend to have long residence times in cells. Dextran-conjugated dyes therefore can be useful in applications in which long-term monitoring of $[Ca^{2+}]_i$ is required. Instances also occur in which cells rapidly transport monomeric dyes into internal organelles (Hepler and Callaham, 1987) but do not do so when dextran conjugates are used. Because the conjugates are made by covalent attachment of monomeric indicators to dextran polymers, individual indicator monomers can reside in slightly different local microenvironments. Therefore, the conju-

---

[3] Quantum efficiency of Ca^{2+}-free form determined relative to carboxy-SNARF-1.

gates, rather than having a unique K_d and identical spectral properties, probably are characterized by a range of microscopic K_ds and a distribution of spectral properties. These characteristics provide a likely explanation for lot-to-lot variations in K_d and spectral characteristics.

III. Loading Indicators into Cells

The common fluorescent indicators for Ca^{2+} are polycarboxylate anions that cannot cross lipid bilayer membranes and therefore are not cell permeant. In the negatively charged form, the indicators can be introduced into cells only by microinjection or by permeabilization, procedures that require some special equipment and skill.[4] By far the most convenient way of loading an indicator into cells is incubating the cells in a dilute solution or dispersion of the AM ester of the indicator. The AM group is used to mask the negative charges on the carboxyl groups present in the indicator molecule. The AM ester form of the indicator is uncharged and hydrophobic. Consequently, this form can cross lipid membranes and gain entry into the interior of cells. The carboxyl groups in the indicator, however, are essential to the ability of the indicator molecule to sense Ca^{2+}; therefore, the AM groups must be removed once the AM ester has entered the cell. Because the AM group is labile to enzymatic hydrolysis by esterases present in the cell, AM esters of the indicator are processed intracellularly to liberate the Ca^{2+}-sensitive polycarboxylate form which, being multiply charged, becomes trapped inside the cell. Trapping of the polyanionic form of the indicator allows cells to accumulate concentrations of up to hundreds of micromolar of the Ca^{2+}-sensitive form of these molecules when incubated with micromolar concentrations of the AM ester in the extracellular medium. Several factors influence the efficiency and quality of indicator loading via the AM ester, and will be discussed subsequently.

A. Limited Aqueous Solubility of AM Esters

AM esters of the common Ca^{2+} indicators have molecular weights in excess of 1000. Being large uncharged organic molecules, these esters have extemely low solubility in aqueous media. For example, at 25°C, the solubility of the AM ester of fura-2 in pure water is only 0.11 μM (Kao *et al.*, 1990). In biological media, in which the ionic strength is typically ~0.15 M, the solubility of fura-2/AM would be even lower. Addition of AM ester in excess of the solubility limit simply would result in precipitation of solid AM ester, which is effectively unavailable for loading cells. In addition, fine particles of solid AM

---

[4] A variety of techniques for loading membrane-impermeant species into cells is discussed by McNeil (1989).

ester often adhere well to the outer surfaces of cells or to the extracellular matrix and can contribute large Ca^{2+}-insensitive fluorescence signals to the measurement.[5] A convenient solution to the solubility problem is the use of Pluronic F-127, a mild nonionic surfactant,[6] as a dispersing agent for AM esters. Typically, aliquots of Pluronic and AM ester stock solutions in dimethylsulfoxide (DMSO) are mixed intimately before dispersal into an aqueous medium. The Pluronic is presumed to sequester the AM ester in micellar form, thus preventing precipitation, and the micelles are presumed to serve as a steady source to replenish AM esters taken up by cells. The net result is significantly improved loading of indicators into cells.

B. Dye Compartmentalization: Loading of Indicator into Subcellular Compartments Other than the Cytosol

1. Minimizing Compartmentalization

In typical experiments, one usually wishes to monitor changes in the concentration of Ca^{2+} in the cytosol; therefore, ideally, one would like the indicator to be loaded exclusively into the cytosol. This ideal situation is almost never realized for two reasons. First, because the AM ester form of the indicator is membrane permeant, it can enter not only the cytosol but all subcellular membrane-enclosed compartments as well. Although this process occurs to a large extent in the cytosol, enzymatic hydrolysis of AM esters also can take place within subcellular organelles. Therefore, some fraction of the indicator molecules tends to be trapped in subcellular noncytosolic compartments. Second, some cell types actively endocytose material from the incubation medium (Malgaroli *et al.*, 1987), including dispersed AM esters, which then are hydrolyzed to release fluorescent indicator molecules within organelles of the endocytotic pathway. Presumably, indicator molecules liberated in this way can end up in a variety of organelles that are connected to the endocytotic pathway by vesicular traffic. Because most subcellular organelles [e.g., endoplasmic reticulum (ER), lysosomes] tend to have high intraorganellar $[Ca^{2+}]$ ($>\mu M$), indicators confined to these organelles would be saturated with Ca^{2+} and would contribute a high-$[Ca^{2+}]$ fluorescence signal that would not vary with changes in cytosolic $[Ca^{2+}]$. Therefore, the net effect of compartmentalized dye is biasing measured cytosolic free Ca^{2+} concentration toward higher values.

The first cause of dye compartmentalization just stated is a reflection of an inherent imperfection of the AM ester loading technique and cannot be remedied easily. The second cause is a cell biological process and can be attenuated. Because endocytosis is a temperature-dependent process, cells loaded at lower

---

[5] This is a problem with AM esters that are fluorescent (e.g., fura-2/AM and indo-1/AM) but not with nonfluorescent AM esters (e.g., fluo-3/AM).

[6] Pluronic F-127 is manufactured by BASF Wyandotte Corporation, Wyandotte, Wisconsin. Free samples accompany AM ester orders from Molecular Probes.

temperatures with AM ester tend to show less compartmentalization of indicator (Malgaroli *et al.*, 1987). The following results illustrate this point. In REF52 cells, roughly 30% of total intracellular fura-2 was in non cytosolic compartments when loading via the AM ester was carried out at 37°C, whereas only 10% was compartmentalized when loading was performed at 23°C.

Although endocytosis is known to be blocked at 10°C in mammalian cells and at 4°C in amphibian cells (J. B. Wade, personal communication), loading at the lowest biologically permissible temperatures does not necessarily yield the best results because AM ester processing in the cytosol is also temperature dependent; at very low temperatures, the concentration of indicator accumulated in the cytosol can be quite low. Optimal loading temperature is determined empirically to be the temperature at which dye compartmentalization is minimized while good cytosolic loading is maintained. In practice, convenience often dictates loading at room temperature as a reasonable compromise.[7]

2. Assessing Extent of Compartmentalization

The extent of indicator compartmentalization can be estimated through a simple semi-quantitative procedure that is based on the observation that micromolar concentrations of digitonin primarily tend to permeabilize the plasma membrane and release cytosolic dye, whereas 1% Triton X-100 can permeabilize and release dye from subcellular organelles (for example, see Kao *et al.*, 1989). The procedure consists of monitoring total fluorescence from a cell or a field of cells, preferably bathed in low-Ca^{2+} medium,[8] and treating the cells first with digitonin and then with Triton X-100. Figure 4 shows the procedure being applied to a cell loaded with fura-2/AM. The fluorescence measured before treatment (F_i) represents contributions from cytosolic and compartmentalized dye plus background. After digitonin release of cytosolic dye, the fluorescence measured (F_d) represents compartmentalized dye plus background. The fluorescence level attained after Triton X-100 treatment is considered the background (F_b). The fraction of total intracellular dye that is compartmentalized is simply $(F_d - F_b)/(F_i - F_b)$.

C. Dye Leakage or Extrusion from Cells

Once an indicator is loaded into cells, it leaks out at a rate that is strongly temperature dependent. For mammalian cells, the loss rate is maximal at 37°C

---

[7] When loading is done in air, the incubation medium should be bicarbonate-free (i.e., some other buffer such as HEPES should be used to maintain pH). Otherwise, steady loss of CO_2 will shift the CO_2–HCO_3^-–CO_3^{2-} equilibrium and rapidly alkalinize the medium.

[8] Nominally Ca^{2+}-free medium or, better yet, medium containing sufficient ethylene glycol bis (β-amino ethyl ether)-N,N,N',N'-tetraacetic acid (EGTA) to make [Ca^{2+}] $< 1 \ \mu M$ (see Section IV,B,1).

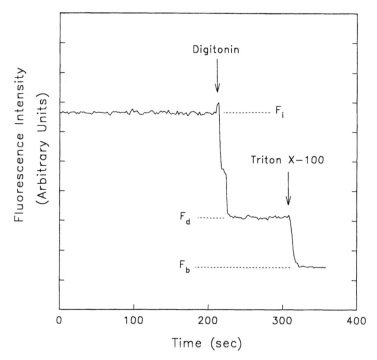

Fig. 4 Procedure for assessing dye compartmentalization. REF52 fibroblast incubated at 1 μM fura-2/AM in Pluronic F-127 dispersion in minimum essential medium (MEM) for 60 min in an air incubator at 37°C. Measurement was done in Hank's basic salt solution (HBSS) containing 2.6 mM EGTA (sufficient to reduce extracellular [Ca^{2+}] to <1 μM), and buffered at pH 7.4 with 10 mM HEPES. The concentration of digitonin was 20 μM; that of Triton X-100, 1% w/v.

and drops off sharply as temperature is lowered,[9] as shown in Fig. 5A. Loss of indicator apparently occurs by an extrusion mechanism for organic anions (Di Virgilio *et al.*, 1988) and can be blocked effectively by inhibitors of uric acid transport such as probenecid and sulfinpyrazone (Di Virgilio *et al.*, 1988, 1990), as illustrated in Fig. 5b. Sulfinpyrazone at tens to hundreds micromolar has been found to be effective, whereas probenecid works at millimolar dosage. High concentrations of inhibitor virtually can stop indicator extrusion but also can induce signs of cellular stress such as blebbing. Therefore, using the minimum concentration of inhibitor necessary to reduce the rate of indicator loss to a tolerable level without attempting to block the process altogether is advisable. In general, slowing indicator loss by lowering the temperature at which experiments are conducted is preferable to using transport inhibitors.

[9] The sharp increase in rate of indicator leakage with rising temperature is another reason that compartmentalization is more severe when cells are incubated with AM esters at high temperatures. At higher temperatures, cytosolic dye is lost rapidly whereas dye in organelles is not, so compartmentalized dye becomes a larger proportion of the total cellular dye content.

Probenecid and sulfinpyrazone are both hydrophobic organic acids and, as such, are practically insoluble in water. These molecules must be neutralized with a stoichiometric amount of base (e.g., NaOH) before an aqueous stock solution can be prepared.

D. Procedure for Loading

Stock solutions of the AM ester of the indicator can be made with dry DMSO as solvent. Typical concentrations can be in the range of 0.1–10 mM. Such DMSO solutions may be stored safely in screw capped polypropylene microcentrifuge tubes in the freezer for many months without apparent degradation.

Dry DMSO is used for making a solution of Pluronic F-127 at a concentration of 15–20% (w/w or w/v is not crucial). This stock solution can be stored at room temperature. When exposed to air, this concentrated stock solution slowly absorbs moisture until Pluronic F-127 begins to precipitate. Because a heterogeneous Pluronic stock is difficult to transfer, preparing a fresh stock is preferable at this stage.

Usually one can load cells by incubation in medium containing (nominally) 0.1 to a few tens micromolar AM ester in a Pluronic dispersion. Typically 0.5–1 μl Pluronic stock is sufficient for 1–10 nmol AM ester. Thus, 1 μl 20% Pluronic stock is adequate for dispersing 1 μl 10 mM AM ester stock into 1 ml medium to yield a 10 μM AM ester solution (1000-fold dilution). Using a dilution of ~1 : 1000 minimizes the DMSO concentration in the final loading medium (to a few parts per thousand). Premixing the requisite volumes of AM ester stock and Pluronic stock is advisable since this minimizes the chances of AM ester precipitation during aqueous dispersal. The Pluronic–AM mixture then is dispersed into aqueous loading medium.

Serum proteins [e.g., bovine serum albumin (BSA)] often have a salutary effect on cells and also can improve loading efficiency, possibly by acting as hydrophobic carriers for the AM ester and preventing its precipitation. Thus, including BSA (0.5–1%) or serum (a small percentage) in the incubation medium may be advantageous.

Enzymatic processing of an AM species to yield the Ca^{2+}-sensitive indicator typically consists of sequential hydrolysis of four to eight AM ester groups, depending on the particular indicator chosen. Only when all the AM groups on a molecule are hydrolyzed does the molecule become properly Ca^{2+} sensitive. For this reason, after removal of AM-containing loading medium, allowing the cells some extra time to complete intracellular processing of the most recently trapped, partially hydrolyzed AM species can be useful in some cases.

Convincing oneself that sufficient AM ester is present in the loading medium, so the amount of indicator taken up by cells is not limited by the amount available, is reassuring. For a typical example, assume that 1 million cells are loaded in 2 ml medium containing 2 μM AM ester. The total amount of AM ester available is 4 nmol. If the cells are 20 μm in diameter, then the total cell volume

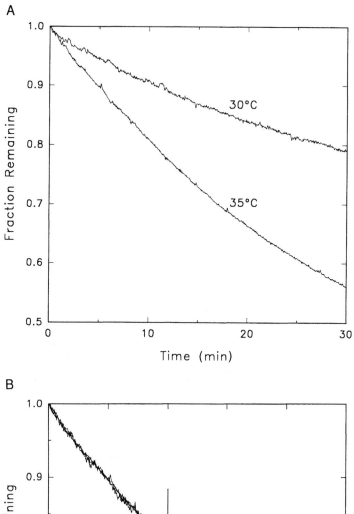

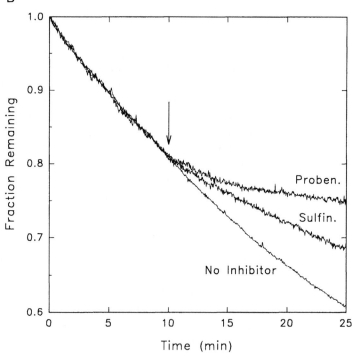

can be estimated at 4.2 μl. Further, if cells are loaded to a final intracellular indicator concentration of 150 μM (a generous estimate), then the total amount of AM uptake by cells is 0.63 nmol, which is still much less than the 4 nmol available in the loading medium.

IV. Manipulation of [Ca^{2+}]

In studying Ca^{2+}-dependent cellular processes, raising or lowering intracellular or extracellular [Ca^{2+}] is frequently desirable. Conventional (slow) techniques for achieving these ends require the use of Ca^{2+} buffers or ionophores and will be discussed in this section. Photorelease (fast) techniques are described in Chapter 2.

A. Using EGTA and BAPTA as Extracellular Ca^{2+} Buffers

Because it is highly selective for binding Ca^{2+} over Mg^{2+},[10] EGTA is the most commonly used Ca^{2+} buffer. However, because two of the ligand atoms in EGTA are tertiary alkylamino nitrogens, the two highest pK_as of EGTA are 8.90 and 9.52,[11] implying that at physiological pH EGTA will exist primarily as protonated species—a fact that is illustrated more quantitatively in Fig. 6. For example, Fig. 6 shows that, at pH 7.2, ~98% of EGTA in solution exists as H$_2$EGTA^{2-}, ~2% as HEGTA^{3-}, and only a negligible fraction is in the EGTA^{4-} form. Therefore, the Ca^{2+}-binding reaction near physiological pH is fairly represented as

$$H_2EGTA^{2-} + Ca^{2+} \rightleftarrows CaEGTA^{2-} + 2\,H^+$$

which immediately suggests that the binding of Ca^{2+} by EGTA should have very steep pH dependence, as a plot of pK'_d(Ca)[12] vs pH shows (Fig. 7). For a

[10] For EGTA, ΔpK_d = pK_d(Ca^{2+}) − pK_d(Mg^{2+}) = 5.58, that is Ca^{2+} is bound 380,000 times more tightly than Mg^{2+} by EGTA. For comparison, ΔpK_d = 1.78 in the case of EDTA—only a 60-fold difference in the strengths with which the two ions are bound. 1,2-Bis(o-aminophenoxy)ethane-N,N,N',N'-tetraacetic acid (BAPTA) has a selectivity similar to that of EGTA; ΔpK_d = 5.20.

[11] At 25°C and 0.10 M ionic strength. Data pertaining to EGTA that are used in this section are from Martell and Smith (1974).

[12] In the metal chelator literature, K_d is used for the "absolute" (or intrinsic) dissociation constant and represents the dissociation constant characterizing the fully deprotonated species of the chelator. K'_d represents K_d that has been corrected for the weakening effect of acidic pH. This convention is not followed consistently in the applications literature.

Fig. 5 (A) Time course of indicator loss at 30° and 35°C from REF52 cells loaded with fura-2. (B) Effect of probenecid (1 mM) and sulfinpyrazone (75 μM) on indicator extrusion from cells at 35°C. For the upper two traces, at the arrow, the specified inhibitor was applied to the cells. For both figures, REF52 cells were loaded with fura-2 by incubation with AM ester/Pluronic dispersion in HEPES-buffered HBSS at 25°C. Experiments were done in HBSS at the specified temperatures. Total fura-2 fluorescence ($F_T = F_{340} + F_{380}$) is monitored over time. Each trace was normalized by dividing by F_T at time 0.

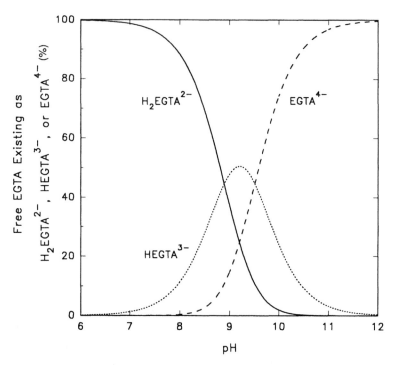

Fig. 6 Percentage of free EGTA existing as H_2EGTA^{2-}, $HEGTA^{3-}$, or $EGTA^{4-}$ as a function of solution pH. Calculations performed with data for EGTA (0.1 M ionic strength, 25°C) tabulated by Martell and Smith (1974).

concrete example, a drop in pH from 7.2 to 7.1 changes the $K'_d(Ca)$ of EGTA by a factor of ~1.6, that is, small errors in pH can lead to significant uncertainties in the dissociation constant. In contrast, knowing that the highest two pK_as of BAPTA are 5.47 and 6.36 (Tsien, 1980), one infers that the ability of BAPTA to bind Ca^{2+} should be only very weakly dependent on pH, as shown in Fig. 7. Comparison of the two traces in Fig. 7 shows that BAPTA has the advantage of being only weakly pH dependent in the physiological pH range. The fact that the two traces cross between pH 7.2 and 7.3 implies that EGTA has the potential advantage of being a progressively stronger binder of Ca^{2+} above the cross-over point (e.g., about 2- and 9-fold stronger than BAPTA at pH 7.5 and 7.8, respectively). The pH insensitivity of BAPTA makes it a less troublesome Ca^{2+} buffer to use, although it is significantly more costly than EGTA (the price differential is between 30- and 70-fold, depending on the supplier and reagent purity).

B. Lowering Extracellular [Ca^{2+}]

In an experiment, lowering extracellular [Ca^{2+}] is often desirable. Depending on how low one wishes to clamp the extracellular [Ca^{2+}], one of the approaches described in the following sections may be adopted. The procedures require

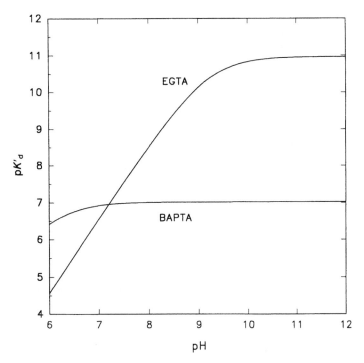

Fig. 7 Plot of pK'_d(Ca) vs pH for EGTA and BAPTA, derived from data from Martell and Smith (1974) and Tsien (1980).

either a stock solution of 1 M Na$_2$H$_2$EGTA at a pH near neutral or a stock solution of 1 M Na$_4$BAPTA.

1. Lowering [Ca^{2+}] to <1 μM but Not Approaching 0

a. Using EGTA

Na$_2$H$_2$EGTA can be added directly to calcium-containing medium at a concentration that is 3–4 times[13] the Ca^{2+} concentration in the medium. Because binding of Ca^{2+} to H$_2$EGTA^{2-} releases protons and acidifies the medium, simultaneously adding tris(hydroxymethyl)aminomethane (TRIS) base at a concentration equal to ~2.2 times the calcium concentration in the medium is also necessary. This amount of TRIS scavenges protons released from H$_2$EGTA^{2-} and maintains the pH of the medium at nearly the level before EGTA addition. This procedure can be applied confidently down to pH ~6.8. At pH levels that are much lower, the strong pH sensitivity weakens EGTA and makes it inconvenient to use as a Ca^{2+} scavenger.

[13] 3 times if pH ≥ 7.0, 4 times if pH < 7.0.

b. Using BAPTA

Add Na_4BAPTA at a concentration equal to 3 times the Ca^{2+} concentration in the medium. Essentially no pH adjustments are necessary. Because of its relative insensitivity to decreasing pH (Fig. 7), BAPTA can be used more conveniently for scavenging Ca^{2+} at lower pH values than EGTA.

2. Lowering $[Ca^{2+}]$ to a Level Approaching 0

If bathing cells in medium in which the free Ca^{2+} concentration approaches "zero" is desired, replacing the high-calcium medium with nominally calcium-free medium[14] that contains EGTA or BAPTA at millimolar concentrations is a simple solution. Again, to avoid problems of rapid weakening of the Ca^{2+} affinity of EGTA with decreasing pH, using BAPTA at pH < 7 is more convenient.

3. Setting Extracellular $[Ca^{2+}]$ to a Precisely Known Value

When the extracellular $[Ca^{2+}]$ must be known precisely, medium containing a well-defined Ca^{2+} buffer system at a fixed pH must be made. The preparation of such solutions is detailed in Chapter 1.

C. Divalent Cation Ionophores

1. Properties of Br–A23187 and Ionomycin

Br-A23187[15] and ionomycin are ionophores that form lipid-soluble complexes with divalent metal cations. These two ionophores are commonly used to increase the permeability of biological membranes to Ca^{2+}. Understanding the differences between the two makes it possible to make a judicious choice during an experiment.

Ionomycin can lose two acidic hydrogens and, as a dianion, can form an uncharged 1 : 1 complex with a divalent metal ion. Br-A23187 can lose a single acidic hydrogen and form an uncharged 2 : 1 complex with a divalent metal ion. This difference makes ionomycin potentially more effective in binding and transporting divalent cations (e.g., two molecules of Br-A23187 are needed to bind and carry a single Ca^{2+} whereas only one molecule of ionomycin is sufficient).

Compared with Br-A23187, ionomycin has somewhat better selectivity for Ca^{2+} over Mg^{2+}; ionomycin prefers Ca^{2+} by a factor of ~2, whereas Br-A23187 shows essentially no preference for one cation over the other (Liu and Hermann, 1978). In addition, these ionophores actually do not bind Ca^{2+} very

---

[14] Because calcium is a ubiquitous environmental "contaminant," nominally calcium-free solutions still can contain micromolar levels of calcium.

[15] The parent compound, A23187, is fluorescent and should be avoided in fluorescence work. The presence of the bromine atom in 4-bromo-A23187 (Br-A23187) effectively quenches the intrinsic fluorescence of the ionophore and makes the molecule useful in fluorescence experiments.

tightly [e.g., $K_d(Ca^{2+}) \approx 100 \ \mu M$ for ionomycin; J. P. Y. Kao, unpublished results[16]]. These factors suggest that the two ionophores would be inefficient in mediating Ca^{2+} transport when relatively low Ca^{2+} concentrations are involved (i.e., at [Ca^{2+}] $\ll K_d$, e.g., $<1 \ \mu M$), because at such low Ca^{2+} concentrations, only a minute fraction of total ionophore actually is engaged in Ca^{2+} binding and transport. This problem becomes evident when either ionophore is used in calibrating Ca^{2+} indicators in cells (see Section V,B).

The most significant difference between the two ionophores lies in the pH dependence of their ability to transport Ca^{2+} (Liu and Hermann, 1978). Transport of Ca^{2+} by Br-A23187 approaches a maximum at pH 7.5, whereas Ca^{2+} transport by ionomycin does not reach a maximum until pH 9.5. The pH at which half-maximal transport is achieved is ~6.4 for Br-A23187 and ~8.2 for ionomycin. Therefore, if one desires to increase transport of extracellular Ca^{2+} into cells in acidic media (pH $<$ 7.0), Br-A23187 is a much better choice than ionomycin.

2. Using Br-A23187 and Ionomycin

Ionomycin can be obtained in either the free acid form or the calcium salt form. Br-A23187 is available as the free acid. All forms are soluble in dry DMSO, which can be used to prepare stock solutions. Because these ionophores are very hydrophobic, they are bound avidly by serum proteins. Serum proteins such as BSA, when present in the medium, greatly reduce the effectiveness of ionophores and, if possible, should be left out of the experimental medium when ionophores are to be used; otherwise, much higher concentrations of ionophore must be used. Br-A23187 and ionomycin have been used at concentrations ranging from 10^{-7} to $10^{-5} \ M$.

In addition to increasing Ca^{2+} flux across the plasma membrane, Br-A23187 and ionomycin also transport Ca^{2+} out of intracellular calcium stores into the cytosol. Therefore, in the continued presence of these ionophores, intracellular calcium stores can remain depleted (Kao *et al.*, 1990).

D. Buffering Changes in Intracellular [Ca^{2+}]

1. Increasing Intracellular Ca^{2+} Buffering Capacity by BAPTA Loading

When changes in [Ca^{2+}]$_i$ are correlated with a biological process, one can ascertain the essentiality of the calcium change in the process by blocking the change in [Ca^{2+}]$_i$ with a calcium chelator. By far the easiest way to introduce extra Ca^{2+} buffering capacity into cells is by incubation with BAPTA/AM in Pluronic dispersion. Compared with the AM esters of common Ca^{2+} indicators, BAPTA/AM has much higher aqueous solubility—15 μM at 25°C (Kao *et al.*, 1990). Therefore, BAPTA can be loaded efficiently into cells via the AM ester.

---

[16] Determined optically at pH 11, at which all ionomycin in solution would be in the dianionic form.

BAPTA/AM is loaded into cells in precisely the same way that AM esters of indicators are loaded. Cells can be loaded with AM esters of BAPTA and an indicator simultaneously.

Figure 8 illustrates the effect of intracellular BAPTA loading on normal changes in $[Ca^{2+}]_i$. Figure 8A shows the changes in $[Ca^{2+}]_i$ in a REF52 cell loaded with fura-2 in response to sequential application of 1 μM bradykinin and 1 μM Br-A23187. Figure 8B shows the responses in a similar cell loaded with fura-2 and BAPTA. Apparently, the presence of sufficient BAPTA practically eliminates the rapid and transient rises in $[Ca^{2+}]_i$ elicited by an agonist. Indeed, even the massive rise resulting from a combination of Ca^{2+} influx and discharging of internal calcium stores mediated by Br-A23187 is suppressed substantially by the buffering action of BAPTA.

2. Possible Controls for the Use of BAPTA

Similar to EGTA, BAPTA is a chelator not only for Ca^{2+} but also for other multivalent metal cations. Thus, insuring that any inhibitory effect observed when using BAPTA is caused strictly by the ability of BAPTA to buffer Ca^{2+},

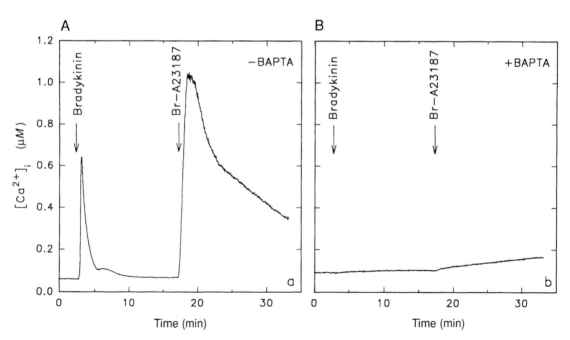

Fig. 8 Buffering action of intracellular BAPTA on changes in $[Ca^{2+}]_i$. (A) $[Ca^{2+}]_i$ of a REF52 cell treated with 1 μM bradykinin and 1 μM Br-A23187. (B) $[Ca^{2+}]_i$ trace of a REF52 cell, preloaded with BAPTA, in response to the same treatments. Cells were loaded with 1 μM fura-2/AM in Pluronic dispersion for 85 min at 25°C. For B, 20 μM BAPTA/AM was also present in the incubation medium. Experiments were conducted in HBSS.

and not because it is scavenging other biochemically important metal ions such as Zn^{2+}, is important. The reagent used to control for heavy metal scavenging by BAPTA is N,N,N',N'-tetrakis(2-pyridylmethyl)ethylenediamine (TPEN; Cal-Biochem, San Diego, California; Fluka Chemical Corporation, Ronkonkoma, New York; Molecular Probes, Eugene, Oregon) (Fig. 9), a membrane permeant metal ion chelator that shows a marked preference for binding heavy metal cations over Ca^{2+} (Anderegg *et al.*, 1977). Whereas the K_d(Ca) of TPEN is 40 μM (Arslan *et al.*, 1985), K_d(Zn) is $2.6 \times 10^{-16} M$ (Anderegg and Wenk, 1967). This enormous selectivity for binding heavy metal ions over Ca^{2+} enables TPEN to scavenge heavy metal ions very effectively, even in the presence of millimolar levels of Ca^{2+}. That TPEN is membrane permeant means it can be applied without using any special procedures. Dry DMSO can be used to prepare stock solutions of TPEN. Typically, TPEN is used in aqueous medium at a concentration of 10^{-6}–$10^{-5} M$.

BAPTA loading through the AM ester is very efficient; high concentrations of BAPTA may be accumulated intracellularly. Thus, ascertaining that observed inhibitory effects are not the result of cytotoxicity arising from the presence of high concentrations of a foreign organic anion may be important. In this case, the control reagent is N-(o-methoxyphenyl)iminodiacetic acid,[17] sometimes referred to as "half-BAPTA." As can be seen from Fig. 9, half-BAPTA is essentially chemically identical to BAPTA except that the molecule is only half of BAPTA. Because the full tetracarboxylate structure of BAPTA is crucial for Ca^{2+} binding, half-BAPTA, lacking such a structure, shows only very weak affinity for Ca^{2+} ($K_d \approx 3$ mM; J. P. Y. Kao, unpublished results). Half-BAPTA thus is expected to mimic BAPTA in all chemical respects except for the ability to buffer Ca^{2+} at physiological concentrations. The AM ester of half-BAPTA is available from Molecular Probes. Cell loading via the AM ester can be done as described previously for other AM esters.[18]

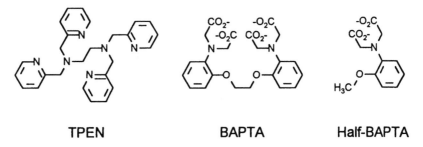

TPEN   **BAPTA**   **Half-BAPTA**

Fig. 9   Structure of BAPTA, half-BAPTA, and TPEN.

---

[17] A trivial name is anisidine-N,N-diacetic acid.

[18] Because it is processed to generate only the Ca^{2+}-insensitive half-BAPTA and yet it is processed intracellularly in the same way that all AM esters are, half-BAPTA/AM also can be used as a control for possible artifacts from AM ester hydrolysis.

V. Conversion of Indicator Fluorescence Signal into Values of $[Ca^{2+}]_i$

Although raw fluorescence signals from intracellularly trapped Ca^{2+} indicators can be informative in a qualitative way, one still must perform some calibration before even semi-quantitative estimates of $[Ca^{2+}]_i$ can be made. Basic principles of, as well as experimental procedures for, calibration for intensity-modulating dyes and for ratiometric dyes are discussed in this section.

A. Calibrating an Intensity-Modulating Fluorescent Indicator

For an intensity-modulating indicator that increases fluorescence emission on binding Ca^{2+}, the free Ca^{2+} concentration is given by

$$[Ca^{2+}] = K_d \left(\frac{F - F_{min}}{F_{max} - F} \right) \tag{1}$$

where F_{min} is the indicator fluorescence intensity at very low $[Ca^{2+}]$ (when all indicator molecules in the sample are Ca^{2+}-free), F_{max} is the corresponding parameter at very high $[Ca^{2+}]$ (when all indicator molecules are converted to the Ca^{2+}-bound form), and F is the measured fluorescence intensity for which we wish to find a corresponding value of $[Ca^{2+}]$. To arrive at a correspondence between measured F and $[Ca^{2+}]_i$, K_d, F_{min}, and F_{max} all must be known. Whereas K_d usually is predetermined *in vitro*, F_{min} and F_{max} must be obtained *in situ*. The most straightforward approach would be to try to equilibrate the indicator-loaded cell with solutions that contain "zero" Ca^{2+} and then high Ca^{2+}. In practice, however, deficiencies inherent in an intensity-modulating indicator makes this approach unattractive. Interpretation of intensity changes is confounded by dye leakage, which causes the total indicator fluorescence from the cell (and therefore F_{min} and F_{max}) to decrease with time. This basic flaw of intensity-modulating dyes suggests that obtaining good quantitative estimates of $[Ca^{2+}]_i$ in cells in which dye leakage or extrusion occurs at significant rates would be difficult. In such cases, a laborious calibration would not be justified. An alternative semi-quantitative calibration procedure developed for fluo-3 is discussed next.

The calibration procedure for fluo-3 depends on the fact that, *in vitro*, $F_{Mn} = 0.2 F_{max}$, where F_{Mn} is the fluorescence intensity when fluo-3 is saturated completely with Mn^{2+} (Kao *et al.*, 1989; Minta *et al.*, 1989). From *in vitro* measurements, that $F_{min} = F_{max}/40$ is also known. Because both F_{max} and F_{min} can be expressed in terms of F_{Mn}, the only parameter that must be determined experimentally is F_{Mn}. *In situ* calibration then consists of

1. applying micromolar levels of ionomycin or Br-A23187 to increase permeability of the cell to divalent metal ions

    2. adding sufficient $MnCl_2$ (typically twice the concentration of Ca^{2+} in solu-tion)[19] to saturate intracellular fluo-3

    3. permeabilizing the cell with digitonin to release fluo-3

Because the fluorescence intensity measured after Step 3 is just the background signal (including cellular autofluorescence), whereas the intensity after Step 2 is (F_{Mn} + background), one can obtain F_{Mn} by subtraction.

The calibration procedure described here is based on the following assump-tions: (1) indicator fluorescence intensity is not diminishing rapidly as a result of leakage; (2) the fluorescence properties (F_{min}, F_{max}, and F_{Mn}) of the indicator are known from *in vitro* measurements and are the same in cells as *in vitro;* and (3) the K_d of the indicator is also the same in cells as *in vitro.*

Quin2 is an example of an indicator the fluorescence of which is quenched completely by heavy metal ions. For calibration of such a system, see the review on quin2 by Tsien and Pozzan (1989).

B. Calibrating a Dual-Wavelength Ratiometric Fluorescent Indicator

A dual-wavelength ratiometric indicator allows excitation spectral intensity or emission spectral intensity of the indicator to be monitored at two different wavelengths. If F_1 is the fluorescence intensity at wavelength λ_1, F_2 is the fluorescence intensity at λ_2, and $R = F_1/F_2$, then the free Ca^{2+} concentration can be shown to be (Grynkiewicz *et al.*, 1985)

$$[Ca^{2+}] = K_d \left(\frac{R - R_{min}}{R_{max} - R} \right) \left(\frac{s_{f,2}}{s_{b,2}} \right) \qquad (2)$$

where R_{min} is the limiting value of the ratio R when all the indicator is in the Ca^{2+}-free form and R_{max} is the limiting value of R when the indicator is saturated with calcium.[20] Experimentally, the factor $s_{f,2}/s_{b,2}$ is simply the ratio of the measured fluorescence intensity when all the indicator is Ca^{2+} free to the intensity measured when all the indicator is Ca^{2+} bound; both measurements are taken at λ_2. Since most of the terms in Eq. 2 are ratios of intensities collected essentially simultaneously at two wavelengths, problems associated with cell shape and dye concentration changes cancel.

Using Eq. 2 to calculate [Ca^{2+}] requires that K_d be known and that R_{min}, R_{max}, and $s_{f,2}/s_{b,2}$ be determined experimentally. A typical calibration entails

---

[19] In Step 2, it is best if the medium contains no carbonate, bicarbonate, or phosphate, which can form insoluble precipitates with Mn^{2+} and thus reduce the concentration of free Mn^{2+}. In addition, the presence of particulate precipitates can add considerable noise to the fluorescence signal because of light scattering.

[20] λ_1 and λ_2 usually are chosen so intensity measured at λ_1 consists mostly of fluorescence emitted by the Ca^{2+}-bound form of the indicator, whereas intensity at λ_2 is mostly generated by fluorescence from the Ca^{2+}-free form. Choosing the wavelength pair in this way increases the difference between R_{min} and R_{max}, making it possible to map [Ca^{2+}] onto a wider range of R values.

1. increasing Ca^{2+} permeability of the cell with ionomycin or Br-A23187 in the presence of "zero" extracellular Ca^{2+} (EGTA or BAPTA in nominally Ca^{2+}-free medium; Section IV,B,2), so all intracellular Ca^{2+} could be depleted

2. increasing extracellular $[Ca^{2+}]$ in the presence of Ca^{2+} ionophore to saturate intracellular indicator

3. permeabilizing the cell with digitonin (at concentrations prescribed in Section III,B,2) to release cytosolic dye so the background signal may be measured[21]

Although the procedure seems straightforward, a few empirical findings are helpful in performing a successful calibration:

1. Many cell types do not tolerate severe calcium deprivation well. During Step 1, these cells often become fragile or leaky or, in the case of adherent cells, detach from the substrate. In many cases, however, one can compensate for the total absence of Ca^{2+} by supplementation with elevated concentrations of Mg^{2+}. Thus, raising the extracellular $[Mg^{2+}]$ to 5–20 mM can help maintain cell integrity. Although Mg^{2+} should bind fura-2 to a limited extent and slightly alter its fluorescence spectrum (Grynkiewicz et al., 1985), in practice R_{min} does not seem to be affected significantly by Mg^{2+} supplementation.

2. Because ionomycin and Br-A23187 become very inefficient at Ca^{2+} transport when intra- and extracellular free Ca^{2+} concentrations are below micromolar levels, depleting the cell of Ca^{2+} entirely is quite difficult. Therefore, one often must wait a long time for true R_{min} to be reached in Step 1. In the example shown in Fig. 10, the interval between ionomycin addition and attainment of R_{min} was in excess of 90 min.

3. Step 2 could be performed in two ways. One could add, in combination with fresh ionophore if desired, sufficient Ca^{2+} to bind to all the EGTA or BAPTA that was introduced in Step 1 and still have a large excess of free extracellular Ca^{2+}, provided that sufficient TRIS base is added to counteract any acidification arising from the Ca^{2+}–EGTA binding reaction. Alternatively, one could replace the medium from Step 1 with nominally Ca^{2+}-free medium and then add a large excess of Ca^{2+} in combination with a fresh dose of

---

[21] At controlled concentrations, digitonin releases primarily cytosolic dye whereas compartmentalized dye remains with the permeabilized cell and would be subtracted out as background. The assumption is that intraorganellar $[Ca^{2+}]$ is significantly higher ($>\mu M$) than cytosolic $[Ca^{2+}]$, so compartmentalized indicator always would be essentially Ca^{2+}-bound and therefore would contribute a constant background to the measured 340- and 380-nm fluorescence signals. This assumption would fail if significant amounts of dye are compartmentalized into organelles that do not have high internal $[Ca^{2+}]$. An alternative approach to obtaining a background reading is adding $MnCl_2$ (at a concentration equal to or greater than the Ca^{2+} concentration in solution) with the digitonin so compartmentalized dye also can be quenched as ionophores transport Mn^{2+} into the organelles. Such an approach assumes that cellular autofluorescence is the true background and ignores the contribution of compartmentalized dye to the background.

ionophore. The aim is to initiate massive Ca^{2+} influx into the cell at a rate that overcomes any cellular mechanism for Ca^{2+} extrusion. In practice, concentrations of ionophore ranging from 10^{-6}–10^{-5} M and external [Ca^{2+}] in the range of one to tens millimolar can be used.

4. A stable baseline is obtained quickly only if the dye released from cells by digitonin permeabilization is swept rapidly away from the region directly above the microscope objective. Otherwise, fluorescence from the released dye still will be captured by the objective and, thus, contribute to the measured background. Once swept away and diluted into the bulk medium, the released dye contributes negligibly to the background.

Elevation of [Ca^{2+}]$_i$ by ionophore can lead to rapid cell lysis and loss of indicator, sometimes before R_{max} can be determined confidently. Almost paradoxically, raising the extracellular [Ca^{2+}] to 10–30 mM in this procedure appears, in some cases, to have a protective effect on cell structure so lysis is deferred and R_{max} can be reached. If high extracellular [Ca^{2+}] is used, the medium should be free of phosphate salts, bicarbonate/carbonate, and even sulfate, since these ions can form precipitates with Ca^{2+}.

Typical data from an experiment performed on a REF52 cell loaded with fura-2 are shown in Fig. 10. Shown in Fig. 10A are the two raw data traces, F'_{340} and F'_{380}, collected when the cell is excited alternately with 340-nm and 380-nm light. The fluorescence signals measured after digitonin permeabilization are considered the background intensities, BG$_{340}$ and BG$_{380}$, that must be subtracted from the respective traces to yield the true F_{340} and F_{380}. The ratio trace is simply a point-by-point division, $R = F_{340}/F_{380}$, and is shown in Fig. 10B. R_{min} is the limiting value of R that is reached during Ca^{2+} deprivation, whereas R_{max} is the limiting value of R reached after treatment with ionophore at high [Ca^{2+}].[22] The factor $s_{f,2}/s_{b,2}$ is essentially $(F'_{f,380}\text{-BG}_{380})/(F'_{b,380}\text{-BG}_{380})$. Using these experimentally derived parameters and a predetermined K_d (224 nM; Grynkiewicz et al., 1985) in Eq. 2, one can convert the F_{340}/F_{380} ratio trace into a plot of [Ca^{2+}]$_i$ as a function of time (Fig. 10C).

This procedure has the advantage that all spectroscopically derived parameters, namely R_{min}, R_{max}, and $s_{f,2}/s_{b,2}$, that are especially sensitive to environmental changes are determined in situ with the indicator residing in the intracellular environment. Only the equilibrium dissociation constant is determined in vitro. R_{min} determined by Ca^{2+} deprivation is assumed to be the true value. Given the ineffectiveness of currently available ionophores at low [Ca^{2+}], con-

---

[22] From Fig. 10B, the ratio values near R_{max} are seen to oscillate significantly because, at saturating concentrations of Ca^{2+}, the fluorescence of the indicator at 380-nm excitation ($F_{b,380} = F'_{b,380} - \text{BG}_{380}$) is weak and cannot be determined with great precision. In forming the ratio, because $F_{b,380}$ is small in magnitude and is in the denominator, noise fluctuations in $F_{b,380}$ become magnified into large-amplitude fluctuations in R_{max}. Therefore, one must average a large number of points to obtain a reliable estimate of R_{max}. Alternatively, the raw fluorescence data (both F_{340} and F_{380}) in the region of interest can be smoothed (digitally filtered) first before a ratio is formed.

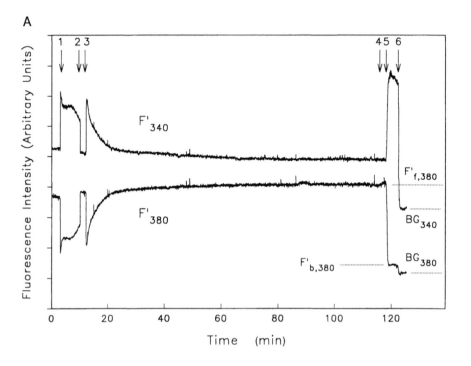

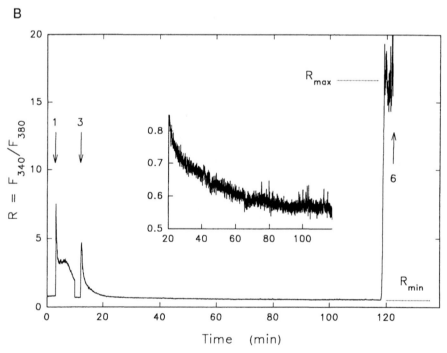

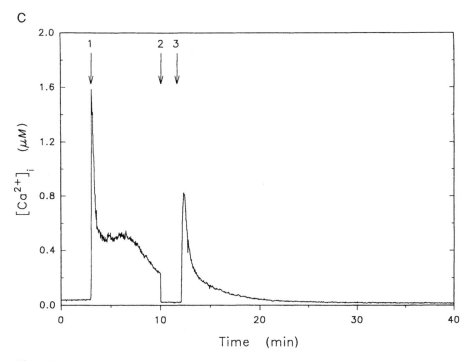

Fig. 10 Procedure for *in situ* calibration of fura-2, a ratiometric indicator. (A) Fluorescence intensity traces collected at 340- and 380-nm excitation. Time marker arrows correspond to (1) addition of 50 nM vasopressin; (2) exchange into Ca^{2+}-free PBS containing 10 mM MgCl$_2$, 2 mM EGTA, pH 7.4; (3) addition of 10 μM ionomycin; (4) exchange into nominally Ca^{2+}-free saline, pH 7.4; (5) addition of 10 μM ionomycin + 20 mM CaCl$_2$; and (6) 20 μM digitonin. Dotted lines mark fluorescence levels corresponding to various parameters discussed in Section V,B. (B) F_{340}/F_{380} ratio trace derived from the data in A. Dotted lines mark R_{min} and R_{max} (0.566 and 16.6, respectively). The parameter $s_{f,2}/s_{b,2}$ is 10.7. *Inset* The portion of the trace from 20 to 118 min at a higher resolution on the vertical scale to reveal the gradualness with which R_{min} is approached. (C) [Ca^{2+}]$_i$ trace derived from the ratio trace by using Eq. 2 in Section V,B. Only the first 40 min are shown. This REF52 cell was incubated with 1 μM fura-2/AM in Pluronic dispersion for 90 min at 25°C in HBSS before being transferred to fresh HBSS for measurement.

cluding that true R_{min} would be difficult to reach[23] and that one could overestimate R_{min} is reasonable. An overestimate of R_{min} results in underestimation of [Ca^{2+}].

Examining the effects of errors in R_{min}, R_{max}, and $s_{f,2}/s_{b,2}$ on the derived value of [Ca^{2+}] is worthwhile. For simplicity, one assumes that errors in the three parameters are independent. Because $s_{f,2}/s_{b,2}$ is related linearly to [Ca^{2+}],

[23] Rather than estimating R_{min} directly from the lowest R values reached in the ratio trace, curve-fitting the portion of the ratio trace that represents the slow descent to R_{min} is also possible. As expected, R_{min} obtained by exponential curve-fitting is somewhat lower than that estimated directly from the ratio trace.

a percentage error in $s_{f,2}/s_{b,2}$ translates into the same percentage error in $[Ca^{2+}]$. Inspection of Eq. 2 reveals that errors in R_{min} should affect primarily low values of $[Ca^{2+}]$ (corresponding to R values near R_{min}). Error in R_{max}, on the other hand, affects the way in which all the R values are scaled and, therefore, should influence all derived values of $[Ca^{2+}]$. These expectations are borne out by calculation.[24]

VI. Concluding Remarks

Fluorescent Ca^{2+} indicators have contributed enormously to our understanding of intracellular calcium regulation. For individuals who are beginning to use these indicators, the technical details can seem bewildering. This compendium of common techniques has aimed to set in order the body of practical empirical knowledge that underlies successful measurements of $[Ca^{2+}]$ through the use of fluorescent indicators.

References

Anderegg, G., and Wenk, F. (1967). *Helv. Chim. Acta* **50,** 2330–2332.

Anderegg, G., Hubmann, E., Podder, N. G., and Wenk, F. (1977). *Helv. Chim. Acta* **601,** 123–140.

Arslan, P., Di Virgilio, F., Beltrame, M., Tsien, R. Y., and Pozzan, T. (1985). *J. Biol. Chem.* **260,** 2719–2727.

Di Virgilio, F., Steinberg, T. H., Swanson, J. A., and Silverstein, S. C. (1988). *J. Immunol.* **140,** 915–920.

Di Virgilio, F., Steinberg, T. H., and Silverstein, S. C. (1990). *Cell Calcium* **11,** 57–62.

Grynkiewicz, G., Poenie, M., and Tsien, R. Y. (1985). *J. Biol. Chem.* **260,** 3440–3450.

Haugland, R. P. (1992). "Handbook of Fluorescent Probes and Research Chemicals." Eugene, Oregon: Molecular Probes.

Hepler, P. K., and Callaham, D. A. (1987). *J. Cell Biol.* **105,** 2137–2143.

Kao, J. P. Y., Harootunian, A. T., and Tsien, R. Y. (1989). *J. Biol. Chem.* **264,** 8179–8184.

Kao, J. P. Y., Alderton, J. M., Tsien, R. Y., and Steinhardt, R. A. (1990). *J. Cell Biol.* **111,** 183–196.

Liu, C.-M., and Hermann, T. E. (1978). *J. Biol. Chem.* **253,** 5892–5894.

Malgaroli, A., Milani, D., Meldolesi, J., and Pozzan, T. (1987). *J. Cell Biol.* **105,** 2145–2155.

Martell, A. E., and Smith, R. M. (1974). "Critical Stability Constants," Vol. 1. New York: Plenum Press.

McNeil, P. L. (1989). *Meth. Cell Biol.* **29,** 153–173.

Minta, A., Kao, J. P. Y., and Tsien, R. Y. (1989). *J. Biol. Chem.* **264,** 8171–8178.

[24] When one uses parameters similar to those for fura-2 in REF52 cells as determined on our instrument ($R_{min} = 0.5$, $R_{max} = 15$, and $s_{f,2}/s_{b,2} = 12$), a 10% overestimation of R_{min} leads to ~19% underestimation of $[Ca^{2+}]$ at 50 nM, ~10% at 100 nM, and ~2% at 500 nM. A 10% overestimation of R_{max} leads to underestimation of $[Ca^{2+}]$ by ~9.5% at 50 nM, ~10.9% at 500 nM, and ~12.5% at 1 μM. A 10% underestimation of R_{max} results in overestimation of $[Ca^{2+}]$ by ~11.8% at 50 nM, ~14% at 500 nM, and ~16.5% at 1 μM.

Tsien, R. Y. (1980). *Biochemistry* **19,** 2396–2404.

Tsien, R. Y., and Pozzan, T. (1989). *Meth. Enzymol.* **172,** 230–262.

Tsien, R. Y., Pozzan, T., and Rink, T. J. (1982). *J. Cell Biol.* **94,** 325–334.

Wong, C. J. H., Lev-Ram, V., Winter, E. L., Ellisman, M. H., Tsien, R. Y., and Bennett, M. V. L. (1992). *Soc. Neurosci. Abstr.* **18,** 1402.

CHAPTER 8

Rapid Simultaneous Estimation of Intracellular Calcium and pH

Stephen J. Morris,* Thomas B. Wiegmann,† Larry W. Welling,‡ and Bibie M. Chronwall§

* Division of Molecular Biology and Biochemistry and
§ Division of Cell Biology and Biophysics
School of Biological Sciences
University of Missouri at Kansas City
Kansas City, Missouri 64110
† Renal Section and
‡ Research Service
Veterans Affairs Hospital
Kansas City, Missouri 64128

I. Introduction

Calcium and hydrogen ions are well-recognized regulators of intracellular activity. The growing interest in the dynamic interregulation of these two ions has sparked interest in measuring their interrelationships. Until recently, their measurement using fluorescent dye indicators was done in tandem fashion. First one ion would be studied with one reporter dye; then the second would be studied with a second dye using a second set of cells. Comparison of effects, therefore, was subject to the criticism that conditions for the two sets of experiments need not be identical. In contrast, loading both dyes into the same cells permits sequential examination of the same sample. If these measurements are made sufficiently rapidly, the kinetics of both ions can be examined at closely related time points.

Over the past 4 years, our laboratories have developed equipment and methodology for rapid acquisition of multiple fluorescence images in real time as well as strategies for subsequent off-line analysis of the data. We are convinced that simultaneous acquisition of the data is a better approach than using alternate acquisition, since simultaneous acquisition solves several problems which appear in multiwavelength experiments by the latter method. In this chapter, we attempt to justify this conjecture. We discuss our imaging system and the associated measurements for calcium and pH in living cells. Since the expense involved is not trivial, we take some time to compare the merits of our system for truly simultaneous measurements with the consecutive measurements approach.

Real-time multiparameter video microscopy is a relatively new field. Like most new research techniques, many pitfalls exist that are not necessarily easily seen but can produce disastrous results if not avoided. This chapter emphasizes techniques and subtle points that usually do not appear in the abbreviated methods sections of research reports but are particularly important when performing simultaneous observations of multiple dyes by fluorescence.

The presentation focuses on simultaneous calcium/pH kinetics. However, the methodology can be applied to any set of probes that passes the basic requirements for truly simultaneous imaging. This technique can be combined with phase contrast or differential interference contrast (DIC) imaging as well as with patch-clamp measurements.

Sections II and III of this chapter describe how we currently make simultaneous calcium/pH measurements. A detailed discussion of the construction of the video microscope that is used for these experiments has been published elsewhere (Morris, 1993), in addition to a discussion of the equipment available for rapid low light imaging and video microscopy. We stress here the quality of the measurements and strategies for fine tuning the system components. Sections IV–VI are a detailed discussion of how to choose filters, dyes, and equipment. An alternative set of dyes is discussed in detail. Section VII presents some of our latest results.

II. Simultaneous Intracellular Calcium and pH Measurements

A. Why Simultaneous Measurements?

1. Speed of Data Collection

We define simultaneous observations as the collection of multiple images at exactly the same time. This definition must be contrasted with definitions used by other investigators that involve loading cells with two or more dyes, then viewing each dye sequentially. We use the simultaneous approach because, as detailed in Section II,B, it provides the most rapid, least distorted picture of changes in the dyes with time and spatial distribution.

2. Need for Correction of Calcium Indicator Dyes Indo and Fura for the pH Dependence of Sensitivity (K_d)

Like the parent compounds on which they are based, the calcium chelators indo and fura have dye–calcium dissociation constants (K_ds) that are pH dependent (Lattanzio, 1990; Roe *et al.*, 1990; Lattanzio and Bartschat, 1991; Martinez-Zaguilan *et al.*, 1991). Calculation of $[Ca^{2+}]$ using these dyes is also pH dependent. Figure 1 shows data from Lattanzio (1990; personal communication) detailing the large change in indo-1 K_d between pH 5.5 and 8. This problem was recognized by the inventors of the dyes in their original publication (Grynkeiwicz *et al.*, 1985). These researchers also noted that this shift was negligible above pH 7.0, leading to the general strategy for calculating $[Ca^{2+}]_i$ of choosing a convenient, published value for K_d and presuming that the pH$_i$ remained stable throughout the experiment.

The danger in the assumption of a fixed K_d is that undetected changes in pH will distort the calcium kinetics artifactually (Roe *et al.*, 1990; Martinez-

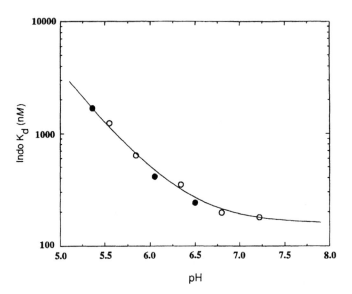

Fig. 1 pH dependence of the dye–calcium dissociation constant (K_d). ○, Data from Lattanzio (1990); ●, data from F. A. Lattanzio (personal communication). The fitted line is Eq. 2.

Zaguilan *et al.,* 1991). Many cells either have resting pH values below 7 or show a shift in pH in the course of the measurements, or both. Thus if small changes in [Ca^{2+}] are to be documented, or if the pH_i changes during the experiment, the calculated calcium level should be based on the prevailing pH. For kinetic experiments, this correction should be made using a pH value obtained at the same time as the calcium data. Since the pH kinetics need not follow the calcium changes, the Ca^{2+} data can be distorted by sequential measurements. Similar considerations apply to distortion from sequential readings for calculation of dye ratios (Section III,A).

B. Strategies for Multi-Ion Measurements

Measurement of calcium using ratiometric dyes has been discussed elsewhere in this volume. We have chosen to employ two ratio dyes, allowing for truly simultaneous measurements. To understand this decision, consider first a simple two-wavelength experiment for a single ratiometric dye such as fura or indo.

1. Consecutive Imaging

The basic assumption of the ratiometric dye method is that the measurements represent a true sample of the two species of dye. This assumption implies that the two measurements are taken (1) at the same point in space and

(2) at exactly the same time. That excitation ratiometric dyes such as fura-2 cannot fulfill the second assumption is immediately obvious; to gather the information needed, one must present the two excitation wavelengths alternately. This presentation allows for errors due to photobleaching and dye leakage, as well as any changes in the concentration of the ion being measured. If the system under investigation is changing very slowly with respect to the sampling speed, then the errors will be minimal.

A typical light path for an excitation ratiometric filter wheel system is presented in Fig. 2. The chief advantage of this type of system is that it is relatively inexpensive, requiring only one light source and one image detector. The vibration problems caused by the rotating filter wheel can be solved by placing the microscope on a separate (vibration-damping) table and connecting the light source and filter wheel to the system with fiber optics. Alternate fluorescence excitation systems that use multiple light sources or a source with multiple ports and mechanical or acousto-optical shutters also will solve this problem. However, these modifications raise the price of the system.

The fastest time resolution for a filter wheel or chopper system forming ratios from a single dye would be half the frame rate or 15 frames/sec. To accomplish this speed, the filters would have to switch wavelength during the short dead time between video frames. The wheel must be timed carefully with stepping motors and its position must be known exactly. Wavelength overlap cannot be tolerated; therefore, the frame transmitted during the change may be lost to the data set.

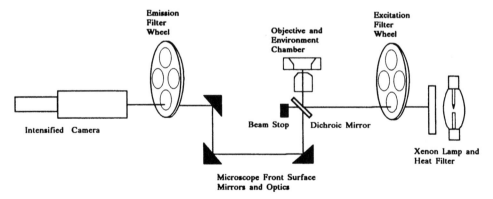

Fig. 2 Typical inverted epifluorescence microscope optical path employing rotating filter wheels for imaging of excitation or emission ratiometric dyes. For an excitation type dye such as fura-2 or BCECF, the excitation wheel would be moved between the two excitation filters and emission at a single wavelength would be imaged. Emission dyes would use a fixed excitation and alternating emission filters. Multiple dyes could be imaged by placing several dichroic mirrors in a revolving wheel (cf. Bright, 1992).

2. Concurrent Imaging

If emission type ratio dyes are used, the experimenter has the choice of consecutive or concurrent data gathering. Images can be created by alternating filters in the emission path (Fig. 2), although one then must face the same problems associated with an excitation wheel. Since the fluorescence emission from dyes such as indo-1 always contains both wavelengths, collecting this information simultaneously using a light path of the type shown in Fig. 3 is possible. The emitted light can be split with a dichroic mirror and images gathered simultaneously from two cameras. With the application of some image acquisition tricks, simultaneous acquisition of both images is possible using standard broadcast video equipment (Morris, 1990).

This design offers the following improvements over filter wheel or monochromator/shutter arrangements: (1) There are no moving parts in the optical train, removing vibration and synchronization problems. (2) By taking two images simultaneously from the same specimen, data collection rates equal camera frame rates. (3) By collecting simultaneous images, the ratios are derived from simultaneous, not sequential, samples. Thus, no temporal sampling error exists. Sampling the spatial distribution of rapid changes in fluorescence is limited only by camera rates, fluorescence emission levels and light losses in the optical train, properties of the dyes themselves, and the sensitivity of the imaging devices. The machine can capture data for 30 ratio images/sec using standard RS170 video equipment. If video field rate cameras are used, this rate can be increased to 60 ratios/sec. The rate will remain the same when multiple dyes are imaged. No special software is required to control or track

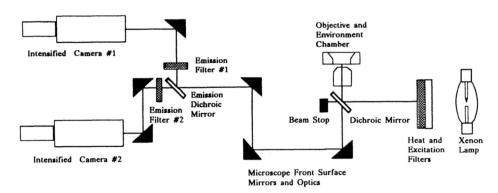

Fig. 3 The optical path for simultaneously imaging both wavelengths of an emission ratiometric dye. Standard epifluorescence is used for a dye such as indo-1. Excitation is via the standard epifluorescence ''cube'' arrangement, except the emission filter has been removed. The two emission wavelengths are divided by a second dichroic mirror and two barrier filters. Images are formed by two intensified charge-coupled device (CCD) cameras. Adjustments in the image splitter (Nikon ''Multi-image Module'') bring the two images into register and focus.

filter wheel positions or to mark the video frames for the prevailing wavelength of the excited light.

The drawbacks of such a system are, first and foremost, that it is expensive. One detector per image is needed. For a four-image system, this setup more than quadruples the cost of the optical train and cameras. Also, as discussed subsequently, to store the concurrent video images on the same RS170 video frame, the images must be made smaller. Thus, one trades image spatial area and resolution for increased sampling speed.

C. Four-Channel Video Microscope for Simultaneous Imaging of Two Ratio Dyes for Calcium and pH

The foregoing discussion presents the need for simultaneous monitoring of both $[Ca^{2+}]_i$ and pH_i. To accomplish this task for two ratiometric days, we use a four-image video microscope system (Fig. 4) that will collect two, three, or four simultaneous low light images. Our system currently is capable of collecting data for 60 ratios/sec at video field resolution and 30 ratios/sec at video frame resolution. A real-time ratio feature of the software allows the experimenter to assess cell viability and response immediately. All images need not be of fluorescence; one can visualize simultaneously a mixture of phase contrast or DIC and fluorescent images (cf. Section II,C,6; Color Plate 2). A detailed description of the construction of the instrument has been published elsewhere (Morris, 1993), so we briefly review the salient features (cf. Section VI for a discussion of equipment).

1. Choice of Dyes for Sensing Calcium and pH

After some deliberation (discussed in Section IV), we settled on indo-1 and SNARF-1 as the dyes of choice for emission ratiometric measurement of intracellular calcium and pH. The emission spectra of the two dyes are shown in Fig. 5. To create the information to image both these dyes, we require simultaneous fluorescence excitation at 350 and 540 nm and images of emission at 405, 475, 575, and 640 nm. A Nikon Diaphot inverted epifluorescence microscope is used as the base (Nikon Instruments, Melville, New York). Cells that have been labeled with both dyes are placed in a thermostatted environment chamber with a thin (#0) glass bottom, of the type used for patch clamp studies. The cells are illuminated at two excitation wavelengths (350 nm and 540 nm). Emitted fluorescence first is divided by a 510-nm dichroic mirror. The fluorescence from indo-1 is split further by a second dichroic mirror at 445 nm and imaged on two microchannel plate (MCP)-intensified charge-coupled device (CCD) cameras at 405 and 475 nm. The emission from SNARF-1 is split by a third dichroic mirror at 605 nm and imaged by two intensified cameras at 575 nm and 640 nm.

A

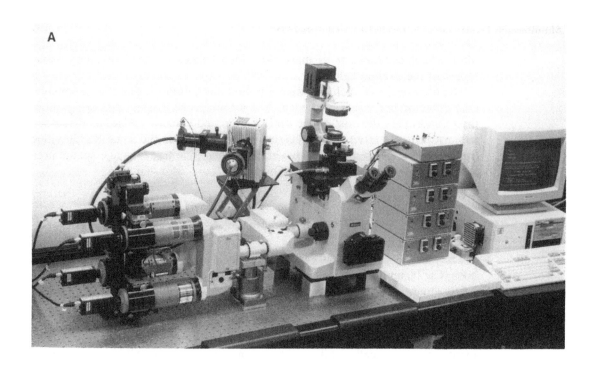

B

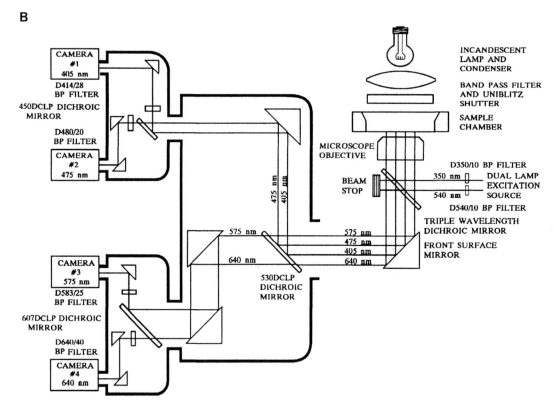

CAMERA
#1
405 nm

D414/28
BP FILTER

450DCLP DICHROIC
MIRROR

D480/20
BP FILTER

CAMERA
#2
475 nm

CAMERA
#3
575 nm

D583/25
BP FILTER

607DCLP DICHROIC
MIRROR

D640/40
BP FILTER

CAMERA
#4
640 nm

475 nm
405 nm

575 nm

640 nm

530DCLP
DICHROIC
MIRROR

MICROSCOPE
OBJECTIVE

BEAM
STOP

575 nm
475 nm
405 nm
640 nm

INCANDESCENT
LAMP AND
CONDENSER

BAND PASS FILTER
AND UNIBLITZ
SHUTTER

SAMPLE
CHAMBER

D350/10 BP FILTER
350 nm   DUAL LAMP
          EXCITATION
540 nm   SOURCE
D540/10 BP FILTER

TRIPLE WAVELENGTH
DICHROIC MIRROR

FRONT SURFACE
MIRROR

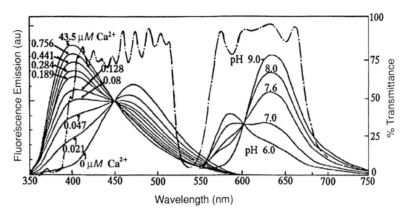

Fig. 5 The emission spectra of indo-1 (*left;* excited at 350 nm) for various concentration of free Ca^{2+} and of SNARF-1 (*right;* excited at 540 nm) at various pH values (cf. Haugland, 1992). The dashed line is the transmittance spectrum of the multiwavelength dichroic mirror used in the video microscope to excite both dyes simultaneously and to image the two dyes at four separate emission wavelengths.

2. Multiple Wavelength Epifluorescence Excitation

A wavelength selectable dual epifluorescence excitation system was built from an Oriel Series Q four-port 75-Watt Xenon lamp housing (Oriel Corporation, Stratford, Connecticut, fitted with two condensers and two parabolic reflectors. Filter holders just in front of the collimating lens allow choice of wavelengths with 5 × 5 cm filters. A second lens focuses the monochromatic light onto the face of a bifurcated Dolan–Jenner (Woburn, Massachusetts) 8-mm diameter fused silica fiber optic bundle. The output from this mixes with a 100-nm × 8-mm diameter liquid light guide (Stenning Instruments, Los Angeles, California). For indo-1 and SNARF-1, we use Chroma Technology (Brattleboro, VT) 350-nm and 540-nm 10-nm width band pass filters. The design is compact and the flexible light guides allow mounting either on or off the optical table. Compared with using mirrors and beam splitters, alignment problems are minimized. Uniblitz shutters (Vincent Associates, Rochester, New York) fitted to the paths of the two fluorescence excitation channels and

---

Fig. 4 Inverted epifluorescence video microscope for simultaneous calcium/pH imaging. The design is similar in concept to the two-image machine in Fig. 3. The microscope has no moving parts. The dual-wavelength excitation system is discussed in the text. Emitted fluorescence is divided first by a 510-nm dichroic mirror housed in an image splitter. The fluorescence from indo-1 is split further by a second dichroic mirror (housed in a second image splitter) at 445 nm and imaged by two intensified cameras at 405 and 475 nm. The emission from SNARF-1 is split by a third dichroic mirror at 605 nm (in a third splitter) and imaged by two intensified cameras at 575 nm and 640 nm. (A) Photograph of the microscope system. (B) Schematic diagram. The designations for the filters and dichroic mirrors are those presently employed.

to the transillumination path can be placed under software control to select wavelengths alternately for excitation dyes such as fura-2 (cf. Section II,C,5).

Multiwavelength epifluorescence illumination is made possible using newly developed "multichroic" mirrors (Marcus, 1988). For this experiment, a Chroma Technology (Brattleboro, VT) multiple band pass dichroic mirror is used, the transmittance spectrum of which is shown in Fig. 5. This mirror strongly reflects light at the two excitation wavelengths (350 nm and 540 nm) while passing almost all emission fluorescent light. This mirror is positioned in the usual place for the epifluorescence dichroic mirror. The three dichroic beam splitters that divide the emitted light are housed in three ganged Nikon multi-image modules. These commercial pieces greatly ease assembly of the microscope. Each module contains a turret that will hold four dichroic mirror/ emission filter combinations. An *x,y* translator and *z* axis focus aid in registration of the images.

Four Chroma Technology band pass filters ensure that no stray light that might have passed through the dichroic splitters enters the cameras. These filters have been designed to maximize emitted photon throughput without compromising the precision of the ratios. They are multicavity filters to maximize transmission within the chosen band and to exclude light on either side. These filters are housed in the dichroic mirror cubes in the image splitters.

3. Simultaneous Collection of Four Video Images of the Same Microscopic Field

We have used the following approach to combine four "quarter images" onto one standard broadcast RS170 video frame (Fig. 6), greatly reducing the cost of the recording equipment involved. An object is placed in the center of the microscope field. The four cameras are positioned off axis so the 256 × 240 pixel rectangular region of interest is in the upper left corner of Camera 1, the lower left corner of Camera 2, and so on. The overlapping images for cameras 2 and 3 are shown at the top of the figure. The outputs from cameras 1 and 2 are fed into video wiper 1 (PVW-1, Primebridge, Singapore); those from cameras 3 and 4 are fed into video wiper II. These outputs are combined by video wiper III, as shown. Some patience is required to set all images to the same size and orientation initially. Once set, there is no drift.

Using the ganged video wipers requires that the cameras be offset from the optical axis of the microscope. We have constructed a tower that holds all four intensified cameras. Each camera has its own set of *x, y,* and *z* translators. Magnification differences are compensated with the coupling lenses between the intensifier and the camera. Color Plate 1 shows images of fluorescence emission at four wavelengths of double-labeled pituitary intermediate lobe melanotrope cells, as well as the ratio images of these cells and a phase contrast image taken before the start of the recording.

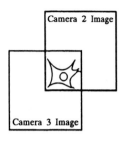

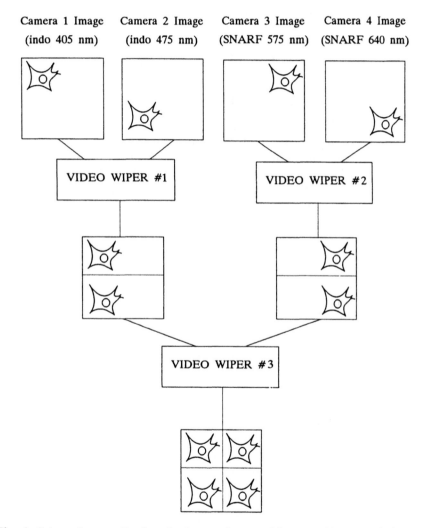

Fig. 6 Scheme for recording four simultaneous images of the same object on a single standard video frame. See text for details.

4. Formation of Low Light Images

After passage through the appropriate dichroic mirrors and barrier filters, the fluorescence emission is focused onto one of four VideoScope KS-1381 image intensifier plates (VideoScope, Washington, D.C.) coupled to Dage/ MTI CCD 1000 cameras (Dage/MTI, Michigan City, Indiana), set at maximum gain and $\gamma = 1.0$. Automatic compensation of gain and pedestal are shut off, producing stable high-contrast images using ultralow light levels with little persistence or bloom. These cameras can be operated at frame or field rates. All cameras are synchronized to the composite video sync output of a ForA FA300 time base corrector (ForA Corporation, Boston, Massachusetts), fed to a Sigma PDA-100A pulse distribution amplifier (Sigma Electronics, East Petersburg, Pennsylvania).

5. Imaging Hardware and Software

The microscope is designed to be modular and independent of the capture and analysis hardware and software; thus, components can be updated as equipment improves or drops in price. We currently use an IBM-compatible 386H/33 PC containing a Matrox MVP-AT imaging board (Matrox Corporation, Dorval, Quebec), running commercial software (MicroMeasure 4000A; Belvoir Consulting, Long Beach, California). The output from the ganged video wipers is presented to the Matrox board. For real-time capture, the combined background-corrected video images are displayed simultaneously on an RGB monitor (in monochrome), and recorded by a 400-line resolution Sony VO 9600 3/4'' U-matic VCR with time code generator/reader(Sony Corporation, Park Ridge, New Jersey) or, when the expense is warranted, directly onto a 450-line resolution Panasonic TQ 3031F Optical Memory Disk Recorder (OMDR) (Panasonic Communications, Secaucas, New Jersey).

Other useful software features include the following:

1. Real-time ratio display and calculation: The software can calculate and display noncorrected ratio images at a rate of 15/sec for alternate full-frame data or multi-image single-frame data. This feature is very useful for qualitative on-line assessment of the viability of a preparation before investing large amounts of experimental time (Tsien and Harootunian, 1990). This initial display can be recorded for analysis if shading error-corrected data is not required, for example, for qualitative quick screening of drug actions. The experimenter can define up to 255 regions of interest (ROI) of any size and shape (cf. Color Plate 1H). The software will integrate the gray levels for these regions and store the results into an ASCII file. 15 ROI samples/frame can be processed in real time. This ability is very handy for previewing experiments or for development work. Background and shading error corrections can be applied off line.

2. Full frame storage: In the configuration described earlier, every video frame contains four 1/4-frame images to be used for off-line analysis. For excitation ratio dyes or analyses at slow data rates, the software will alternate among two, three, or four camera inputs, taking successive frames or averaging 2–999 frames before changing to the next channel. Thus, for relatively slow experiments, one can collect two half-screen images for one dye, followed by two for the other dye, or one an alternate full-screen images of all four wavelengths. The software will control several Uniblitz shutters, either for excitation ratio dyes or for intermittent observations to save the samples during long experiments. Data can be stored on the VCR or the OMDR (especially handy for averaged data) or directly on the fixed disk (useful for experiments with low total images). The array processor and software can perform real-time background subtraction if desired.

6. Other Configurations: Combination of Phase Contrast and Fluorescence

This imaging system is not fixed in the four-image/single-frame format of Color Plate 1, nor must all the images be of fluorescence. Color Plate 2 shows images of cells being injected via an intracellular pipette (Eppendorf North America, Madison, Wisconsin). The cells were imaged at three fluorescence wavelengths (405 nm, 475 nm, and 640 nm); a phase contrast image at 587 nm was collected simultaneously. The optical system was as described in Fig. 4, with the following changes. The phase contrast image was created by placing a 587 DF10 band pass filter in the transillumination path and setting the Camera 3 intensifier at lowest gain. This adjustment greatly reduces the relative contribution of 575-nm Indo-1 fluorescent light. Madin-Darby canine kidney (MDCK) cells were loaded with indo-1 AM and placed in the bath. (These cells have begun to collect dye in subcellular compartments and ordinarily would not be used for experiments.) The injector pipette was loaded with 2.5 mM 6-carboxy-x-rhodamine in 140 mM KCI, 10 mM HEPES at pH 7.2. Several cells already have been injected successfully, as judged by retention of the rhodamine dye. The image represents a 16-frame average taken from the video tape during the 500-msec injection. Analysis of the video record showed no appreciable change in intracellular Ca^{2+} during or immediately after injection. In other experiments, some indo-labeled MDCK cells were injected with a mixture of rhodamine and GTPγS or GDPβS. The noninjected cells in the field served as controls for subsequent exposure to *Escherichia coli* ST$_B$ enterotoxin (cf. Section VII,A).

The "windows" created by the video wipers are completely adjustable (Fig. 6), allowing the format to be changed. We have used a format of three equalsized 170 × 480 rectangular windows to capture simultaneously two fluorescence dye images and one phase contrast image of cells undergoing fusion (Morris *et al.*, 1993).

III. Correction Procedures and Calculations

Accurate ratio calculations require that several corrections be applied to the images. In addition to the basic need for appropriate correction for background and shading error (Inoue, 1986; Tsien and Harootunian, 1990), simultaneous excitation of both fluorophores creates some special problems.

A. Calculation of pH_i and $[Ca^{2+}]_i$

Calculation of pH and Ca^{2+} concentration follows the scheme put forward by Martinez-Zaguilan *et al.* (1991). Depending on the required precision of the data, $[Ca^{2+}]_i$ can be corrected for pH_i on a region-by-region or pixel-by-pixel basis as follows. All images are corrected first for background, shading error, spillover, and geometric distortion. Fortunately, SNARF-1 is not sensitive to either the calcium concentration or the presence of indo-1 (Martinez-Zaguilan *et al.*, 1991; unpublished observations) and no correction is required. Indo-1 spillover correction is described in a subsequent section. For each region or pixel of the calcium image to be corrected, the corresponding region in the pH image is defined. The pH ratio for each region is calculated and converted to a pH value, either by interpolation from a standard curve of SNARF-1 at several pH values or from the relationship

$$pH = pK_a + \log (S_{f2}/S_{b2}) + \log [(R - R_{min})/(R_{max} - R)] \qquad (1)$$

where R, R_{max}, and R_{min} are the experimental, maximum, and minimum pH ratios, (S_{f2}/S_{b2}) is the ratio of fluorescence values for free and bound forms of the dye at the R denominator wavelength, and $pK_a = 7.30$.

The indo-1 K_d corresponding to this pH ($K_{d,corr}$) is calculated from the relationship

$$K_{d,corr} = [K_{d,max} + 10^{(pH - pK_a)} \cdot K_{d,min}]/[10^{(pH - pK_a)} + 1] \qquad (2)$$

$K_{d,max}$, $K_{d,min}$, and pK_a were calculated by least squares fit of Eq. 2 to the data of Lattanzio (1990) and F. A. Lattanzio (personal communication) for indo-1/EGTA at 37°C as 80.9 μM, 141 nM, and 3.65. These values are revised slightly from Morris (1993) because of the inclusion of new data (cf. Fig. 1). Finally, the pH-corrected $[Ca^{2+}]_i$ is calculated using the relationship (Grynkiewicz *et al.*, 1985):

$$[Ca^{2+}] = K_{d,corr} \cdot (S_{f2}/S_{b2}) [(R - R_{min})/(R_{max} - R)] \qquad (3)$$

Since performing these calculations is relatively time consuming for repetitive kinetic data, when analyzing whole-cell Ca^{2+}/pH, we define ROIs, extract integrated gray-level values for these ROIs from uncorrected images, and store them in an ASCII file at video rates, using the MicroMeasure software. The ROI values then are corrected off line for total background and shading error before graphing. Color Plate 1 E,F shows the ratio maps for the data of Color Plate 1A–D after correction for background, shading error, and spillover. (See Fig. 22 for the integrated calcium and pH data for the cell ROIs marked in Color Plate 1H.)

B. Special Considerations

1. Spillover

"Spill over" of emitted light from indo-1 into the 575-nm and 640-nm images of the SNARF-1 dye can be seen in the spectra of the two dyes (Fig. 5) and will contribute unwanted fluorescence to the SNARF-1 images, which must be subtracted before calculating the ratio. This adjustment can be handled in a straightforward manner, similar to scintillation-counting spillover adjustments.

Each experimental image first is corrected sequentially for background, shading error, and geometric distortion. The spillover is calculated by determining the intensity of fluorescence for indo-1 standards at 475 nm (I_{475}), 575 nm (I_{575}), and 640 nm (I_{640}) at various calcium concentrations. To correct for spillover into the 575-nm image, a two-parameter straight line fit of these data is performed for the equation

$$I_{575} = C_1 \times I_{475} - C_2 \qquad (4)$$

The spillover at 575 nm for any pixel (or ROI) then can be calculated as

$$I_{spillover,575} = C_1 \times I_{475,experimental} - C_2 \qquad (5)$$

A similar correction procedure is applied to the 640-nm image.

Figure 7 shows data and fitted lines for a typical spillover correction. The slopes and intercepts would be substituted into Eq. 5 to calculate the spillover for any given experimental pixel or ROI measured at 475 nm. The calculated value then is subtracted from the experimental value for the 575-nm or 640-nm image. Alternatively, linear look-up tables (LUTs) for the 256 gray level values can be constructed and applied to the 575- and 640-nm images.

The slopes and intercepts of the lines will vary with the relative gains of the detector systems. In the example presented in Fig. 7, the gain for the 640-nm image was about 5-fold higher than for the 575-nm image. The values for C_1 and C_2 must be recomputed each time the gains are changed and should be calculated for every experimental day.

2. Multiwavelength Excitation of a Given Dye

Like its parent compound fluorescein, SNARF-1 is excited by the UV light used to excite indo-1 (Fig. 8), adding fluorescence to both SNARF-1 emission images. We have chosen a UV excitation wavelength at 350 nm, for efficiency of indo-1 excitation and because this wavelength is a relative minimum point

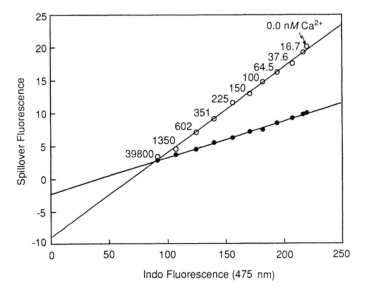

Fig. 7 Graph of levels of indo-1 spillover at 575 nm (○) and 640 nm (●) compared with the 475-nm peak. The calcium concentrations for the measurements are noted on the graph. Units are in gray levels. Thus, the values at $x = 0$ and $x = 255$ could be used to construct straight line look-up tables (LUTs) for image correction. Additional details are found in Section III,B.

for SNARF-1 excitation (Fig. 8B). Contributions from 350-nm excitation are still on the order of 15–25% of the excitation of 540 nm; these values will skew the pH calculations if not included in the standardization, which can be handled easily by calibrating the pH dye in the microscope using excitation at both 350 nm and 540 nm.

3. Energy Transfer and Other Special Considerations

Special problems of energy transfer and excitation of SNARF-1 by emission photons of indo-1 probably can be neglected. The concentrations of the dyes in the cytoplasm are usually too low for resonance energy transfer to take place. Emission levels from indo-1 are negligible compared with the excitation light supplied for SNARF-1.

C. Calibration of Dyes

Accurate calibration of the calcium dye has been discussed elsewhere in this volume. Errors in calibration of the calcium dye will cause the same problems in these experiments as they would normally. Miscalibration of the pH dye creates a special problem. Since the pH value will be used to choose the "correct" indo-1 K_d value for the calcium calculation, errors in the pH stan-

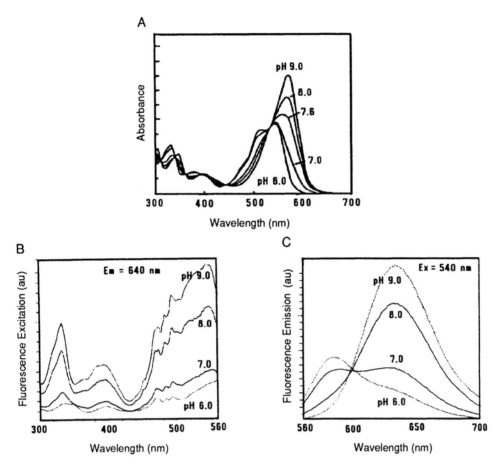

Fig. 8 (A) Absorption spectrum, (B) fluorescence excitation spectrum, and (C) fluorescence emission spectrum of SNARF-1 at various pH values.

dardization can have drastic effects on the apparent intracellular calcium value.

As for the calcium dyes, external standards may not reflect the sensitivity of the pH dye in the cell cytoplasm. Therefore, we also obtain intracellular SNARF-1 standard curves. Gathering a complete internal standard curve for each cell is nearly impossible. The ionophores that must be added to the cells make them unfit for further experimentation, requiring that the calibration be done at the end of the experiment. When ionophores are applied, the cells often detach and float away before standardization can be completed. As a result, internal standardization is performed on separate sets of cells, as follows. Double-labeled cells on cover slips are placed in the microscope environment chamber and treated with nigericin (5 μM) and monensin (5 μM) in

140 m*M* KCl buffered with 10 m*M* HEPES at pH values between 6.0 and 8.0 (Grinstein *et al.*, 1989; Rosario *et al.*, 1991). Images of 5–20 cells are collected at each pH. Ratios are extracted and the averaged data from 3–5 repetitions (20–100 cells) are plotted against pH and fit to a straight line.

Comparison of a typical internal and external standard curve for MDCK cells can be seen in Fig. 9. The two curves have different slopes and intercepts; they intersect at ~pH 6.5. Differences between assigned pH values would be small in the range from pH 6.5 to 7.0, where a pH difference would produce the greatest errors in calculated [Ca^{2+}]. Above pH 7.0, changes in K_d can be neglected. The choice of standard curve was found to have only a small effect on eventual calculation of calcium levels (Fig. 10). The pH data for the cell in Fig. 10 were calculated using both standard curves. The calcium data were corrected using both these pH curves as well as a fixed K_d for pH 7.05. The intracellular standardization produces consistently lower pH values than the extracellular curve. Despite these differences, the pH-corrected values for intracellular calcium were virtually identical. Both corrected curves show more prominent changes in [Ca^{2+}]$_i$ than the uncorrected (fixed K_d) calcium curve. From this discussion, obviously pH calibration for each cell type must be checked carefully.

IV. Choosing Calcium and pH Indicator Dyes

To take advantage of the truly simultaneous approach to multiple ion measurements, two ratiometric dyes are necessary, the spectra of which allow unambiguous data collection of the information needed to create the two ratio

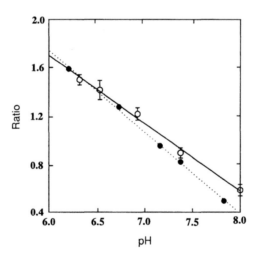

Fig. 9 SNARF-1 intracellular (○) versus external (●) pH calibration curves. (Reprinted with permission from Wiegmann *et al.*, 1993.)

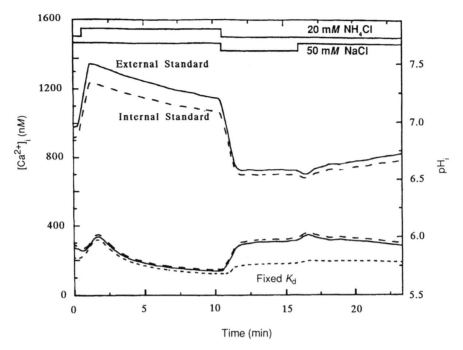

Fig. 10 Effects of using the internal vs external calibration on the calculated pH and calcium levels. The effect of changing intracellular pH on intracellular $[Ca^{2+}]$ is shown. At 1 min, the standard balance salt solution was exchanged for one containing 20 mM NH_4Cl, which alkalinized the cell cytoplasm. After the cell returned to baseline pH, the solution was exchanged for an NH_4Cl-free, Na^+-free medium. The cell pH dropped suddenly and recovered slowly. Both NH_4Cl addition and removal produced a release of intracellular Ca^{2+} (cf. Section VII,B). The two upper traces are the intracellular pH using the external standard curve (———) and the internal standard curve (— —). The three lower traces are the intracellular Ca^{2+} calculated from a fixed K_d for pH 7.05 (----), and the K_d calculated from the external standardized pH curve (— —), and the K_d calculated from the internal standardized pH curve (———).

maps. The combination of indo-1 and SNARF-1 forms nearly an ideal pair for Ca^{2+} and H^+. However, situations will arise in which one or both dyes are not suited to the experimental system, or other combinations of ions are to be measured. To aid in choosing dye pairs, we discuss the properties of the dyes that must be examined. The possibility of combining a single wavelength dye with a ratiometric dye is presented.

A. Dye Spectra, Significance of the Isosbestic Point, and Excitation Problems

A fluorescence spectrum consists of two parts: a scan of excitation wavelengths recorded at a fixed emission wavelength (usually the emission peak) and an emission wavelength scan recorded at a fixed excitation wavelength (again, usually the excitation peak). Figure 11 shows the excitation and emis-

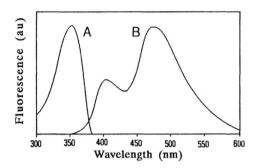

Fig. 11 (A) Fluorescence excitation and (B) fluorescence emission of indo-1 in 50 n*M* calcium.

sion spectra of indo-1 in 50 n*M* Ca^{2+} at pH 7.2. The sample excitation was scanned first from 300 nm to 450 nm using an emission setting of 475 nm. Then emission was scanned from 360 nm to 600 nm, using a fixed excitation of 350 nm. The spectrum shows that the fluorophore has two emission peaks. A series of emission scans at varying free calcium concentrations (Fig. 5) reveals that the 400-nm peak predominates at high $[Ca^{2+}]$ and the 475-nm peak predominates at low $[Ca^{2+}]$.

In Fig. 5 the intersection of the scans at about 440 nm for indo-1 and 600 nm for SNARF-1 is called the isosbestic point. This point is the wavelength at which the emission intensity is invariant with concentration of the ion to be tested. (A similar point will be found in the excitation spectrum of excitation ratiometric dyes such as fura.) Such a point is a certain indication that the dye is a two-state dye that will show a spectral shift when it changes from the free to the bound form. The isosbestic point often is chosen to monitor the relative concentration of the dye. Dye leakage or photobleaching will register as a reduction in intensity.

Whenever possible, obtain fluorescent excitation rather than absorption spectra. All absorbing wavelengths do not produce fluorescence, as illustrated in Fig. 12, which shows that, for the excitation pH indicator BCECF, no isosbestic point exists in the absorption spectrum at 439 nm whereas one appears in the excitation spectrum (Haugland (1992)). Presuming that the peak at 495 nm is chosen for one excitation wavelength for a ratiometric determination with BCECF, monitoring at 439 nm for the second wavelength also will give an estimate of the dye concentration in the cells. If an estimate of dye concentration is not needed, then use of 400 nm, which shows an almost 2-fold change in fluorescence excitation for pH 8 versus 6, would double the dynamic range of the assay. This information is not available from the absorption spectrum. Figure 8 shows that a number of isosbestic points exist in the SNARF-1 absorption spectrum, but none in the fluorescence excitation spectrum. Thus, SNARF-1 cannot be used in the excitation ratio mode.

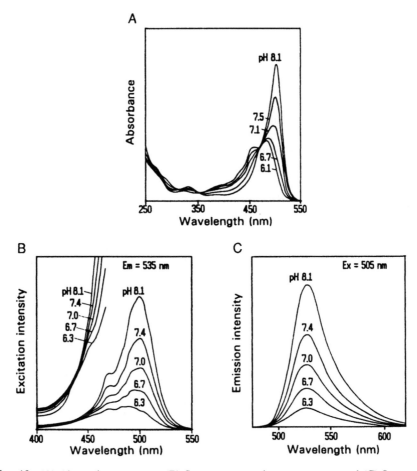

Fig. 12 (A) Absorption spectrum, (B) fluorescence excitaton spectrum, and (C) fluorescence emission spectrum of BCECF at various pH values.

B. Spectral Requirements for Simultaneous Imaging of Two Ratiometric Dyes

To image two different dyes simultaneously it must be possible to separate the excitation and emission peaks involved. The dye pair must have certain fluorescence properties.

1. Emission

Ideally, the emission spectra of the two dyes should not overlap. Some slight overlap can be tolerated but requires appropriate correction. The indo-1/SNARF-1 combination shows a tolerable amount of emission overlap (cf. Section III,B; Fig. 5).

2. Excitation

In the ideal situation, both dyes have the same excitation maximum wavelength, which removes the need to supply a second excitation wavelength simultaneous with the first. This expectation is unreasonable; the dyes are more likely to have different excitation maxima.

The second best possibility is for the excitation maximum for the second dye to appear where it will not interfere with reading the emission of the first. This is the case for the indo-1/SNARF-1 combination (Fig. 5). The problem that arises when the dyes have different excitation peaks and one is excited weakly at the peak excitation for the other is discussed in Section III,B.

Numerous ratio-type dyes for pH (reviewed by Haugland, 1992) are available. SNARF-1, an emission ratio-type fluorescein analog, has a near ideal fluorescence spectrum for use with indo-1 (Fig. 5), as well as a pK of 7.30. The pH-sensing properties of this dye are relatively insensitive to other ions including calcium and to other dyes such as indo or fura, making SNARF-1 a good choice to pair with indo. This dye is loaded easily into cells as the AM ester, in conjunction with indo-1/AM (Morris *et al.*, 1991).

C. Checking for Spectral Problems

For a dye pair such as indo/SNARF, we refer to the dye with the lower excitation peak and lower emission peaks as dye 1 and to the other as dye 2. As discussed earlier, the problems most often encountered are (1) dual excitation of dye 2 and (2) spillover of emission from dye 1 into the emission of dye 2. Experimental testing for these common problems is straightforward.

1. Find or collect spectra that will warn of possible problems (overlaps, etc). Collecting these yourself, using water-based solutions similar to the intracellular milieu, is a good idea. Most published spectra are collected in dimethylsulfoxide (DMSO), ethanol, or other hydrophobic organic solvents, which will not be used in actual experiments. These spectra will be blue-shifted compared with those in water-based solutions. Likewise, extinction coefficients should be collected from buffer solutions.

2. To check for excitation of dye 2 by the wavelength for dye 1, load cells with dye 2 only, place them in the microscope, and, recording dye 2 emission, excite first with one wavelength and then with the other. If significant excitation occurs at the wavelength for dye 1, then calibrations for dye 2 will need to be made with both excitation wavelengths applied.

If isosbestic points occur in the excitation spectrum, a situation can arise in which the two excitations contribute subtractively to the emission of dye. If possible, arrange to shift the excitation wavelength for dye 1 to an isosbestic point for dye 2. Then, the dye 2 emission caused by dye 1 excitation will be independent of ion concentration.

3. To check for spillover of dye 1 emission into the images for dye 2, load cells with dye 1 only, place them in the microscope, excite with the dye 1 wavelength, and record images at all four emission wavelengths. Spillover of more than a small percentage should be removed from dye 2 images (Section III,B,1). The relative spillover can be controlled by the gain settings of the four image intensifiers and cameras. Lowering the gains on the image intensifiers and cameras for dye 2 will reduce the apparent spillover.

V. Possible (but Untried) Dye Combinations

We planned to review a number of other possible combinations of ratiometric dyes that could be used in the simultaneous four-wavelength mode we employ for our experiment. Only one emission ratio dye generally is available for calcium measurements: indo-1. Except for the various other SNARF derivatives (reviewed by Haugland, 1992), we have found only one other emission ratio dye for pH: 1,4-dihydroxyphthalonitrile (DHPN). The properties of this dye are reviewed in a subsequent section. Other calcium and pH ratiometric dyes may exist in the literature, but our exhaustive searches have failed to uncover them. Thus, for the time being, only the indo/SNARF combination is available for the dual ratiometric dye approach. Other strategies are reviewed in Section IV,B.

A. Emission Ratio pH Sensing Dyes

1,4-DHPN has been used as a ratiometric dye for intracellular pH by flow cytometry (Valet *et al.,* 1981). The use of this dye has been extended by Kurtz and Balaban (1985) as a ratio dye for pH measurements by microfluorimetry. The dye can be loaded as the membrane permeant derivative 1,4-diacetoxyphthalonitrile (1,4-DAPN) by diffusion into the cell cytoplasm followed by esterase cleavage. The dye has a relatively weak extinction coefficient compared with SNARF-1 (~6.3 and 7.2 for the acid and base forms vs. 26 and 53). Three excitation peaks occur at ~345 nm, 375 nm, and 405 nm, corresponding to the fully protonated, monoprotonated, and deprotonated forms. Emission peaks vary depending on the excitation wavelength. Using a band pass filter centered at ~390 nm (375–407 nm) for excitation of the monoprotonated and deprotonated forms, Kurtz and Balaban produced emission spectra with a peak at 450 nm for pH 6.0 shifting to 490 nm at pH 9.0. Judging from their spectra, one would expect a dynamic range of about 4 using the 450/490 ratio between these pH values. This range contrasts with a range of ~15 for SNARF-1 between pH 6.0 and 9.0. The emission spectrum of DHPN almost completely overlaps that of indo-1; since currently no other choice for the calcium dye is available; DHPN is of little use to this type of experiment.

B. Other Strategies

Since other dual-wavelength dye pairs are not available, we discuss the application of ratioing a single-wavelength indicator dye against a second inert reference dye, then extend this application to reporting two different ions using a single-wavelength and a dual-wavelength dye.

1. Ratiometric Determination Using a Single-Wavelength Indicator Dye and a Single-Wavelength Reference Dye

The reason for the popularity of ratiometric dyes is the relative ease of calibration. The ratio is dimensionless. Knowing the concentration of the dye in the cell is not essential to knowing the absolute ion concentration. Dye leakage or photobleaching is no longer a problem. However, if one can justify the assumptions involved, ratioing a single-wavelength indicator dye against an inert reference dye, for example, a pH-sensitive dye such as 6-carboxyfluorescein and a pH-insensitive dye such as Texas Red, is quite straightforward. Both dyes can be excited at the fluorescein maximum (493 nm); fluorescence emission would be recorded at 530 nm for fluorescein and 640 nm for Texas Red. The ratio of the two emissions is proportional to the pH. This ratioing will work if the following assumptions are true:

1. Both dyes are in the same intracellular compartment.
2. Both dyes are taken up by the cells to the same extent.
3. Neither dye changes concentration during the measurements or both dyes change to the same extent, that is, any change in the ratio is caused by changes in indicator dye fluorescence; therefore no differential leakage or photobleaching of either dye is allowed.
4. The reference dye is really inert; it will not be affected by any changes in the intracellular milieu. Thus, carboxyfluorescein would not be a good choice as a reference for a calcium indicator dye such as Fura-Red™, since it is pH sensitive.

Many of the assumptions could be met if the two dyes were both coupled to a polymer such as dextran. The polymer would have to be loaded by injection, scrap loading, or osmotic shock; however, once in place, problems of differential loss from leakage disappears. Derivatized dextrans are available for fluorescein, rhodamine, indo, fura, SNARF, and other indicators from Molecular Probes (Eugene, Oregon). A mixture of indo- and rhodamine-labeled polymers would be easy to prepare. A more elegant solution would be making a dually derivatized dextran, either oneself or by custom synthesis. Further, dextrans are not compartmentalized as easily as free dyes. The problem of differential photobleaching remains, but results might be adjusted using measured bleaching rates for the dextran-coupled derivatives.

2. Extension to One Single-Wavelength Indicator and One Ratiometric Indicator

Ratioing a single-wavelength indicator dye against the isosbestic point of a dual-wavelength dye is quite possible. For example, the single-wavelength calcium indicator fluo-3 has been combined with SNARF-1 by Rijkers *et al.* (1990) for flow cytometry calcium measurements. These investigators excited both dyes with a 488-nm laser, split the emission with a 600-nm dichroic mirror, and read fluorescence at 525 nm for fluo-3 and 610 nm, the isosbestic point of SNARF-1. By choosing the isosbestic point, Rijkers and colleagues avoided the problem of the pH dependence of SNARF-1. These researchers made no attempt to extract cell pH from the SNARF-1 fluorescence.

We suggest the combination Calcium Green™ and SNARF-1. (See Haugland, 1992, for a discussion of the Calcium Green family of calcium indicators.) The spectra for this combination are presented in Fig. 13. Cells loaded

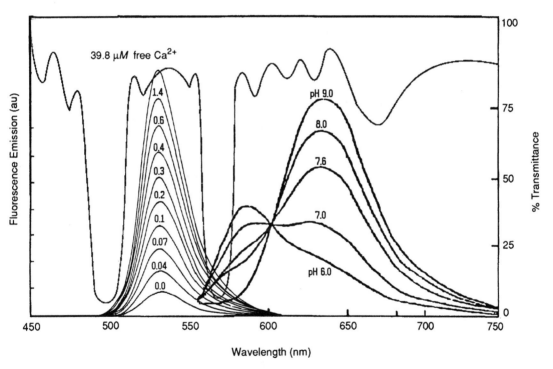

Fig. 13 The emission spectra of Calcium Green™ (*left;* excited at 488 nm) at various concentrations of free Ca²⁺ and of SNARF-1 (*right;* excited at 540 nm) at various pH values (cf. Haugland, 1992). Also included is the transmittance spectrum of a commercial multiwavelength dichroic mirror used to excite fluorescein isothiocyanate and rhodamine isothiocyanate and to separate their emission fluorescence. See Section V,B,2 for additional details.

with both dyes would be excited at 488–495 nm for Calcium Green and 550 nm for SNARF-1. Calcium Green emission would be imaged at 525–530 nm. SNARF-1 emission images would be formed at 600 nm (the SNARF-1 isosbestic point measured in 140 mM KCl, 10 mM HEPES at pH 7.2) and 640 nm (cf. Fig. 8). The complete optical path including designations for the dichroic mirrors is laid out in Fig. 14. The multiwavelength epifluorescence dichroic mirror chosen for this experiment is a fluorescein/rhodamine mirror from Chroma Technology. Using their dual-wavelength fluorescein/rhodamine excitation filter may be possible, which would remove the need for a two-wavelength excitation source. Calcium Green could be replaced with fluo-3, if desired.

If, for some reason, one wanted two independent reports of $[Ca^{2+}]i$, it might be possible to load cells with both indo-1 and a single-wavelength calcium indicator such as Calcium Crimson™. Indo-1 would be excited at 350 nm and

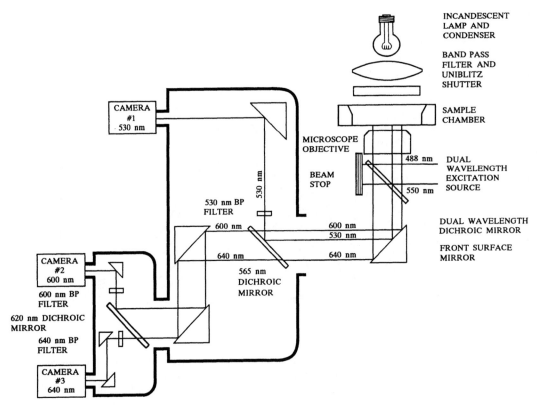

Fig. 14 Optical path for simultaneous excitation of Calcium Green™ and SNARF-1 and imaging of their emission fluorescence at three wavelengths. See Section V,B,2 for details.

imaged at 400 and 435 nm. Calcium Crimson would be excited at 495 nm and imaged at 660 nm. This procedure would reduce the sensitivity of the indo-1 assay by a factor of ~2 (the dynamic range of the change at the indo-1 475-nm peak).

Using this approach, performing a two-dye experiment in a ratiometric mode with three cameras rather than four becomes possible, reducing the cost of the experimental apparatus by that of one intensified camera and the Nikon image splitter. Also, this approach moves the excitation into the visible range, where it will cause less photodynamic damage. Caged compounds released by UV excitation could be employed. The fourth camera could be used to form a phase contrast or DIC image (cf. Section II,C,6) or a fourth fluorescence image from a third dye. Ratioing two single-wavelength indicator dyes against a third reference dye is also possible. The assumptions concerning differential changes in dye concentrations must be met for all three dyes. Again, many of the assumptions for the single dye/ratio dye combination could be met by labeling the cells with dextran that is derivatized with all three dyes.

VI. Equipment

A. Choosing Optical Filters and Dichroic Mirrors

The successful separation of the various emission fluorescence peaks is critically dependent on the choices of optical filters and dichroic mirrors in the microscope system. Use of poor quality components easily may render an expensive microscope system useless. Components must be chosen that do not allow unwanted light to enter the cameras.

1. Characterization of Dichroic Mirrors

Note that dichroic mirrors are not completely effective at transmitting or reflecting any given wavelength. Figure 15 shows the spectrum of the dichroic mirror we use to split the emission fluorescence for SNARF-1 in our system. This mirror *tends* to reflect light below 610 nm and pass light above this wavelength, but note that the cutoff is not completely efficient for values just above or below 610 nm. The midpoint of the split can be shifted up or down by 10–15 nm depending on the orientation of polarized excitation light. Spring (1990) has used this shift effectively to "tune" the wavelength split of the excitation dichroic mirror. The other property to note is that dichroic mirrors (like band pass filters) can become suddenly transparent to light at wavelengths removed from the characteristic "split" wavelength. This property allows the manufacture of multiple-wavelength dichroic mirrors (cf. Figs. 5, 13). The mirror in our example is almost completely efficient in the ranges needed to divert the low and high emission peaks of SNARF-1 to the intensified cameras. To ensure

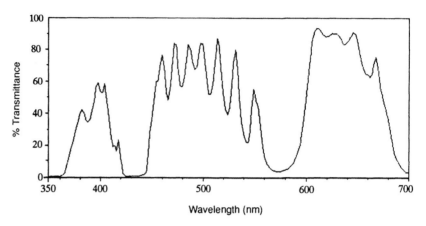

Fig. 15 The transmittance spectrum of the dichroic mirror used in Fig. 6 to separate the 575-nm and 640-nm fluorescence peaks of SNARF-1. See Section VI,A,1 for details.

that no stray unwanted light enters the image intensifiers, band pass filters are placed between the intensifiers and the mirror (Fig. 4B). When choosing mirrors for multiwavelength experiments, ask the manufacturer for a spectrum covering the entire range of wavelengths of the experiment.

2. Interference–Type Band Pass Filters

These band pass filters typically consist of a three-element sandwich (Fig. 16). A simple interference filter will have high transmittance at a characteristic wavelength with a Gaussian distribution of transmittance falling to a low value (not necessarily zero) at some distance from this center point. At wavelengths removed from the median, the filter will again pass light (Fig. 16A). To stop high and low wavelengths "leaking" through and to narrow the band pass, the interference filter is sandwiched between a low pass filter and a high pass filter (Fig. 16B), giving the final spectrum shown in Fig. 16C. These filters are characterized by a median wavelength and the peak width at half height, for example, 550 nm and 20 nm in Fig. 16C. However, about 30% of the total light passed will fall outside these boundaries. Obviously such an arrangement still suffers from unwanted light from the wide distribution around the characteristic wavelength, as well as from loss of light due to decreased sensitivity as one moves away from the median (peak) wavelength. Improvements in coating materials and techniques have, in turn, produced vastly improved band pass filters with greater transmittance efficiency and rejection of unwanted light (Marcus (1988)). Figure 16D shows such a band pass filter. This filter, which has the same median wavelength and peak width at half height as the filter in Fig. 16C, is easily seen to give a >50% increase in fluorescence intensity of

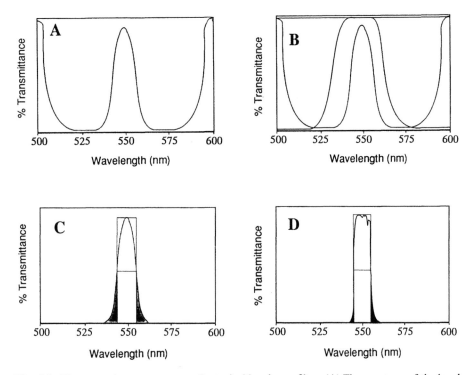

Fig. 16 The transmittance spectrum of a typical band pass filter. (A) The spectrum of the band pass interference filter. (B) The spectrum of A with the addition of a low pass filter and a high pass filter. (C) The spectrum resulting from the addition of the three curves in B. The horizontal bar shows the width of the filter's half-maximal transmission. The shaded areas show the amount of light passed both above and below the half-maximal points. (D) A multicavity band pass filter with the same half-maximal transmission spread as in C. Less light is passed beyond the half-maximal points and more light is transmitted within them.

light. Further, we will have far less stray light from wavelengths beyond the set boundaries. These multicavity filters are commercially available.

3. Choosing Band Pass Filters

One of the great benefits of the ratio method is the large dynamic range of response, on the order of $20\times$ for indo-1 and $15\times$ for SNARF-1. To take best advantage of this characteristic, filters should be chosen that maximize fluorescence peak sensitivity for one of the two species of dye (free or bound) while excluding light from the other. Figure 17A shows the spectrum for the two-state ratiometric dye indo-1 at its K_d, the ion concentration at which the free and bound species are at equal concentrations. Also included are the

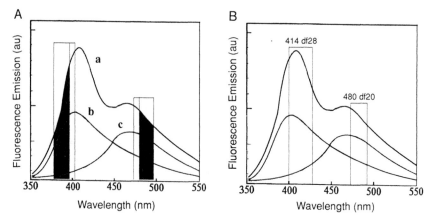

Fig. 17 (A) Indo-1 emission spectrum along with two choices for band pass filters to form the images for the ratio. Curve a represents the emission fluorescence for indo in 120 nM free Ca^{2+}, which is approximately its K_d at pH 7.2. Curves b and c are the contributions from the calcium-bound and calcium-free forms of the dye, respectively. The sets of rectangles represent the ranges for two series of band pass filters discussed in Section VI.A.3. (B) The same spectrum with the band pass filters presently used for indo-1.

contributions of the free and bound species to the spectrum. As one moves from either peak toward the isosbestic point, more and more of the total fluorescence comes from the opposite species. Choosing two narrow-range band pass filters (e.g., 5 nm) at the two peak wavelengths (400 nm and 475 nm) would give a large dynamic range (Fig. 17A); however, sensitivity as well as total light level can be increased by choosing broader band widths (e.g., 20 nm) with a range for each filter that favors the low side of the lower peak and the high side of the higher peak.

Another problem to consider is generated by the response characteristics of the detector. Figure 18 shows the response spectrum of a typical KS1381 microchannel plate image intensifier used in the imaging system. Sensitivity is low in the area around 400 nm and nonexistent at 390 nm. If we choose a narrow band pass filter at the low peak for indo-1, throughput will drop to near zero. Therefore, we use the filter in Fig. 17B, with a range from 400 to 428 nm, sacrificing some sensitivity in exchange for increased image brightness and an increased signal-to-noise ratio.

B. Multiwavelength Light Sources

Our experimental setup requires that we simultaneously present two excitation wavelengths to the cells in the microscope. Drawing both excitation wavelengths from the same light source will remove errors caused by differential brightness changes, such as drift or flicker, occurring in two separate light

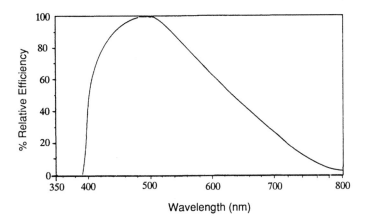

Fig. 18 Spectral response of a typical VideoScope KS 1381 microchannel plate image intensifier. Absence of response below 400 nm dictates the use of the D414/28 filter.

sources. The least expensive way to achieve this goal is to use an appropriate dual-wavelength band pass filter (Marcus, 1988). Such filters can be purchased for a few hundred dollars for dye combinations such as fluorescein/rhodamine or fluorescein/Texas Red. However, making such a filter to pass two wavelengths, one of which is below 400 nm and the other above, is very difficult.

We currently use a light source consisting of an Oriel Series Q four-port lamp housing (Oriel Corporation), a 75-Watt Xenon bulb, two condensers, two parabolic rear reflectors, two filter holders, and two lenses to focus output onto the ends of our bifurcated fiber optic bundle (Fig. 5). This system can be assembled with off-the-shelf parts for about $3000. The problem we have encountered is that positioning of the arc to illuminate both light paths brightly is very critical.

C. Cure for Shading Inhomogeneities in Bifurcated Fiber Optic Bundles

It is easy to create two excitation wavelength inputs by placing filters in front of the inputs to a bifurcated fiber optic bundle, then leading the output of the bundle into the epifluorescence path on the microscope. However, this arrangement often results in highly inhomogeneous distribution of the two wavelengths of light from the end of the bundle, because of highly nonrandom placement of the fibers in the bundle. Even the so-called randomized bundles seem to have this problem. This placement problem creates "hot spots" for one excitation wavelength over the other. To mix the two wavelengths, we use an 8-mm diameter liquid light guide (Stenning Instruments), coupled with silicon grease of appropriate refractive index to the output of our 8-mm fused silica fiber optic bundle (Dolan–Jenner Industries). Output from the light guide is homogeneous. The problems we have encountered are as follows. (1) The liquid in the guide has about the same transmission properties as glass; thus, the 350-nm light is attenuated about 50% compared with transmission through quartz or fused silica. (2) The windows of the liquid light guide are not optically flat; the image must be defocused to remove inhomogeneities.

D. Advantages and Disadvantages of Using Oil Immersion Objectives

We routinely use the Nikon 40× DL Phase Fluor 1.30 NA oil immersion objective for most of our experiments (Nikon Instruments). Experimenters have their choice of oil or dry objectives. The oil objectives have far more light gathering power because of their larger numerical aperture (Inoue, 1986), which translates into brighter fluorescence emission for the same level of excitation. An oil objective allows for lower excitation levels and, thus, less photobleaching and photodynamic damage. The major disadvantages are the following. (1) The objective is coupled by the immersion oil to the chamber containing the cells and becomes a heat sink. If absolute temperature regulation is required, then the objective will need to be wrapped with a thermostatted blanket. (2) Oil immersion objectives have smaller working distances than dry objectives. This feature is designed into the objective by the manufacturer, who expects the objective to be used on an upright microscope to observe specimens through a standard (#1) thickness cover slip. (Designs may change if investigators generate a demand for longer working distance oil immersion objectives.) To create greater working distances for our experiments, we use a 25-mm diameter #0 glass cover slip (Carolina Biological Supply, Burlington, North Carolina) for the bottom of our thermostatted chamber. We grow the cells on 18 × 18-mm #00 glass cover slips (Corning Glass Corporation, Corning, New York). The thickness of the two pieces of glass is less than that of the standard #1 cover slip for which the objective was designed. Alternative approaches are growing the cells in chambers with thin glass bottoms and placing these chambers into thermostatted holders or growing the cells on thin, round cover slips and sandwiching these into special adapters that fit into the thermostatted holder.

VII. Results

We have completed a number of studies using the video microscope to follow calcium and pH changes in cells. The first study demonstrates the use of the machine to look for pH changes to be used to correct the Ca^{2+} measurements. The last two studies show changes in one ion driven by the other.

A. *Escherichia coli* Heat-Stable Enterotoxin B (ST$_B$) Opens a Plasma Membrane Calcium Channel

The heat-stable enterotoxin B of *E. coli* (ST$_B$) is a 48-amino-acid extracellular peptide that induces rapid fluid accumulation in animal intestinal models. Unlike other *E. coli* enterotoxins which elicit cAMP or cGMP responses in gut (heat-labile toxin and heat-stable toxin A, respectively), ST$_B$ induces fluid loss by a hitherto undefined mechanism that is independent of cyclic nucleotide elevation. We studied the effects of ST$_B$ on intracellular calcium concentration ([Ca^{2+}]$_i$ (Dreyfuss *et al.*, 1993), another known mediator of intestinal ion and fluid movement.

Ca^{2+} and pH measurements were performed on MDCK cells, HT-29 intestinal epithelial cells, and primary pituitary intermediate lobe melanotrope cells. ST$_B$ treatment induced a slow dose-dependent rise in [Ca^{2+}]$_i$ with virtually no effect on internal pH in all three cell types (Fig. 19A,B). ST$_B$-mediated calcium elevation was not inhibited by drugs that block L-, T-, or N-type voltage-gated calcium channels. The rise in [Ca^{2+}]$_i$ was dependent on a source of extracellular Ca^{2+} and was not affected by prior treatment of cells with thapsigargin or cyclopiazonic acid, agents that deplete internal calcium stores by blocking the calcium-translocating ATPase. In contrast to these results, somatostatin and pertussis toxin pretreatment of cells completely blocked the ST$_B$-induced rise in [Ca^{2+}]$_i$ (Fig. 19C). Collectively, these data suggest that ST$_B$ opens a G$_i$- or G$_o$-protein-linked receptor-operated calcium channel in the plasma membrane of the cell. The nature and distribution of the ST$_B$-sensitive calcium channel is presently under investigation.

Cell pH occasionally showed a slight drop in response to ST$_B$-promoted calcium entry (Fig. 19B). Often no change was seen, especially if the Ca^{2+} rise was relatively slow. The significance of Ca^{2+}-driven pH changes is discussed in a subsequent section.

B. pH-Driven Release of Intracellular Calcium in Kidney Cells

The interrelationships between changes in intracellular calcium concentration ([Ca^{2+}]$_i$) and intracellular pH (pH$_i$) in MDCK cells, kidney glomerular epithelial cells, and pituitary melanotropes exposed to NH$_4$Cl loading were analyzed simultaneously in the fluorescence video microscope (Wiegmann *et al.*, 1993). Addition of NH$_4$Cl produced the expected alkalinization and was accompanied by a concurrent rise in [Ca^{2+}]$_i$. When the NH$_4$Cl was removed in a Na$^+$-free medium and the cells became acidic, a second rise in [Ca^{2+}]$_i$ was recorded. Both changes in [Ca^{2+}]$_i$ were from intracellular stores, since they persisted in the absence of extracellular calcium. The mechanism(s) coupling the pH and calcium changes is currently under investigation. (See Fig. 10.)

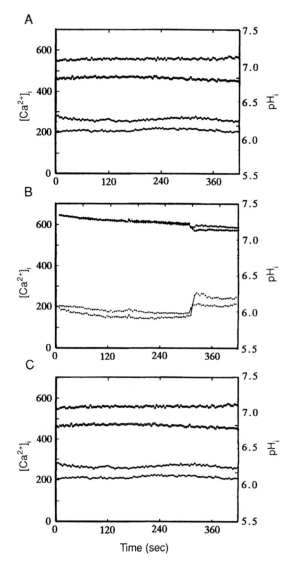

Fig. 19 Response of HT-29 cells to ST_B treatment. Integrated Ca^{2+} and pH levels from two cells are shown for each experiment. Cells are initially in Ca^{2+}-free balanced salt. ST_B toxin (50 nm) was added at 60 sec. Ca^{2+} was added at 300 sec to a total concentration of 3 mM. (A) Control (no ST_B added). (B) ST_B added at 60 sec. (C) Somatostatin (1 μM) was added to the bath 5 min before the start of the recording.

C. Calcium-Driven Cytoplasmic Acidification in Secreting Cells

For a number of electrically excitable secreting cells, depolarization by high potassium or treatment with secretagogues has been shown not only to promote a rise in intracellular calcium levels, but also to cause a drop in intracellular pH (Kuijpers *et al.*, 1989; Rosario *et al.*, 1991; Törnquist and Tashjian, 1992).

Melanotropes, the predominant cell type of the pituitary intermediate lobe, synthesize and store the proopiomelanocortin (POMC)-derived peptides α-melanocyte stimulating hormone (α-MSH) and β-endorphin in secretory vesicles. These peptides are released by exocytosis in response to K^+ depolarization or β-adrenergic agonists. We used the imaging system to study calcium and pH responses of primary rat melanotropes to K^+ depolarization, as well as to ionophores and calcium-transport ATPase blockers (Beatty *et al.*, 1993).

Changes in $[Ca^{2+}]_i$ produced changes in pH_i. K^+-induced depolarization of melanotropes produced increases in $[Ca^{2+}]_i$ by activation of L-type Ca^{2+} chanels. Ca^{2+} entry was coupled closely to reductions in pH_i. We examined a large number of recordings of the type presented in Fig. 20. In some cases,

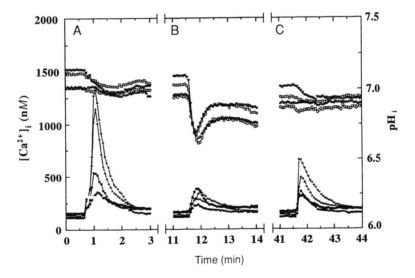

Fig. 20 Effect of cyclopiazonic acid (CPA) treatment on cultured primary melanotropes. The calcium and pH responses of the four cells identified in Color Plate 1H are graphed. (A) After recording 60 sec of baseline, the bath K^+ concentration is raised iso-osmotically to 66 mM. The resulting depolarization opens L-type voltage-activated calcium channels. Both intracellular Ca^{2+} and H^+ rise (pH falls) rapidly. The calcium level recovers more rapidly than the pH. (B) The bath has been exchanged for the standard balanced salt solution (BSS). Addition of 20 nM CPA to the bath causes the depletion of the intracellular calcium stores. Again, both intracellular Ca^{2+} and H^+ rise rapidly. (C) The bath is exchanged for BSS. After recovery, the cells are depolarized again with K^+. The responses are qualitatively similar to those seen in A.

before the calcium values were corrected for the prevailing pH, the rise in $[Ca^{2+}]_i$ appeared to respond more slowly than the drop in pH_i. This effect disappears after pH correction (Fig. 21). In no case does the pH begin to fall before the Ca^{2+} level rises. Thus, we can accept the null hypothesis that pH decline does not activate the L-type Ca^{2+} channel. The findings demonstrate the need for pH correction of indo-1 recordings.

The potassium effects were dependent on entry of extracellular Ca^{2+} rather than on direct release of Ca^{2+} from intracellular stores. No changes in $[Ca^{2+}]_i$ or pH_i were seen in the presence of either 50 μM extracellular ethylene glycol bis(β-aminoethylether)-N,N,N',N'-tetraacetic acid (EGTA) or 2 mM 1,2-bis(o-aminophenoxy)ethane-N,N,N',N'-tetraacetic acid (BAPTA). The responses were blocked by the L-type channel blockers verapamil, nitrendipine, and nifedipine, but not by T-type or N-type channel blockers. A secondary Ca^{2+} elevation, sometimes followed by oscillations, often was produced from K^+ depolarization and appeared to result from Ca^{2+} released from intracellular stores. A rise in $[Ca^{2+}]_i$ was coupled to release of Ca^{2+} from intracellular stores by thapsigargin or cyclopiazonic acid (Fig. 20B), or by ionophore-mediated influx across the plasma membrane. After thapsigargin or cyclopiazonic acid treatment, the primary Ca^{2+} elevation in response to K^+ depolarization persisted but the secondary Ca^{2+} peak disappeared (Fig. 20C), suggesting that the second peak was generated by Ca^{2+} release from intracellular stores.

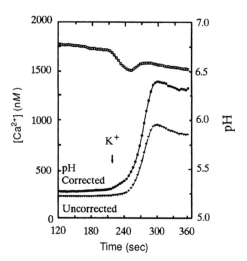

Fig. 21 Effects of correcting the indo-1/$[Ca^{2+}]$ K_d for the prevailing pH_i compared with using a fixed K_d to calculate $[Ca^{2+}]_i$. The pH kinetic data (○) for one cell are shown. The $[Ca^{2+}]_i$ kinetic data were calculated using a fixed K_d for a pH of 7.0 (■) and by first calculating the $K_{d_{corr}}$ from the prevailing pH (●) and Eq. 3.

Our data support the following mechanism for the calcium-driven pH changes. Close association between increases in intracellular Ca^{2+} from any source and increased intracellular H^+ were observed for all treatments (K^+ depolarization, Ca^{2+} ionophore treatment, or release from intracellular stores) suggesting that the observed pH_i changes were caused by release of H^+ on binding of Ca^{2+} to intracellular calcium buffers. Although the pH drop is "passive," because of the poor buffer capacity for H^+, the change will be sensed by all cytoplasmic components. Unlike Ca^{2+}, the diffusion of which in cytoplasm is slowed by high buffer capacity (Allbritton *et al.*, 1992), H^+ changes are transmitted rapidly throughout the cell. Thus, local changes in $[Ca^{2+}]_i$ can affect a larger volume of cytoplasm through the secondary H^+ signal. This signal becomes a second messenger pathway akin to the generation of cAMP or inositol trisphosphate and cannot fail to influence numerous pH-dependent cell activities. Therefore, the calcium-driven pH change provides a direct coupling of a sudden large influx of calcium into the cytoplasm to pH-sensitive processes which may, in turn, be linked to processes such as exocytosis. The calcium may come from extracellular or intracellular sources. We expect the elucidation of this pathway to be of significant future interest.

Acknowledgments

This work was supported by grants from the Kansas Affiliate of the American Heart Association, the Loeb Charitable Trust, National Science Foundation Grant IBN-9211912, and National Institutes of Health (NIH) Grant GM44071 (S. J. Morris); NIH Grant NS28019 (B. M. Chronwall); and the Department of Veterans Affairs (T. B. Wiegmann and L. W. Welling).

References

Allbritton, N. L., Meyer, T., and Stryer, L. (1992). Range of messenger action of calcium ion and inositol 1,4,5-trisphosphate. *Science* **258**, 1812–1815.

Beatty, D. M., Chronwall, B. M., Howard, D. E., Wiegmann, T. B., and Morris, S. J. (1993). Calcium regulation of intracellular pH in pituitary intermediate lobe melanotropes. *Endocrinology* **133**, 972–984.

Bright, G. R. (1992). Multiparameter imaging of cellular function. *In* "Fluorescent Probes for Biological Function of Living Cells—A Practical Guide" (W. T. Matson and G. Relf, eds.). San Diego: Academic Press.

Dreyfus, L. A., Harville, B., Howard, D. E., Shaban, R., Beatty, D. M., and Morris, S. J. (1993). Calcium infux mediated by the *Escherichia coli* heat-stable enterotoxin B (ST_B). *Proc. Natl. Acad. Sci. U.S.A.* **90**, 3202–3206.

Grinstein, S., Cohen, S., Goetz-Smith, J. D., and Dixon, S. J. (1989). Measurements of cytoplasmic and cellular volume for detection of Na^+/H^+ exchange in lymphocytes. *Meth. Enzymol.* **173**, 777–790.

Grynkiewicz, G., Poenie, M., and Tsien, R. Y. (1985). A new generation of Ca^{2+} indicators with greatly improved fluorescence properties. *J. Biol. Chem.* **260**, 3440–3448.

Haugland, R. (1992). "Molecular Probes Handbook of Fluorescent Probes and Research Chemicals," 1992–1994 edition. Junction City, Oregon: Molecular Probes.

Inoue, S. (1986). "Video Microscopy." New York: Plenum Press.

Kuijpers, G. A. J., Rosario, L. M., and Ornberg, R. L. (1989). Role of intracellular pH in secretion from adrenal medulla chromaffin cells. *J. Biol. Chem.* **264,** 698–705.

Kurtz, I., and Balaban, R. S. (1985). Fluorescence emission spectroscopy of 1,4-dihydroxyphthalonitrile: A method for determining intracellular pH in cultured cells. *Biophys. J.* **48,** 499–508.

Lattanzio, F. A. (1990). The effects of pH and temperature on fluorescent calcium indicators as determined with Chelex-100 and EDTA buffer systems. *Biochem. Biophys. Res. Commun.* **171,** 102–108.

Lattanzio, F. A., and Bartschat, D. K. (1991). The effect of pH on rate constants ion selectivity and thermodynamic properties of fluorescent calcium and magnesium indicators. *Biochem. Biophys. Res. Commun.* **177,** 184–191.

Marcus, D. A. (1988). High performance optical filters for fluorescence analysis. *Cell Motil. Cytoskel.* **10,** 62–70.

Martinez-Zaguilan, R., Martinez, G. M., Lattanzio, F., and Gillies, R. J. (1991). Simultaneous measurement of intracellular pH and Ca^{2+} using the fluorescence of SNARF-1 and fura-2. *Am. J. Physiol.* **260,** C297–C307.

Morris, S. J. (1990). Real-time multi-wavelength fluorescence imaging of living cells. *BioTechniques* **8,** 296–308.

Morris, S. J. (1993). Simultaneous multiple detection of fluorescent molecules: Rapid kinetic imaging of calcium and pH in living cells. *In* "Optical Microscopy: Emerging Methods and Applications" (B. Herman and J. J. Lemasters, eds.), pp. 177–213. New York: Academic Press.

Morris, S. J., Beatty, D. M., Welling, L. W., and Wiegmann, T. B. (1991). Instrumentation for simultaneous kinetic imaging of multiple fluorophores in single living cells. *SPIE Proc.* **1428,** 148–158.

Morris, S. J., Sarkar, D. P., Zimmerberg, J., and Blumenthal, R. (1993). Kinetics of cell fusion mediated by viral proteins. *Meth. Enzymol.* **221,** 42–58.

Rijkers, G. T., Justement, L. B., Griffioien, A. W., and Cambier, J. C. (1990). Improved method for measuring intracellular Ca^{2+} with fluo-3. *Cytometry* **11,** 923–927.

Roe, M. W., Lemasters, J. J., and Herman, B. (1990). Assessment of Fura-2 for measurements of cytosolic free calcium. *Cell Calcium* **11,** 63–73.

Rosario, L. M., Stutzin, A., Cragoe, E. J., Jr., and Pollard, H. B. (1991). Modulation of intracellular pH by secretagogues and the Na^+/H^+ antiporter in cultured bovine chromaffin cells. *Neuroscience* **41,** 269–276.

Spring, K. R. (1990). *In* "Optical Microscopy for Biology" (B. Herman and K. A. Jacobson, eds.), pp. 513–522. New York: Wiley-Liss.

Törnquist, K., and Tashjian, A. H., Jr. (1992). pH homeostasis in pituitary GH_4C_1 cells: Basal intracellular pH is regulated by cytosolic free Ca^{2+} concentration. *Endocrinology* **130,** 717–725.

Tsien, R. Y., and Harootunian, A. T. (1990). Practical design criteria for a dynamic ratio imaging system. *Cell Calcium* **11,** 93–109.

Valet, G. A., Moroder, R. L., Wunsch, E., and Ruhenstroth-Bauer, G. (1981). Fast intracellular pH determination in single cells by flow cytometry. *Naturwissenschaften* **68,** 265–266.

Wiegmann, T. B., Welling, L. W., Beatty, D. M., Howard, D. E., Vamos, S., and Morris, S. J. (1993). Simultaneous imaging of intracellular $[Ca^{2+}]$ and pH in single MDCK and glomerular epithelial cells. *Am. J. Physiol. Cell Physiol.* **265,** C1184–C1190.

CHAPTER 9

Improved Spatial Resolution in Ratio Images Using Computational Confocal Techniques

Thomas J. Keating and R. John Cork

Department of Biological Sciences
Purdue University
West Lafayette, Indiana 47907

I. Introduction

Light microscopy has undergone a renaissance in recent years because of advances in digital video technology and the development of new fluorescent probes. The increased availability of low light level cameras and inexpensive computers has allowed the acquisition and analysis of fluorescence images to be performed with greater quantitative accuracy. As a result, monitoring and quantifying many physiological variables, including $[Ca^{2+}]_i$, in individual cells with relatively high spatial resolution is now possible.

Nonetheless, as with all images acquired with a microscope, digital video images are degraded by out-of-focus information and by the limited resolving power of the microscope in the plane of focus. Both these effects combine to limit the resolution and quantitative accuracy of fluorescence images by producing a characteristic blur that obscures details and smears light from one region of an image into neighboring areas. This problem may be especially significant in measurements of $[Ca^{2+}]_i$ and other physiological parameters because the variations in intensity among different regions of the sample may be only slight. Blurring of the light can make a localized increase appear to be distributed more broadly but to be less intense and, in some cases, could obscure localized increases entirely.

If information about how the microscope affects images is available, it is possible to use it to compensate for such effects and undo much of the degradation that occurs during image formation. In this chapter, we describe how these adjustments can be made and demonstrate that such processing improves the quantitative accuracy of these images, particularly when using a ratiometric dye.

II. Computational Confocal Image Processing

In a light microscope, the objective lens is focused on only a thin section of the sample. The resulting image, however, also contains out-of-focus light that comes from other regions of the sample, above and below the plane of focus. This out-of-focus information appears as a cloudy blur that obscures the fine details of the in-focus image. In addition, even the information in the image that arises from the in-focus parts of the sample are blurred to some extent by diffraction of light as it travels through the aperture of the objective (Spencer, 1982).

The way in which a microscope distorts and blurs an image is described by its point spread function (PSF). The PSF is a mathematical description of how the image of a point source will appear in the in-focus plane and at all levels out of focus. For any microscope, the PSF is almost completely dependent on the particular objective used. This function is affected by the numerical aperture (NA) and any aberrations of the objective lens, as well as by the index of refraction of the immersion medium. PSFs can be calculated from theoretical approximations or can be determined empirically. The details of how to obtain a PSF are discussed in a subsequent section.

Once a PSF has been determined or approximated, the physical blurring or distorting of an image can be described mathematically by convolving the PSF with the intensity distribution in the real sample (see Inoué, 1986, for an explanation of convolutions). Treating each focal plane separately, so the in-focus plane of sample light intensity is convolved with the in-focus PSF of the microscope and light from the out-of-focus parts of the sample are convolved with the out-of-focus PSFs of the microscope, is the easiest method. The observed image is essentially the sum of the information from the various planes of focus, after each has been convolved with its respective PSF (Fig. 1).

In a confocal microscope, out-of-focus information is rejected physically and the observed image comes essentially from one focal plane. The essence of the computational confocal technique is to use the PSF to reverse computationally some of its effects and to produce an image that more accurately portrays the in-focus portion of the sample than the original observed image.

If we consider a simple example in which only two neighboring focal planes contribute to the out-of-focus blur, an observed image, o_j, can be represented in mathematical terms (Agard *et al.,* 1989)

$$o_j \sim i_j {}^* s_0 + i_{j-1} {}^* s_{-1} + i_{j+1} {}^* s_{+1} \tag{1}$$

where i_j, i_{j+1}, and i_{j-1} are the parts of the sample that are in-focus, above focus, and below focus, respectively; s_0, s_{+1}, and s_{-1} are the respective in-focus and out-of-focus PSFs; and the asterisks represent convolutions. Since we do not have access to the real in-focus information in each focal plane, we assume that it is approximately equal to the observed image from the out-of-focus plane (i.e., $o_{j+1} \sim i_{j+1}$ and $o_{j-1} \sim i_{j-1}$). The deblurred in-focus image of the in-focus plane (i_j) then can be obtained from

$$i_j \sim [o_j - c\,(o_{j+1} {}^* s_1 + o_{j-1} {}^* s_1)] {}^* g \tag{2}$$

where c is an empirical constant, g is a filter that corrects for the in-focus PSF, and s_1 is the out-of-focus PSF, which is assumed to be equivalent above and below the in-focus plane. This processing scheme is called the "nearest-neighbors" algorithm (Agard *et al.,* 1989). An additional simplification is the "no-neighbors" algorithm (Monck *et al.,* 1992)

$$i_j \sim [o_j - 2c\,(o_j {}^* s_1)] {}^* g \tag{3}$$

in which, for the purposes of processing, we assume that $o_j \sim o_{j+1} \sim o_{j-1}$. The nearest-neighbors and no-neighbors processing algorithms will be discussed in more detail in a subsequent section.

III. Point Spread Function

To reverse the distorting effects of the microscope, the PSF first must be determined directly or calculated from optical theory. The PSF can be found empirically by imaging a fluorescent bead that is smaller than the resolution of the system, so it acts as point source of light. If the microscope were a perfect imaging system, the bead would appear as a single point of light (contained in a single pixel on the video screen) when in-focus and would not be seen when

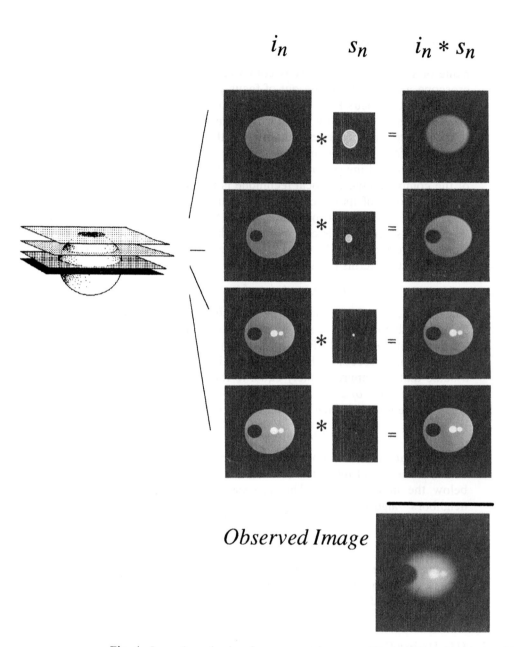

$$i_n \qquad s_n \qquad i_n * s_n$$

Observed Image

Fig. 1 Image formation in a fluorescence microscope. The formation of an observed image can be thought of as the sum of the actual information in each plane of the sample, convolved with its appropriate PSF. On the left is a schematic representation of the sample, showing the positions of four of the infinite number of planes that make up a real sample. For each of these planes, the in-focus portion of the sample, the appropriate PSF, and the result of the convolution are shown. The top row of images corresponds to the portion of the sample that is 6.5 μm above the in-focus plane, the second row to 3.0 μm, the third row to 1.0 μm, and the fourth row to the in-focus plane. Finally, the single image in the lower right is the observed image, which is the sum of the blurred information from the entire sample using planes that were 0.1 μm apart.

out-of-focus. In reality, however, the in-focus image of the bead consists of a bright center around which the intensity decreases as one moves away from the center. Thus, one can see directly the effect that the in-focus PSF has on images as they are collected with the microscope.

A. Empirical

To observe the out-of-focus PSF, one should focus the microscope above or below the bead. One feature to note in the out-of-focus images of the beads is their radial symmetry, or perhaps their lack thereof. Although the empirical PSF may not be radially symmetrical, the theoretical PSF (discussed subsequently) is perfectly circular. Since imperfections such as this that must be accounted for may exist in the real PSF, an empirically measured PSF is preferable to one based on theoretical calculations.

It is important to collect the PSF images under conditions that resemble as closely as possible the manner in which the images to be processed were collected. One should, of course, use the same objective lens since the objective is the major determinant of the PSF (Agard *et al.,* 1989). Also, the wavelength of light should be similar to the emission wavelength of the indicator that was used (e.g., fluorescein-type beads would suffice for images collected with fura-2, since the emission peaks for both fluorescein and fura-2 are 500–520 nm). Other factors to keep constant are the thickness of the cover slip, the type of immersion oil used, and the magnification of the final image.

The beads we use are latex microspheres impregnated with fluorescent indicators and are available from several companies (e.g., Polysciences, Warrington, Pennsylvania; Molecular Probes, Eugene, Oregon). The bead used should be smaller than the resolution limit of the system so it can act as a point source; if the actual shape and structure of the bead are detectable in the PSF, the images that are convolved with those bead images will be distorted. If the beads cluster together, they might form an asymmetric image, leading one to believe that the PSF is not circular. This issue can be resolved by examining other beads and noting their asymmetries, since finding other bead clusters having the same orientation is unlikely.

As a practical matter, extremely small beads will have a dim signal and will produce noisy images, especially in the out-of-focus planes. Therefore, using the largest size bead that is still below the resolution limit is a good idea. To reduce noise, one can average the images of several different beads at each plane of focus. Care must be taken, however, to align the images accurately before averaging them (Hiraoka *et al.,* 1990). Alternatively, averaging can be performed more easily in the Fourier domain, as is explained in Appendix I.

To change the focus, a stepping motor can be attached to the fine focus mechanism of the microscope. Alternatively, the fine focus knob can be turned by hand, using the 1-μm gradations that generally are marked on the knob to guide the movement. Regardless of the method chosen, a series of sections should be collected by traveling in the same direction since a considerable amount of free play exists in the focus mechanisms of most microscopes.

Images of the bead should be collected at regular intervals starting 5–10 μm below focus and moving an equal distance above the plane of focus of the bead. The distance between adjacent images and the number of images collected will be determined by how the processing will be performed later (see subsequent discussion), as well as by the method used to move the focus mechanism. The deblurring techniques discussed in this chapter only require at most three PSF images: one in focus, one a certain distance below focus, and another the same distance above focus. However, since one does not know exactly where the in-focus plane will be or which out-of-focus image PSFs will be used to deblur the images, collecting as many sections as possible and having them available later is useful.

B. Theoretical

Another method for estimating the PSF is to calculate it based on theoretical models of diffraction-limited optical systems. These calculations, which are explained by Castleman (1979) and Agard (1984), form the basis of the PSFs used by many of the commercial image deblurring software packages currently available. However, as discussed by Agard *et al.* (1989), the theoretical PSF differs substantially from the actual PSF, aside from the fact that the theoretical PSF cannot account for aberrations in individual lenses. Despite these shortcomings, however, the theoretical PSF can be used to deblur images, producing dramatic improvements in resolution (Castleman, 1979; Agard, 1984; Monck *et al.*, 1992).

A C language program for calculating the theoretical PSF can be obtained by contacting the authors of this chapter. This program is an implementation of the equations of Castleman (1979) and Agard (1984) and can be adjusted for differences in numerical aperture of the objective, index of refraction of the immersion oil, size of the pixels in the image, dimensions of the image, and wavelength of the emission light. The specific equations used in this subroutine are presented in Appendix II.

IV. Deblurring and Inverse Filtering Algorithms

Once the PSF is obtained, it can be used to produce more detailed images in a number of ways. These methods differ in the number of focal planes that are included and in the techniques used to do the image processing; however, all the methods use the same basic two-step process discussed for Eq. 2. An observed image can be improved first by subtracting information from out-of-focus parts of the sample, a process that we refer to as deblurring. A deblurred image then can be corrected for the effects of the in-focus PSF, a procedure we call inverse filtering. In some cases, only one of these procedures may be possible, or necessary, but all the techniques discussed in this section fall into the general category of one or the other.

A. Deblurring

To implement a deblurring algorithm, a method of convolving images with corresponding PSFs is first necessary. Two methods exist for performing convolutions of digital video images. These convolutions may be calculated directly, convolving each pixel in the image with the values from the convolution mask or kernel (Inoué, 1986, p. 367). Alternatively, convolutions may be done in the Fourier domain, in which they become a simple point-by-point multiplication of the image transform and the PSF transform, the contrast transfer function (CTF).

Many general-purpose image-processing programs can perform convolutions and allow the user to specify the convolution kernel. This calculation is fairly computer-intensive and will be slow unless the kernel is kept relatively small (3×3 pixels, 5×5, or 7×7). Also, depending on the computer available, the kernel may accept only integer values. A suitable kernel can be obtained from a PSF by taking the appropriate number of pixels from the center of the bead image (for example, taking the 9 pixels in the 3×3 region in the center of the bead). If the PSF gray values are used directly in the convolution kernel, the resulting convolved image is most likely to have many pixels with the maximum gray value (255). Therefore, the values in the kernel should be scaled down so the overall intensity of the image being convolved is not changed. Typically, the sum of all the terms in such a filter kernel is 1.0; however, if only integer values are allowed, use the smallest integers possible and divide the convolved image by the sum of the kernel values.

A small kernel obtained from an out-of-focus PSF is most likely to have terms that all have about the same value, as does a low pass, or smoothing, filter kernel (Inoué, 1986). Larger kernels can describe the actual PSF of the system more accurately; however, at focal planes further from the in-focus plane, larger and larger kernels are needed to describe the PSF. For example, on our system, an in-focus PSF can be approximated closely with a 13×13 kernel, but a 35×35 kernel is required to approximate the PSF 7 μm out of focus. As the kernel becomes larger, the convolution becomes slower; at some point, performing the calculations in the Fourier domain instead makes sense (Appendix I).

Having chosen a method for performing the convolutions, you must select a scheme for deblurring the image. Different schemes have been proposed for deblurring images. Two that have been employed for calcium images are the "nearest-neighbors" scheme (Yelamarty et al., 1990) and the "no-neighbors" algorithm (Monck et al., 1992). Other schemes can be envisioned, for example, including more out-of-focus planes, each one convolved with its corresponding PSF.

In the nearest-neighbors scheme, one collects three images: one from the plane that will be deblurred and the other two from neighboring sections, equal distances above and below the plane of the original image. The two neighboring sections are convolved with their corresponding out-of-focus PSFs. The resulting blurred images are subtracted from the in-focus image, producing a deblurred image. As mentioned earlier, one can simplify these calculations by assuming that the out-of-focus PSFs from above and below the in-focus plane are equivalent. If this assumption is known not to be true, convolving each out-of-focus image with the correct PSF makes sense. Alternatively, using an immersion oil with a different index of refraction, one should be able to make the out-of-focus PSFs more or less equivalent (Hiraoka et al., 1990).

The amount of out-of-focus information to be subtracted from the observed image is determined by the constant c in Eq. 2. Ideally, if all the observed out-of-focus images did not contain any information from their neighboring planes (i.e., $o_n = i_n$, $n = \pm 1$), then c should equal 1.0, but because each out-of-focus image is a combination of information from many focal planes, c must be reduced. Theoretically, the most appropriate value for c would be 0.5 (Agard et al., 1989); however, this value would also remove all the low frequency information from the in-focus plane, resulting in a deblurred image with greatly exaggerated edges and no regions with constant intensity. In general, the constant must be determined empirically to provide the most deblurring without removing useful information from the required image. Maximum values for c tend to be in the range 0.45–0.5, but the actual value chosen depends to a large extent on how much signal is present in the original image. One possible way to determine the optimum value is to plot profiles of the gray values across the deblurred image using different values for c. Ideally, optimal deblurring should be when the intensity profiles of the edges of objects in the image are as square as possible. As we have found in practice, however, the values of c that best restore the ratio values in test images do not always produce square intensity profiles for the structures in the individual images (Fig. 2; see also Monck et al., 1992).

In the nearest-neighbors algorithm, the PSFs from above and below usually are assumed to be equivalent, which may not always be the case. If an experimental PSF has been determined, any variations can be included in the deblurring scheme. If more than two out-of-focus planes are collected, some combination of weighting constants must be derived to determine how much of each blurred image to subtract from the observed image. One way to do this might be to measure how the total intensity of the bead images drops off as the focus is changed.

A simple deblurring scheme that requires less time collecting out-of-focus images is the no-neighbors scheme. This method is similar to the nearest-

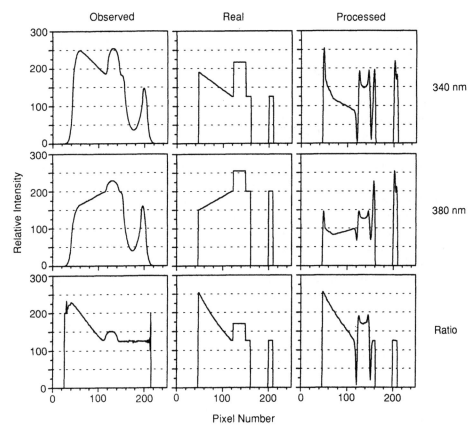

Fig. 2 Intensity profiles from the test cell. A model cell was generated and simulated observed images were calculated as described in the text with parameters that match our system, including a 40×/1.3 NA objective lens; pixel size was 0.3 μm by 0.27 μm. The profiles are shown as relative intensity against pixel number. Pixels numbers represent positions along the line AA in Color Plate 3. Profiles are shown for the simulated observed images, the real in-focus data, and the processed images. For each of these, profiles are shown for the 340-nm and 380-nm images, as well as for the 340/380 ratio. The intensity values are arbitrary and were stretched to fill the available gray scale (0–255). The three ratio profiles on the bottom row are plotted on the same intensity scale and can be compared directly. The three ratio profiles correspond to the images in Color Plate 3.

neighbors scheme, except only a single image is collected. This image is convolved with an out-of-focus PSF, and the resultant image is substracted from the original observed image to correct for out-of-focus blur. The choices that must be made to implement this scheme are described by Monck *et al.* (1992); the choice of interplanar spacing between in-focus plane and out-of-focus PSF, and the amount of blurred image to subtract are the most critical to the results.

B. Inverse Filtering

After deblurring an image, it may be improved further by correcting for the in-focus PSF of the system, using an appropriate filter for the term g in Eq. 2. Again, this calculation can be done in real space or in the Fourier domain. In real space, one must convolve the in-focus image with a filter kernel that corrects for the in-focus PSF; this generally will be some form of edge sharpening filter (Inoué, 1986; Marangoni et al., 1992). An $n \times n$ filter kernel based on the specific in-focus PSF can be constructed by taking the center $n \times n$ pixels from the in-focus PSF and multiplying all but the central pixel by -1. The central pixel should have such a value that the total of all the terms is equal to 1.0. All values may have to be scaled to keep the sharpened image within the usable gray scale; different weightings applied to the center value will adjust the amount of edge sharpening relative to the requirement for keeping areas of uniform intensity unchanged.

An alternative to using sharpening kernels is creating a Fourier domain function that corrects for the in-focus PSF, and multiplying the in-focus image by this filter. At its simplest, such a filter would be the inverse of the in-focus CTF. However, such a filter would accentuate the high frequency noise in the image, probably making the sharpened details indistinguishable from the background. A filter that optimizes the image restoration while minimizing the noise contribution is the Wiener filter (Berriel et al., 1983; Monck et al., 1992). The equation for the Wiener filter we have used is

$$G = S_0/(S_0^2 + \alpha) \tag{4}$$

where G and S_0 are the Fourier transforms of g and s_0, respectively, and α is an empirical constant that is a function of the signal-to-noise ratio of the image.

The different methods of deblurring can be used with either type of sharpening process; also, sharpening can be performed on images without deblurring them. For example, an image of a flat tissue culture cell may not have much out-of-focus information to remove and may be improved by sharpening alone. Conversely, in an image of a large sample taken at a lower magnification, most of the image degradation may be due to out-of-focus blur whereas the in-focus PSF may not be a large factor.

V. Ratios of Processed Images

Since computational confocal image processing requires that the out-of-focus information in an image be removed, one must begin with images that have a high signal-to-noise ratio to have a usable signal left after processing. On a video imaging system, one should perform frame averaging to minimize noise in the original image. Also, since image processing relies on detecting subtle differences between gray values in different regions of the sample, using as much as possible of the available gray scale by adjusting the gain and offset of the camera and the digitizing board is important. The system we use produces 8-bit images, giving 2^8 or 256 different gray values. Slow-scan cooled charge-coupled device (CCD) cameras usually are equipped to produce images with 10–16 bits per pixel of gray values, allowing finer resolution of intensities and a wider dynamic range.

As for other corrections applied to ratio images, for example, background subtraction, the most important rule to follow is to process the two images in an equivalent manner before calculating the ratio. This requirement includes using the same values for the empirical constants as well as using the same PSF images to process the sections. The problem that remains, then, is choosing the values for the empirical constants c and α, and choosing the out-of-focus PSF to use to recover the proper ratios after correcting the images.

To investigate this question, we took an approach similar to that of Monck *et al.* (1992) and generated a computer model of a cell, as though it were loaded with a ratiometric Ca^{2+} dye. We then calculated pairs of images as they might be observed through a microscope, using the theoretical PSF to blur the sections. The theoretical cell (Fig. 2; Color Plate 3) contained a gradient of calcium, the peak of which is at the left side of the in-focus image and drops off linearly over 20 μm in all directions. The model cell also contains a spherical organelle that has a higher concentration of dye and a higher Ca^{2+} level inside. Finally, another spherical organelle excludes the dye (but is otherwise transparent) and therefore has a ratio of zero.

Computer-generated "observed" images were created that represented the two wavelengths of a Ca^{2+} ratiometric dye such as fura-2 or indo-1; the ratios of the pairs of observed images were calculated. The ratios in the observed images were different from the "real" in-focus ratios in several respects (Fig. 2; Color Plate 3). First, in regions in which the gray levels differed in the neighboring planes, for example, in the region of the gradient, the ratio values were underestimated. In addition, ratio values extend outside the boundaries of the cell because of blurring at the edges of the individual observed images. Finally, the organelle that excludes dye should contain only zeros for its ratio values, yet it has the same ratio as the surrounding cytoplasm, again because of information from neighboring areas scattered into the dark region.

The images then were processed with the theoretical PSFs and the ratios of the processed images were calculated (Fig. 2; Color Plate 3). Since we know what the ratios in the various regions of the image should be, we used this model to derive some general rules about how to choose the empirical constants to guide us when we processed images with unknown ratios.

We started by performing only deblurring on the observed images using different values of c and different out-of-focus PSFs. We then calculated the ratios of the deblurred images and compared these to the actual in-focus ratios. For a given out-of-focus PSF, as the value of c increased, the amount of light remaining in inappropriate areas of the image, for example, in the dark organelle or outside the edge of the cell, decreased. In the ratios of these images, the values outside the cell and in the dark organelle decreased until, at a certain value of c, they were equal to zero. The value of c that eliminated the ratios outside the cell was lower than the value required to eliminate the ratios inside the dark organelle because out-of-focus light was smeared into the organelle from all directions whereas the out-of-focus blur outside the cell arose from a much more limited area and therefore was not as extensive.

If the images were processed with the minimum value of c that was required to eliminate the ratios in the dark organelle, the ratio values in the rest of the image were close to their correct values. In fact, increasing or decreasing c from this value generally made the ratio values in the rest of the image less accurate. Similar results were obtained when we adjusted c to obtain the correct ratio in the bright organelle: the correct value of c gave ratios that were closest to the actual values in the rest of the image. Therefore, if a reference point inside the cell has a known ratio, the pairs of images can be processed with different c values until the reference structure has the correct ratio. From the results with the model cell, one can conclude that this value is the best value for c and that the ratio values are much closer to their actual levels than in the ratio of the unprocessed images. This improvement is confined mainly to regions in which the gray values differ in the neighboring focal planes.

When images were processed with the optimal c value, the edges of structures in the images tended to be exaggerated. In the intensity profiles (Fig. 2), the gray values at the edges of structures became very high, producing spikes. In most cases, the spikes were equivalent in the two images and were cancelled out in the ratio. In the case of the bright organelle, however, the edges of the structure were sufficiently different in the two images that the spikes were not cancelled out in the ratio. Although the values in the center of the bright organelle were accurate in the ratio of the processed images, the values nearer the edge of the structure were completely inaccurate. A situation such as this might arise in a real cell if the dye becomes concentrated in organelles inside the cell, particularly if the calcium level in the organelle is much higher than that in the cytoplasm. One way to reduce the occurrence of this problem in practice is to avoid cells in which the dye has become compartmentalized into the organelles.

The out-of-focus PSF that is used puts a limit on the amount of out-of-focus blur that can be subtracted from the in-focus image. As the PSFs get further from the in-focus plane, they blur the image more and leave less of the sharp details in the image, meaning only lower spatial frequencies can be removed when this blurred image is subtracted from the in-focus image. Using an out-of-focus PSF from a plane closer to the in-focus plane blurs the image less and therefore allows higher frequency information to be subtracted away. In the theoretical model, and when processing images of real cells, we use the PSF from 1 μm out of focus to deblur the images. This value effectively removes the out-of-focus blur but still leaves sufficient in-focus information to produce a clear image.

Changing the value of α in the Wiener filter (Eq. 4) did not alter the values of the ratios of the processed images greatly, except when α was extremely large (>5.0) or small (<0.01). Although the Wiener filter is supposed to be the equivalent of dividing by the in-focus PSF, the α factor is required to prevent higher frequencies from being overemphasized in the final image. As α approaches zero, the Wiener filter is closer to the inverse of the in-focus PSF and brings out the features in the image that have higher spatial frequencies. When α is too small, however, the images become noisy, since the highest frequency information is noise. When α is very large, the Wiener filter acts more like a low pass filter, blurring the images with which it is convolved. Since values between 0.01 and 5.0 do not change the values of the ratios very much, the choice of an optimal value is somewhat arbitrary; choosing a value that makes the images look the best seems reasonable. With real images, the choice of α will be dictated mainly by the amount of noise that is present in the images.

To provide an example of how these techniques work on real cells, we collected images of undifferentiated PC-12 cells loaded with fura-2/AM, exciting with 360-nm and 380-nm light (Color Plate 4). Shown is a cluster of three cells, one of which has a large organelle that excludes the dye, as in the model cell. The images were deblurred with a 1-μm out-of-focus PSF that was generated by imaging a fluorescent bead. The value of c was increased from 0.45 until the dark organelle in the cell on the right was completely black. However, because of limitations in the signal-to-noise ratio in these images, processing the images with the optimal value of c removed too much information and made the ratio of the processed images very noisy. As a compromise, we used a value for c (0.475) that made most of the ratio values in the dark organelle equal to zero without removing too much of the signal. The value of α in the Wiener filter was 5.0, causing it to act like a low pass filter. Lower values of α made the ratio appear noisy but did not change the overall ratio values significantly. After processing the 360-nm and 380-nm images, several changes occurred in the ratio compared with the ratio of the unprocessed images. One difference was that the ratio values inside the dark organelle were zero after processing. In addition, the boundaries of the cells were defined more clearly, including where the cells bordered each other.

VI. Hardware and Software Requirements

The most basic requirement for performing these image processing procedures is a computer-based image acquisition system that allows the collection of low light level fluorescent images with minimal noise. An essential component of such a system is a sensitive video camera such as a silicon-intensified (SIT) camera, an intensified SIT (ISIT), an intensified CCD

(ICCD), or a cooled slow-scan CCD camera. Although we have not tried the ISIT or ICCD, we were able to acquire good images with the Hamamatsu C2400-08 SIT camera (Hamamatsu Corporation, Bridgewater, New Jersey), images that showed substantial improvement on processing. The extent to which images can be processed depends on the signal-to-noise ratio, the dynamic range, and the spatial resolution of the detection device(s). The choice of camera must be made with these criteria in mind, aiming for the lowest noise and highest resolution possible for the money available.

In addition to the camera, a frame grabber/digitizer board and a graphics card and monitor are required to display the acquired images. All these components can be controlled by a number of off-the-shelf software packages that run on Macintosh- or MS-DOS-based personal computers as well as on UNIX workstations. The best idea is to pick a software package and determine which hardware will work with it. The software should be able to control the camera and perhaps some peripheral devices such as shutters or a filter wheel. In addition, it should be able to perform at least basic image processing operations, such as frame averaging and arithmetic manipulations of images, so a background could be substracted or one image divided by another. Image data can fill up even a large hard disk quickly. Therefore, an additional piece of equipment that is extremely useful is an optical disk drive. These devices can store 500–800 megabytes of data on inexpensive removable disks.

As described earlier, the no-neighbors deblurring scheme can be performed with small convolution kernels, processes that usually can be performed by inexpensive image-processing packages. Alternatively, the source code for the programs described in this chapter that perform Fourier domain image processing can be obtained by contacting the authors of this chapter. The programs are written in a generic format that should make them usable on any computer that has a C language compiler. However, we cannot be certain that the programs will run on every machine. In particular, problems may occur in running the programs, unmodified, on computers that use the MS-DOS operating system because data arrays in that environment cannot be larger than 64 kilobytes (kB) each. This size is too small to accomodate an array of 256×256 floating-point numbers (256 kB), a size that is required to calculate the Fourier transform of a reasonably large image.

Fourier transforms are time-consuming operations, especially on personal computers. If one has access to a powerful mainframe computer, the processing programs can be run on that machine instead, using image files downloaded from the acquisition computer. This approach also circumvents the MS-DOS memory limitation just discussed. To connect to the mainframe computer, use the fastest data transfer rate available, for example, that provided by an Ethernet connection. Another way to increase the computing power of a personal computer is to add a floating-point accelerator or an array processor (see Monck *et al.*, 1992).

The image processing software we use is from G. W. Hannaway and Associates (Boulder, Colorado) and runs on a Sun SPARC Station II (Sun Microsystems, Mountain View, California). This system can perform a wide variety of image processing tasks and includes a library of C language-callable routines that allows us to write our own custom functions. Since this machine runs UNIX, some knowledge of that operating system is required to use the Hannaway package effectively. The system accepts input from any camera that puts out an RS-170 signal, for example, the SIT camera used to acquire the images in Color Plate 4. Video acquisition and display operations are performed by several IT-150 series boards (Imaging Technology, Woburn, Massachusetts), including a frame buffer, analog-to-digital converter, a real-time convolver, and an arithmetic logic unit. Since the processing is performed off line, this system is sufficiently fast for our needs; to process a 256×256 image completely with the no-neighbors algorithm takes about 10 sec.

VII. Conclusions

In this chapter, we have described image processing techniques that can be used to improve the quality of ratio images. This information, in addition to the source code for the image processing programs, should enable anyone with a basic image acquisition system to process images with convolutions or Fourier domain techniques. Readers should consult the references, especially Castleman (1979), Agard (1984), and Inoué (1986), for additional information.

We have shown, in theory and in practice, that improving the spatial resolution of ratio images is possible by processing the individual images with computational confocal techniques. Using the no-neighbors deblurring scheme, one can reduce the amount of out-of-focus blur that is present in the two images. The ratios of these processed images more accurately depict the features of the cell, in the sense that non-zero ratio values no longer appear in regions in which no dye was present, for example, outside the cell or inside organelles that exclude the dye.

In the model cell, the ratio of the "observed" images contained incorrect values in regions in which the ratios were different in the neighboring planes. However, it is possible to choose parameters for processing the images that almost completely recovered the correct ratio values in these regions of the cell. Based on the results with the model cell, we might conclude that, in the ratio image of the PC-12 cells in Color Plate 4, the values are more accurate than in the unprocessed ratio image.

Although we focused on the no-neighbors technique, in principle the nearest-neighbors deblurring algorithm should be equally useful. In fact, taking into account neighboring sections may reduce the ripples that appear around certain objects, as they did around the bright organelle in the test cell. Using the nearest-neighbors technique has the obvious drawback that one must collect six images (three at each wavelength) to produce a single ratio image.

Appendix I: Fourier Transforms

Often solving mathematical problems is easier by transforming them and finding the solution in a different domain. For example, when multiplying large numbers (without the aid of a calculator) taking the logarithms of the numbers and adding those, and taking the inverse logarithm of the sum to obtain the answer, is easier than performing a straight multiplication. In a similar way, transforming images to the Fourier domain simplifies certain operations one might perform on those images. Convolutions are time-consuming operations when the two arrays are large; however, when the image arrays are transformed into the Fourier domain, the convolution is achieved simply by multiplying the two arrays together. Of course, performing the Fourier transform is itself time-consuming, but at a certain point the time required for the Fourier transforms on the images is shorter than the time required to convolve the two image arrays directly, especially when one uses the fast Fourier transform (FFT) algorithm discussed in this appendix.

This brief discussion of Fourier transforms and the Fourier domain will be conducted primarily in qualitative terms; however, more thorough and quantitative treatments can be found in articles by Brigham (1974) and Press *et al.* (1988). The purpose of this section is to enable someone who has not had experience using Fourier transforms to perform the Fourier domain image processing techniques discussed in this chapter. Additionally, since many commercially available image processing packages now include Fourier domain deblurring functions (essentially the same as the one presented here), a basic understanding of what these programs do will help the users of such systems interpret the results these programs produce.

A fluorescence image which, on a computer screen or video monitor, is represented by a two-dimensional array of intensity values is said to be in real space. When the image is Fourier transformed, that information is then in the Fourier, or frequency, domain or in reciprocal space. The Fourier transform converts the real space image into an array of spatial frequencies. The term "spatial frequency" refers to how close together the intensity changes are; for example, an image that is sharp and in focus will contain a greater proportion of high frequency values than an image that is blurry and out of focus. Given this information, one can see that, for the most part, sharpening or blurring an image can be accomplished by enhancing or attenuating the appropriate spatial frequencies.

Whereas a fluorescence image consists of an array of numbers with one value used to represent each point, the Fourier transform of that image produces an array of complex values in which a pair of numbers, one real and one imaginary, represents each spatial frequency. Because the frequency domain values are complex numbers, operations on them must be performed accordingly.

If $(a + bi)$ and $(c + di)$ are complex numbers, where a and c are the real parts and b and d are the imaginary parts, the following rules hold.

Addition: $(a + bi) + (c + di) = (a + c) + (b + d)i$
Subtraction: $(a + bi) - (c + di) = (a - c) + (b - d)i$
Multiplication: $(a + bi)(c + di) = (ac - bd) + (ad + bc)i$

Also, complex quantities can be described by polar coordinates:

Magnitude of $a + bi$: $\|a + bi\| = (a^2 + b^2)^{\frac{1}{2}}$
Phase of $a + bi$: $\Theta = \tan^{-1}(a/b)$

Addition and subtraction produce the same results whether they are performed in real space or in the Fourier domain (Brigham, 1974). As discussed earlier, however, multiplication of two images in the Fourier domain is equivalent to convolving them in real space, one of the reasons Fourier transforms are so useful.

The Fourier transform algorithm that is used normally is the FFT. The differences between the basic and fast transform routines are explained in Brigham (1974) and Press *et al.* (1988). The FFT routine only operates on images with dimensions that are powers of 2, for example, 128×128 pixels, 256×256, and 512×512, or even rectangular dimensions such as 256×512. If the image dimensions are not powers of 2 in one or both directions, one can simply fill in zeros in the extra spaces until the dimensions of the image are powers of 2.

At this point, however, another property of the Fourier transform must be mentioned, namely, the fact that it is "circular." This description means that, for the purposes of the transform, the opposite edges of the image are next to each other. In practical terms, this characteristic implies that, if two opposite edges of the image have significantly different brightness levels, this area will seem like a rather sharp edge, containing much higher spatial frequencies than the rest of the image. When this image is processed with Fourier domain filters, it will contain "ripples" that emanate from the edges, often obscuring details in the image. This effect also occurs when spots occur on the images that are unusually dark or bright; these can be caused by dust on the camera or by highly fluorescent particles near the sample, respectively.

The best way to deal with this problem is to collect images that have similar intensities at the edges. If this is not possible, the opposite edges of the images should be brought into register with each other in a manner that eliminates the sharp transition, a process called windowing. The simplest way to window the data is to perform a linear interpolation between the intensity values at the opposite edges, using perhaps 5 or 10 of the rows or columns along the edges to allow a gradual transition. Several other windowing techniques are discussed by Press *et al.* (1988).

As discussed earlier, the way in which a microscope alters image information, both in focus and out of focus, is called the PSF. The frequency domain equivalent of the PSF is the CTF, which is obtained simply by calculating the Fourier transform of the PSF. To convolve the out-of-focus PSF with an image, as is required to perform nearest-neighbors or no-neighbors image deblurring, one can multiply the Fourier transform of the image by the CTF.

After transforming the out-of-focus image of a bead to the Fourier domain, two steps are required before the Fourier domain image can be used as an out-of-focus CTF. First, the CTF function must be adjusted so it does not alter the phases of the images with which it is convolved. Second, the magnitudes must be adjusted so the average intensity of the image that is multiplied by the CTF is not changed.

When two complex numbers are multiplied together, the phase of the product is different from the phases of either of the two original numbers. The exception to this rule is when the imaginary part of one of the factors is zero, in which case the phase of the result is the same as the phase of the other factor. Therefore, to avoid changing the phases in the images that are multiplied by the CTFs, the real parts (R) at each frequency of the CTF should equal the magnitude of the complex number:

$$R^* = (R^2 + I^2)^{\frac{1}{2}}$$

$$I^* = 0$$

where the asterisks are used to distinguish the new values from the old ones, and I represents the imaginary part of the number. If the phase information in an image is altered in the Fourier domain, the image will not look the same when it is reverse-transformed to real space.

The other correction to be applied to the CTF is even easier to perform, since it simply involves dividing each element in the array by the first value. The first value in the array corresponds to the "zero-frequency" information (sometimes called the DC value) of the image and is the sum of the intensities in real space. Dividing all the frequencies by this value normalizes the image, with the zero-frequency component equal to 1. Since the zero-frequency component, which is the sum of the real space intensities, is 1, multiplying by the normalized CTF will not change the total intensity of the input image. This correction should be performed on the CTF after setting the real components equal to the magnitudes.

The CTF filter now can be multiplied, using complex arithmetic, by a Fourier domain image to simulate the convolution of the image with the PSF.

Then the result is reverse-transformed and displayed. When subtracting blurred sections for the nearest-neighbors and no-neighbors routines, one can perform the subtraction in the Fourier domain since addition and subtraction operate the same in real space and in the frequency domain. The same rule applies to multiplying the blurred images by the constant c, which is also independent of the domain in which it is performed. Before displaying the reverse-transformed results, each pixel value must be divided by the total number of pixels in the image, since the FFT routine increases each value by this constant factor (Press *et al.*, 1988). When a processed image is reverse-transformed to real space, it sometimes will contain negative values. Since a fluorescence image cannot have negative intensities, the negative values should be set to zero (this rule is sometimes called a nonnegativity constraint). Performing all the calculations using floating-point numbers until the final image is determined is a good idea, to maximize gray-scale resolution.

To construct a Wiener filter in Fourier space of the form given in Eq. 4

$$G = S_0/(S_0{}^2 + \alpha)$$

one first must process the in-focus CTF, S_0, so the real component equals the magnitude and the values are normalized to the zero-frequency component, as just described. At this point, each frequency component of the Wiener filter can be calculated directly using the appropriate α value. Since the imaginary components of S_0 are zero, one need not use complex arithmetic to calculate G. The Wiener filter and the CTF functions can be stored on disk as floating-point arrays to avoid having to recalculate them each time images are processed.

Appendix II: Theoretical CTF Equations

The following equations, published by Castleman (1979) and Agard (1984), give a numerical description of the CTF. The CTF is the Fourier domain equivalent of the PSF.

$$S(q) = 1/\pi(2\beta - \sin 2\beta) \, \text{jinc}[8\pi w/\lambda \, (1 - q/f_c)q/f_c]$$

Where:

$\text{jinc}(x) = 2J_1(x)/x$

$J_1(x)$ = Bessel function of x; a C language version of this function is presented by Press *et al.* (1988) and usually will be included in the math library of a C language compiler

$\beta = \cos^{-1}(q/f_c)$

$w = -d_f - \Delta z \cos \alpha + (d_f^2 + 2d_f\Delta z + \Delta z^2 \cos^2 \alpha)^{\frac{1}{2}}$

d_f = focal distance of the objective lens in μm

Δz = change in focus in μm; for the in-focus CTF, $\Delta z = 0$ and, as a result, $w = 0$; these equations assume that the out-of-focus PSF is equivalent above and below the in-focus plane so, for example, $\Delta z = 1$ whether one is 1 μm above or 1 μm below the plane of focus

$\alpha = \sin^{-1} (NA/\eta)$; α is the widest angle of incident light rays, relative to the optical axis, that can be captured by the objective lens (see Fig. 2 in Agard, 1984); NA is the stated numerical aperture on the objective lens and η is the index of refraction of the immersion medium recommended by the manufacturer of the objective lens

$f_c = 2\,NA/\lambda = 2\,\eta \sin \alpha/\lambda$; f_c is the inverse of the resolution limit, which is the highest spatial frequency that can be passed by the objective

η = index of refraction of the immersion medium

λ = wavelength of emission light in microns

$q = (u^2 + v^2)^{\frac{1}{2}}$; u, v = spatial frequency in the horizontal and vertical directions, respectively, in units of μm^{-1}

Acknowledgments

We thank Kenneth R. Robinson, in whose laboratory this work has been carried out, for his continued support. We thank Heather Harper Keating (Department of Biology, Purdue University) for supplying the PC-12 cells used in the studies in this chapter. This work has been supported by grants from the National Institutes of Health and the National Science Foundation.

To contact the authors about the CTF program, contact Mr. Keating through Internet: tom @ video. bio. purdue. edu. and Dr. Cork through Internet: john @ video. bio. purdue. edu.

References

Agard, D. A. (1984). Optical sectioning microscopy: Cellular architecture in three dimensions. *Annu. Rev. Biophys. Bioeng.* **13**, 191–219.

Agard, D. A., Hiraoka, Y., Shaw, P., and Sedat, J. W. (1989). Fluorescence microscopy in three dimensions. *Meth. Cell Biol.* **30**, 353–377.

Berriel, L. R., Bescos, J., and Santisteban, A. (1983). Image restoration for a defocused optical system. *Appl. Opt.* **22**, 2772–2780.

Brigham, E. O. (1974). "The Fast Fourier Transform." Englewood Cliffs, New Jersey: Prentice-Hall.

Castleman, K. R. (1979). "Digital Image Processing." Englewood Cliffs, New Jersey: Prentice-Hall.

Hiraoka, Y., Sedat, J. W., and Agard, D. A. (1990). Determination of three-dimensional imaging properties of a light microscope system. *Biophys. J.* **57**, 325–333.

Inoué, S. (1986). "Video Microscopy." New York: Plenum Press.

Marangoni, R., Colombetti, G., and Gualtieri, P. (1992). *In* "Image Analysis in Biology" (D.-P. Häder, ed.). Boca Raton, Florida: CRC Press. pp. 55–74.

Monck, J. R., Oberhauser, A. F., Keating, T. J., and Fernandez, J. M. (1992). Thin-section ratiometric Ca^{2+} images obtained by optical sectioning of fura-2 loaded mast cells. *J. Cell Biol.* **116**, 745–759.

Press, W. H., Flannery, B. P., Teukolsky, S. A., and Vetterling, W. T. (1988). ''Numerical Recipes in C.'' Cambridge: Cambridge University Press.

Spencer, M. (1982). ''Fundamentals of Light Microscopy.'' Cambridge: Cambridge University Press.

Yelamarty, R. V., Miller, B. A., Scaduto, R. C., Jr., Yu, F. T. S., Tillotson, D. L., and Cheung J. Y. (1990). Three-dimensional intracellular calcium gradients in single human burst-forming units—erythroid-derived erythroblasts induced by erythropoietin. *J. Clin. Invest.* **85,** 1799–1809.

CHAPTER 10

Confocal Imaging of Ca^{2+} in Cells

Pamela A. Diliberto, Xue Feng Wang, and Brian Herman

Department of Cell Biology and Anatomy
University of North Carolina School of Medicine
Chapel Hill, North Carolina 27599

I. Introduction

Confocal fluorescence microscopy provides a powerful method for examining intracellular free calcium ion distribution and dynamics within the single intact living cell. Confocal microscopy, used in conjunction with Ca^{2+}-sensing fluorescent indicator dyes, facilitates the visualization of thin optical sections through the cell, rejecting out-of-focus light arising from the excited fluorophore. These images are in contrast with images produced by conventional nonconfocal fluorescence microscopy, which contain information from light emitted from the entire thickness of the cell. The resultant high vertical and horizontal spatial resolution of the confocal technique affords one the ability to image Ca^{2+} within the context of detailed intracellular structure.

Major variables to be considered when establishing an appropriate confocal Ca^{2+}-imaging methodology for one's specific needs and applications include the choice and proper use of both the Ca^{2+} indicator dye and the confocal instrumentation itself. Considerations including required spatial and temporal resolution, quantitative ability, and monetary investment necessarily must be taken into account. In many instances, sacrifices or compromises must be made in one aspect to obtain the desired end-point in another. In this chapter, we provide a brief but useful outline of (1) currently available indicator dyes, instrument systems, and their appropriate uses; (2) qualitative and quantitative methods for evaluating the confocal Ca^{2+}-imaging data; and (3) illustrative examples of applications of this rapidly evolving technology to the study of intracellular Ca^{2+} metabolism and signaling in living cells.

II. Ca^{2+} Indicator Dyes

A. Types of Dye

Two major types of fluorescent Ca^{2+} indicator dye are currently available for use in confocal Ca^{2+} imaging: (1) quantitative ratiometric dyes, (2) single visible wavelength dyes. The familiar quantitative ratiometric dyes fura-2 and indo-1, extensively used in conventional fluorescence microscopic Ca^{2+} imaging, are excited by ultraviolet (UV) wavelength light with a resultant emission in the visible spectrum (Grynkiewicz *et al.*, 1985). Ca^{2+}-dependent spectral shifts in either the excitation (fura-2) or emission (indo-1) properties of these dyes permit determination of dual-wavelength ratio values that are dependent on the Ca^{2+} concentration but are independent of the intracellular dye concentration and specimen pathlength.

Single-wavelength dyes, on the other hand, all are excited by visible wavelength light and show no spectral shift on binding Ca^{2+} (thus the alternative names "visible wavelength" and "nonratiometric" dyes). A partial list of available visible wavelength Ca^{2+} indicator dyes includes fluo-3, rhod-2, Calcium Green™, Calcium Orange™, Calcium Crimson™, and Fura-Red™ (Minta *et al.*, 1989; Haugland, 1992). These dyes may exhibit a decrease (Fura-Red) or an increase (all others) in fluorescence emission intensity on Ca²⁺ binding, and vary from one another in their fluorescence spectra, quantum efficiencies, and Ca²⁺-binding properties. These dyes are therefore not ratiometric Ca²⁺ indicators; the emitted fluorescence intensity is dependent not only on the Ca²⁺ concentration but also on the amount of excited dye (concentration and pathlength).

Advantages and disadvantages of the two types of dye center around differences in the ability to calibrate fluorescence signals quantitatively to absolute Ca²⁺ concentration (ratiometric vs nonratiometric measurements) and in the instrumentation required (UV vs visible wavelength light source and optics; dual vs single excitation or detection capabilities). These factors are discussed in greater detail in subsequent sections.

B. Cell Loading

Intracellular introduction of Ca²⁺ indicator dyes into living cells can be accomplished by a variety of methods. The main goals of successful dye loading are (1) to achieve an intracellular dye concentration that is sufficient to provide adequate signal strength for detection while remaining low enough not to disturb normal intracellular physiology, and (2) to achieve a homogeneous or known distribution of the dye within the cell. This latter point is especially important in the case of visible (single) wavelength dyes, for which data cannot be ratioed and thus are dependent on the dye concentration. Two commonly used loading techniques are (1) bath application of the cell-permeant acetoxymethyl (AM) ester derivatives of the Ca²⁺ indicators, which subsequently are de-esterified intracellularly by endogenous enzymes and trapped within the cells, and (2) microinjection of the cell-impermeant salt forms of the dyes into single cells.

We have found AM-loading of a variety of dyes to be an efficient and useful method for dye introduction. This method is much less labor intensive than microinjection and produces very reproducible results within a given cell population; these cells can be imaged quickly and simultaneously with minimal trauma. Cells cultured on glass cover slips initially are prechilled for 5–10 min at 4°C to prevent endocytotic uptake of the AM esters into intracellular organelles from the incubation medium (Roe *et al.*, 1990). Cells then are incubated for 20–45 min at 37°C with 2–10 μM AM ester (either in the presence or absence of the dispersing agent Pluronic-127™; BASF Wyandotte Corporation, Wyandotte, Wisconsin). The duration, dye concentration, and requirement for Pluronic are determined empirically for the loading incubation. We have found these parameters to be dependent on both the cell type and the specific dye to be used. For example, in BALB/c 3T3 fibroblasts, we found that shorter incubation times (20–30 min) and higher concentrations (10 μM)

of fluo-3/AM (Color Plate 5) resulted in similar signal intensity but much more homogeneous intracellular dye distribution than longer times (60 min) and lower concentrations (4–5 μM), as assessed by laser scannig confocal microscopy (LSCM; Bio-Rad MRC600; Bio-Rad Laboratories, Richmond, California; Olympus LSM-GB200; Olympus, Lake Success, New York). This loading did not require the presence of a dispersing agent. Fura-Red/AM loading in these same cells, however, was absolutely dependent on the presence of Pluronic and required long incubation times (45–60 min) with high concentrations (10 μM) of dye. (A major factor thought to influence the loading conditions required for Fura-Red/AM is the low quantum efficiency of this indicator, necessitating high intracellular concentrations to obtain detectable signals). After loading, excess extracellular AM ester is washed from the cells and the cover slips are mounted in a chamber containing 37°C medium on the stage of an inverted fluorescence microscope. Using this technique, we have loaded cells under similar conditions successfully with the AM esters of Calcium Green-1 and Calcium Crimson for LSCM imaging of intracellular Ca^{2+}.

C. Compartmentalization

As alluded to in the previous section, AM-loaded dyes can become sequestered preferentially in specific intracellular organelles if the proper precautions are not taken or if the loading conditions are not optimized for the specific cell type and Ca^{2+} indicator. Recognition and knowledge of the extent of this compartmentalization of the dye becomes especially critical when working with the nonratiometric visible wavelength dyes. Quantitative comparison of Ca^{2+} changes in different intracellular compartments becomes extremely difficult, if not impossible, in the case of inhomogeneous spatial distribution of the dye. In addition, compartmentalization artifacts can arise with ratiometric as well as nonratiometric dyes because of compartment-specific environmental effects on the fluorescence spectra of the dye. Differences in ionic strength, pH, viscosity, polarity, and intracellular proteins in the microenvironment all can alter the spectral properties of the dye (Roe *et al.*, 1990).

We have assessed the intracellular location of Ca^{2+} indicator dyes in LSCM-imaged cells loaded with the AM esters of fluo-3, Calcium Green, or Fura-Red. Digitonin, applied in progressively increasing concentrations, was used to permeabilize and release dye from the different intracellular compartments sequentially (Roe *et al.*, 1990). Our results indicate that the major fraction of the three dyes is localized to the cytoplasm and nucleus of the cell. All three dyes exhibited minimal localization to the mitochondrial and insoluble compartments (≤5%, total) under similar loading conditions (10 μM, 30 min). Differences between the dyes were apparent in the degree of localization to the lysosomal/endosomal compartment, the extent of which was dependent on the specific loading conditions employed. In BALB/c 3T3 cells, fluo-3 localization in particular appeared to be highly sensitive to changes in the loading conditions. This finding emphasizes the point that determination of intracellular dye compartmentalization is required for each specific dye, set of loading conditions, and cell type if meaningful interpretation of the data is desired.

D. Ca²⁺ Binding Affinity

Binding affinities of the fluorescent indicator dyes for Ca^{2+} vary significantly from one another when considered in relation to the normal physiological concentration range of intracellular Ca^{2+}. Values for dye–Ca^{2+} dissociation constants (K_ds) for the various dyes extend over more than a 10-fold concentration range (133 nM, 316 nM, and 1.0 μM for Fura-Red, fluo-3, and rhod-2, respectively; Haugland, 1992), and can be influenced by physical parameters such as pH and temperature. Basal or resting cytoplasmic Ca^{2+} concentration in most cells is ≤100 nM, whereas Ca^{2+} agonist or electrical stimulation may produce intracellular Ca^{2+} levels that approach the low micromolar range. Thus, the dynamic range of the intracellular Ca^{2+} concentration changes to be examined must be taken into account when selecting the appropriate indicator dye for single-cell confocal Ca^{2+} studies. High affinity dyes (low K_d values) such as fura-2, indo-1, and Fura-Red, if present in sufficient quantity, may buffer or "clamp" Ca^{2+} levels in the same range as resting intracellular Ca^{2+}. The high affinity dyes also become saturated below micromolar concentrations of Ca^{2+}, and thus may not provide the sensitivity needed to monitor high magnitude Ca^{2+} alterations that may occur in localized subcellular regions (nucleus, neuronal plasma membrane) as a result of agonist stimulation or the artificially induced Ca^{2+} increases resulting from caged Ca^{2+} release. Several of the newer visible wavelength dyes (fluo-3, rhod-2, Calcium Greens, etc.) have weaker affinities for Ca^{2+}, improving the ability to monitor changes in higher Ca^{2+} levels. For fluo-3, we have not found that this extension of the upper range of Ca^{2+} sensitivity has compromised our ability to monitor changes in the low Ca^{2+} concentration range (see subsequent discussion).

E. Bleaching

Decreased signal intensity resulting from fluorophore bleaching is a common occurrence and a concern in confocal imaging. This artifact arises from the fact that, as the confocality is improved, the signal is collected from a smaller and smaller volume of the specimen and thus, from fewer molecules of excited fluorophore. Higher intensity light sources provide the needed increase in output signal but also increase the risk of photobleaching. Bleaching can be minimized by decreasing the duration and/or intensity of exposure to the lowest possible levels that still provide adequate signal-to-noise characteristics in detection. Of course, this minimization is aided by high sensitivity detection capabilities.

In confocal Ca^{2+} imaging, the problem of bleaching becomes most critical with the single-wavelength dyes, for which interpreting whether a decrease in signal intensity originates from a decrease in Ca^{2+} or from bleaching of the dye is difficult. Note that the ratiometric Ca^{2+} dyes are not exempt from concerns of bleaching, since the extent of bleaching is dependent on the degree of excitation and, therefore, may occur to different extents in localized regions of the cell with spatially heterogeneous intracellular Ca^{2+} responses. In living cells loaded with Ca^{2+}-binding indicator dyes, increasing the amount of indicator present while decreasing exposure intensity may help solve the bleaching problem, but bear in mind that intracellular Ca^{2+} buffering effects increase with increased dye concentrations. We have found minimal bleaching effects in BALB/c 3T3 cells loaded with fluo-3 as described and imaged by LSCM (Bio-Rad and Olympus systems), provided the laser intensity is attenuated sufficiently (~1% transmission neutral density filtering).

III. Instrumentation

A. Laser Scanning Confocal Microscope

The LSCM is the most popular confocal instrumentation in use today. LSCM has been employed successfully to image intracellular Ca^{2+} by a number of laboratories (e.g., Hernandez-Cruz *et al.*, 1990; Lechleiter *et al.*, 1991; Lechleiter and Clapham, 1992; Stricker *et al.*, 1992). These systems provide great improvement in horizontal and vertical spatial resolution over conventional fluorescence microscopy. In LSCM, the laser beam is scanned across the specimen and emitted light is collected through a pinhole aperture. The pinhole provides the necessary mechanism for the rejection of out-of-focus light, but also necessitates the collection and detection of signal from only one point of the specimen at a time. Generation of two-dimensional cell images therefore are accomplished by scanning the laser beam across the specimen and recording the single-point detector output in a two-dimensional array. Standardly, laser beam scanning is accomplished using vibrating mechanically positioned mirrors, which limit the speed at which the beam can be moved across the specimen and therefore reduce the temporal resolution of image acquisition. Temporal resolution can be improved by scanning a smaller area of the specimen, including the commonly used one-dimensional line scan.

Our laboratory has been using LSCM (Bio-Rad MRC600; Olympus LSM-GB200) to image Ca^{2+} in living cells. We have found the optical performance of this type of system in Ca^{2+} imaging to be excellent. Vertical spatial resolution of approximately 0.5–0.7 μm is obtained using 40× or 60× high numerical aperture objectives. Maximum temporal resolution of full-frame cell image acquisition is, however, limited to the range of one frame per second. This rate often is not fast enough to observe intracellular Ca^{2+} dynamics of interest such as Ca^{2+} oscillations and wave propagation. The other inherent limitation of the LSCM systems is the requirement for a laser light source, which in many cases (including the two just mentioned) is restricted further to specific visible light wavelength output by the use of a certain laser type (i.e., Kr/Ar, He/Ne). This restriction necessitates the use of the visible wavelength Ca^{2+} dyes, with their associated quantitation problems, for intracellular Ca^{2+} imaging. On the other hand, visible wavelength dyes are preferred over UV dyes for caged compound release studies. UV Ca^{2+} indicator dyes are not highly compatible with these studies because of the UV absorbance by many caged compounds and their by-products, which can lead to "inner filter effects" on dye excitation, in addition to the concerns of dye bleaching by the photolytic pulse and, conversely, the possible undesired photolytic release of caged compound during the Ca^{2+} monitoring process.

The availability of UV lasers in LSCM systems is increasing rapidly. UV lasers, however, are expensive and require expensive optics. Conventional microscopes usually are not designed to be achromatic from UV through the visible wavelengths, and therefore should not be used in conjunction with UV confocal light sources and other confocal attachments not designed explicitly for use with UV.

B. Spinning Disk

An alternative illumination design to LSCM is the scanning of the light beam across the specimen through the use of a rotating disk containing pinholes (Nipkow disk). This design offers the advantages of nonlaser light sources and real-time imaging with direct viewing. Arc lamp light sources provide a more flexible and inexpensive choice of excitation wavelengths, including UV. Conventional microscope systems with video camera detection can be modified by adding a spinning disk attachment (such as the K2S-BIO; Nikon, Inc., Melville, NY) to construct an affordable confocal system with imaging capabilities for both visible and UV Ca^{2+} indicator dyes. The significant disadvantage of this type of system that we have found in living-cell Ca^{2+}-imaging applications is the very poor efficiency of emitted light collection. Thus, high intensity illumination (with the associated problems of dye bleaching, fluorophore saturation, and cellular photodamage) results in only low level signal intensity, producing very dim cellular images.

C. Slit Scanner

In slit scanning confocal microscopy, the illuminating light source is scanned across the specimen through a narrow slit rather than a pinhole aperture. The resultant emitted light is descanned and detected after passage through a variable aperture detection slit. Slit scanning, by design, provides less confocality than spot scanning but offers the advantage of nonlaser illumination and direct viewing, as does the spinning disk design. Slit scanning improves the scan rate of acquisition of a given specimen area and requires less intense illumination, with potentially less dye bleaching and photodamage to living cells, than does spot scanning. Low cost attachments for conventional microscopes are available and can be used in combination with standard arc lamp light sources (with their UV capabilities) and video camera detection devices.

D. Real Time

The use of acousto-optical deflector (AOD) systems with LSCM technology can improve the temporal resolution of confocal image acquisition significantly. In conventional laser scanning systems, the movement of mechanical galvanometer-driven mirrors used to position and scan the laser beam proves to be the rate-limiting step. In real-time AOD systems, the laser beam is scanned rapidly across the specimen by nonmechanical deflection through an arrangement of prisms and AO devices in two dimensions. This type of modification can result in increased scanning speeds to video rates or higher. We have found the commercially available AOD LSCM, the Noran Odyssey (Noran Instruments, Middleton, WI), to be capable of obtaining discernible full-frame scan images of fluo-3-loaded BALB/c 3T3 cells with a temporal resolution of ~125 msec. The trade-off for increased time resolution is a significant increase in the signal-to-noise ratio of the output. The optional capability of averaging a series of rapidly acquired images over a user-defined period of time (e.g., 0.25–1 sec) appears to provide an acceptable compromise between ''real-time'' temporal resolution and signal noise in terms of image quality, provided the Ca^{2+} dynamics of interest can be examined under these conditions.

E. Fluorescence Lifetime

Confocal fluorescence lifetime imaging is currently in the early stages of development. The combination of conventional lifetime measurement techniques with confocal microscopy will allow quantitation of intracellular Ca^{2+} by single excitation wavelength dyes with high spatial resolution. Fluorescence lifetime measurements entail the determination of fluorescence decay parameters of intracellularly loaded fluorescent Ca^{2+} indicators based on time-domain or phase-domain parameters of the emitted light. Lifetime measurements enable calculation of the fraction of Ca^{2+}-bound vs Ca^{2+}-free indicator molecules, thus permitting Ca^{2+} quantitation that is independent of the indicator concentration and specimen pathlength. Two-dimensional fluorescence lifetime imaging techniques have been developed in conjunction with conventional fluorescence microscopy (Wang *et al.*, 1992). Fluorescence lifetime imaging of Ca²⁺ in solution has been reported for the visible wavelength Ca²⁺ indicator dyes Calcium Green, Calcium Orange, and Calcium Crimson (Lakowicz *et al.*, 1992b). Application of this technology to living cell Ca²⁺ imaging using phase-domain lifetime determinations has been performed using quin-2 as the Ca²⁺ probe (Lakowicz *et al.*, 1992a). With the implementation of time-domain fluorescence lifetime measurements in combination with LSCM to image fluorescent molecules in algal cells (Buurman *et al.*, 1992), this technology soon will be employed in confocal lifetime imaging of cellular Ca²⁺. The major limitation perceived in the application of current lifetime technology to confocal Ca²⁺ imaging in living cells appears to be one of limited temporal resolution. Future improvement and development of high-sensitivity high-speed detection and collection capabilities should resolve these problems.

IV. Calibration

Calibration of confocal fluorescence intensity measurements to intracellular Ca²⁺ values in living cells can be performed by *in vitro* or *in vivo* calibration techniques, each method having its own limitations. In general, these methods are the same as those employed for Ca²⁺ calibration using conventional fluorescence microscopy. The common use of nonratiometric single-wavelength Ca²⁺ dyes in confocal imaging, however, introduces additional challenges to the problem of accurate Ca²⁺ calibration. The inability to perform ratio measurements with these dyes leaves one with a fluorescent signal that is a function not only of the Ca²⁺ concentration, but also of the pathlength and the dye concentration. Fortunately, confocal microscopy offers an advantage over conventional microscopy in this regard. Pathlength variations are minimal because of the relatively small and constant volume of the cell that is imaged from the thin optical section in the focal plane. Variations in signal intensity are, therefore, primarily independent on Ca²⁺ concentration and dye concentration.

A. *In Vitro*

Intracellular fluorescence intensity measurements can be calibrated by constructing a Ca²⁺ calibration curve with solutions of known free Ca²⁺ concentrations. These solutions are formulated to simulate the intracellular environment (ionic strength and composition, pH, viscosity) and are imaged through the microscope optics. Inner filter effects encountered when performing measurements on high intensity solutions containing high concentrations of indicator dyes can be avoided by decreasing the pathlength of the solution. We have found it most convenient to image solutions contained in small diameter (~0.5 μm) glass capillary tubes. Calibration measurements should be performed at

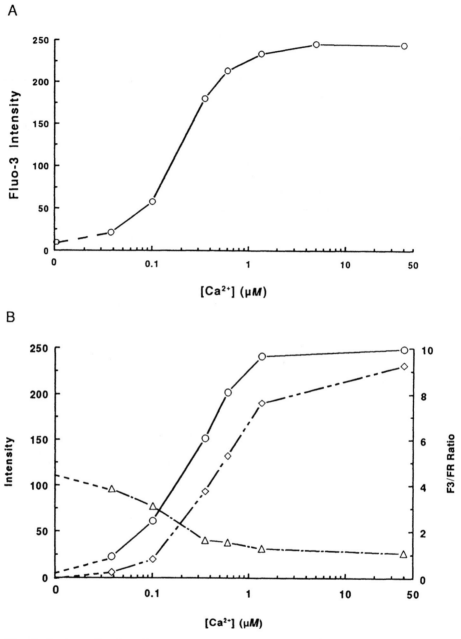

Fig. 1 *In vitro* calibration of fluo-3 and Fura-Red™ intensity to Ca²⁺ concentration. Fluo-3 and Fura-Red salts were dissolved in MOPS-buffered (pH 7.2) high K⁺ solutions containing varying concentrations of free Ca²⁺ (adjusted with EGTA: CaCl₂). Solutions were placed in capillary tubes and imaged by LSCM (Bio-Rad MRC600, 488-nm excitation) at 37°C. Instrument settings were essentially the same as those employed for live-cell imaging experiments. Images were

the same temperature used for living cell measurements (37°C) because of the temperature dependency of the Ca^{2+}–dye binding affinities.

Calibration of ratiometric dyes performed in this manner is fairly straightforward because the ratio values obtained are independent of the dye concentration. Therefore knowledge of the intracellular dye concentration is not required. Calibration of single-wavelength dyes does require knowledge of the intracellular dye concentration if a direct comparison of signal intensities derived from calibration solutions and cell measurements will be performed. Average intracellular dye concentration can be determined in a population of dye-loaded cells by quantitating the amount of dye released from suspensions of disaggregated cells on lysis by digitonin. Using this method and the dye loading conditions described earlier, we have determined intracellular dye concentrations in AM-loaded BALB/c 3T3 cells to approximate 75 μM, 150 μM, and 25 μM for fluo-3, Fura-Red, and Calcium Green, respectively. Once the intracellular dye concentration is known, the Ca^{2+} calibration curve can be constructed using this known dye concentration as a constant (Fig. 1A). The obvious limitation of this type of calibration is the necessary assumption that all cells in the population contain the same concentration of loaded dye and that, within a single cell, the dye is distributed homogeneously. An additional limitation of *in vitro* calibration, in general, lies in the assumptions that must be made about the intracellular environment of the indicator dye. Again, subcellular enviromental heterogeneity that may influence the spectral properties of dyes cannot be considered in this method.

B. *In Situ*

In situ Ca^{2+} calibration methods address the inter- and intracellular heterogeneity issues by measuring the fluorescence characteristics of the indicator dyes in the same cells in which the Ca^{2+}-imaging experiments are performed. Manipulations of intracellular Ca^{2+} are accomplished using a Ca^{2+} ionophore such as ionomycin to raise intracellular Ca^{2+} levels in the presence of a known extracellular Ca^{2+} concentration. Additional assessment of the amount of dye present and the amount of signal arising from cellular autofluorescence can be made by sequential treatment of cells with heavy metals (Mn^{2+} or Zn^{2+}) to quench Ca^{2+}-dependent fluorescence, and by release of intracellular dye by digitonin. *In situ* calibration of fluo-3-loaded cells by these methods has been performed successfully (Kao *et al.*, 1989). *In situ* calibration offers the advan-

corrected for background and channel-crossover contributions. (A) Emission intensity of fluo-3 (75 μM) solutions as a function of free Ca^{2+} concentration. (B) Fluo-3/Fura-Red emission ratios as a function of free Ca^{2+} concentration. Solutions containing a mixture of fluo-3 (75 μM; 0) and Fura-Red (150 μM; △) were imaged with dual-channel PMT detection. Emission intensities in the two channels (fluo-3, Fura-Red) are represented on the left axis; the ratio values (fluo-3/Fura-Red; ◇) calculated from these data are on the right.

tage over *in vitro* methods that the potential variations in intracellular environmental factors that may influence the spectral properties of the dye (i.e., local ionic strength, the presence of heavy metals, polarity, viscosity of the local environment) may be taken into account. However, this method has a number of limitations. In many cases, when using an ionophore, equilibrating the Ca^{2+} concentration throughout multiple intracellular compartments of a living cell is difficult. In addition, when using low affinity Ca^{2+}-binding dyes (such as fluo-3), raising the intracellular Ca^{2+} concentration to the point of dye saturation using an ionophore may be difficult without lysing the cell.

C. Fluorescence Lifetime

Determination of the fluorescence lifetimes of intracellularly loaded Ca^{2+} indicator dyes provides direct calibration of the output of single-wavelength dyes to Ca^{2+} concentration that is independent of the amount of dye present. The future development of confocal fluorescence lifetime imaging will facilitate intracellular Ca^{2+} calibration efforts greatly. Fluorescence lifetime imaging of cellular Ca^{2+} using conventional microscopy and phase-domain measurements of Ca^{2+} indicator lifetimes has been reported (Lakowicz *et al.*, 1992a). Calibration of predetermined fluorescence lifetime values of Ca^{2+}-bound and -unbound dye to confocal lifetime imaging data will permit precise quantitation of Ca^{2+} concentrations at the subcellular level.

V. Ratiometric vs Nonratiometric Measurements

Although the ability to quantitate intracellular Ca^{2+} through the use of ratiometric Ca^{2+} indicator dyes makes this technique highly attractive for confocal as well as conventional microscopic Ca^{2+} imaging, the requirement for UV excitation of these dyes imposes strict demands on instrumentation design that are only now beginning to be realized in commercially available products. Currently, the UV-compatible confocal systems can be prohibitively expensive. Visible wavelength laser scanning systems, on the other hand, are becoming quite common since they are relatively inexpensive and easy to use. Thus, the current question is not one of the theoretical merits of ratiometric vs nonratiometric measurements, but the practical and realistic one of how to best use the currently available technology.

In addition to the limitations in instrumentation imposed by the use of ratiometric UV wavelength dyes, these dyes also possess certain disadvantages in living cell applications compared with visible wavelength indicators: (1) increased photodamage to living cells by UV light; (2) increased cellular autofluorescence with UV excitation; and (3) a decreased compatibility with the simultaneous use of caged compounds, which are cleaved photolytically or released by UV light. The ideal, and hopefully forthcoming, solution to the problem of ratiometric confocal Ca^{2+} imaging would be the development of a visible wavelength ratiometric Ca^{2+} indicator dye. The only dye currently available that has some potential in this respect is Fura-Red (Molecular Probes, Eugene, Oregon). Like its UV counterpart fura-2, this indicator exhibits a Ca^{2+}-dependent spectral shift in excitation wavelength. Fura-Red ratioing, which therefore requires dual-wavelength excitation (rather than dual-wavelength emission like indo-1), introduces the added complexities associated with the mechanism and speed of changing excitation wavelength. These potentially rate-limiting factors result in a loss of time resolution with the Fura-Red ratio imaging system, time resolution that may prove to be critical in the study of fast intracellular Ca^{2+} dynamics. These concerns, added to the low quantum yield and high Ca^{2+} affinity (133 nM at 22°C) of Fura-Red, do not support this dye as ideal.

An alternative approach to solving the quantitation problem inherent in the use of single-wavelength Ca^{2+} dyes is the simultaneous excitation and monitoring of two different indicator dyes, coloaded into the same cells. Signal output from the two dyes with different spectral properties is expressed as a ratio value, thus imitating the principles involved in single-indicator ratiometric techniques. We have been using this technique with LSCM Ca^{2+} imaging to ratio fluo-3 and Fura-Red coloaded as the AM esters into BALB/c 3T3 cells (Diliberto and Herman, 1993). Both Ca^{2+} dyes are excited with the 488-nm line of a standard argon ion laser; emission is split to dual-channel photomultiplier tube detectors (emission maxima for fluo-3 and Fura-Red of 525 nm and 645 nm, respectively). At this excitation wavelength, Fura-Red emission intensity decreases with increased Ca^{2+} binding whereas fluo-3 emission increases with increased Ca^{2+} (see calibration curve, Fig. 1B). Resting coloaded cells exhibited an inverse relationship between the intensities of fluo-3 and Fura-Red (Color Plate 6), indicative of low cytoplasmic Ca^{2+} levels and even lower nuclear Ca^{2+} levels. Stimulation of these cells with a Ca^{2+} agonist (e.g., platelet-derived growth factor, PDGF) resulted in an increase in fluo-3 and a decrease in Fura-Red (increase in fluo-3/Fura-Red ratio), whereas addition of an extracellular Ca^{2+} chelator (ethylene glycol bis(β-aminoethylether)-N,N,N′,N′-tetraacetic acid, EGTA) resulted in a reversal of these changes to levels less than the original basal Ca^{2+} level (Fig. 2). Calibration of the cellular fluo-3/Fura-Red ratio values was performed by *in vitro* methods, imaging solutions containing a mixture of known concentrations of both indicators through the microscope optics under the same conditions used experimentally for cell imaging (i.e., temperature, laser settings, detector settings). Under these conditions, a 10-fold increase in fluo-3/Fura-Red ratio was obtained over the Ca^{2+} concentration range of 34 nM to ≥1.3 μM, at which point fluo-3 fluorescence appeared to reach saturation (Fig. 1B). We have estimated nuclear free Ca^{2+} concentrations in these cells, calibrated by this method, to range from 50 to 75 nM at rest to ≥700 nM in cells maximally stimulated with BB-PDGF. Unfortunately, we have found that Fura-Red, coloaded with fluo-3, inhibits the nu-

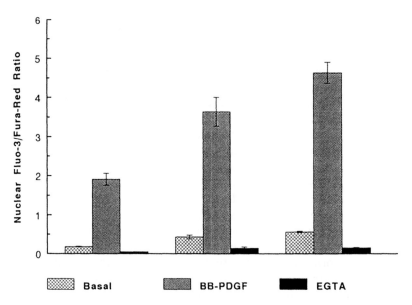

Fig. 2 Nuclear fluo-3/Fura-Red™ ratios in resting and BB platelet-derived growth factor (BB-PDGF)-stimulated cells. Cells coloaded with fluo-3/AM and Fura-Red/AM under three different sets of loading conditions were monitored for nuclear fluorescence intensity changes, as illustrated in Color Plate 6. Loading conditions: (*left*) Fluo-3/AM (4 μM, 30 min), Fura-Red/AM (10 μM, 45 min); (*center*) Fluo-3/AM (4 μM, 30 min), Fura-Red/AM (10 μM, 30 min); (*right*) Fluo-3/AM (7.5 μM, 30 min), Fura-Red/AM (10 μM, 30 min). Fluo-3/Fura-Red intensity ratios were calculated from the mean ± SEM values obtained from 7–9 individual mapped nuclei imaged on each of the three coverslips. The three ratio values shown for each of the three loading conditions represent resting cells, BB-PDGF-stimulated (10 ng/ml) cells, and the same cells exposed to 2.5 mM EGTA 12.5 min after BB-PDGF treatment.

clear Ca^{2+} oscillations induced by PDGF that are observed in cells loaded with fluo-3 alone (see subsequent discussion). We hypothesize that this effect of Fura-Red on the temporal characteristics of the Ca^{2+} response is caused by its high Ca^{2+} binding affinity, high intracellular concentration, and Ca^{2+} buffering effects.

The limitations involved in the use of this method center mainly around the uncertainty of the intracellular dye concentrations and of the spatial homogeneity of dye distribution. We have found that, by varying the coloading conditions of the two indicators, cells will exhibit fluorescence patterns indicative of different intracellular amounts of one dye in relation to the other (i.e., the dye concentration ratio is not a constant, but can be manipulated experimentally). This result is reflected in a dependency of the ratio values associated with changes in intracellular Ca^{2+} levels on the indicator loading concentrations (Fig. 2). Therefore, to calibrate ratio values to Ca^{2+} concentration, the intracellular dye concentrations, or minimally the ratio of these concentrations, first must be determined before a meaningful calibration curve can be constructed, using these dye concentrations as constants and varying the amount of free Ca^{2+} (Fig. 1B).

Questions of spatial homogeneity of the distribution of a single-wavelength dye within a cell are difficult to answer. Theoretically, dual-indicator ratioing, using two different Ca^{2+} indicators or a single Ca^{2+} indicator in combination with a second Ca^{2+}-insensitive dye, makes it possible to compensate for inter- and intracellular heterogeneity in dye concentration. This theory holds, however, only if the simplyfing assumption is made that differences in dye uptake between different cells or different compartments within a single cell are the same for both indicators employed. Experimental conditions must favor the establishment and maintenance of a constant dye ratio. Differential de-esterification (if loaded as AM esters), compartmentalization, bleaching, and/or leakage of the two dyes will lead to artifactual changes in the fluorescence ratio values that are not accurate reflections of Ca^{2+} alterations. Additionally, when a Ca^{2+}-independent dye is used as the second dye, one should confirm that signal intensity does not vary under the conditions used experimentally to alter cellular Ca^{2+}. This confirmation is not difficult, but not many compounds are currently available that meet this criterion. The pH dependency of many intracellular probes rules out their use as inert indicator dyes in many living cell systems in which Ca^{2+} alterations are to be assessed (i.e., growth factor-induced changes in intracellular pH as well as Ca^{2+}). One potentially effective indicator pair for use in visible wavelength confocal Ca^{2+} ratio imaging is that of fluo-3 (Ca^{2+} dependent) and SNARF-1 (Ca^{2+} independent). With 488-nm excitation, SNARF-1 possesses a pH-independent isoemissive point at 610 nm. Successful use of this combination of indicators, loaded as the AM esters, has been reported for flow cytometry measurements of Ca^{2+} alterations in blood peripheral mononuclear cells (Rijkers *et al.*, 1990).

VI. Image Processing

Because of the dynamic nature of intracellular Ca^{2+} alterations, image processing in living cell confocal Ca^{2+} measurement applications places a large emphasis on temporal resolution. Three-dimensional image processing and reconstruction, although possible for a static system, is currently not feasible for the majority of intracellular Ca^{2+} alterations that commonly occur on a time scale of milliseconds to seconds. Confocal Ca^{2+} image acquisition and processing, instead, concentrate on those Ca^{2+} alterations that occur in a defined one- (line) or two- (plane) dimensional region of the cell as a function of time. Analyses of experimental data, therefore, usually require the processing and manipulation of large numbers of image and/or data files. One of the greatest practical challenges to confocal Ca^{2+}-imaging experimentation today is devising an efficient method for qualitative and/or quantitative data evaluation, as well as a clear and concise way in which to convey this information to other investigators.

Multiple image sets of digitized two-dimensional Ca^{2+} image data obtained as a function of time can require massive amounts of data storage space and time, depending on the size of the images and the time resolution and duration of the set. Less time and space consuming are one-dimensional line scanning data that also can be stored as "images" in which time represents the second dimension. One common image processing concern for these types of data is related to the ease of retrieving the quantitative information of interest from the images. Commercially available confocal microscope systems often come equipped with a number of image processing options, including such manipulations as image averaging, background subtraction, ratio processing, and a variety of pictorial and graphical representations of temporal and spatial image parameters. Note, however, that the different imaging systems and software do not necessarily employ a common or standard image format. In our experience, this often means that processing or file conversion software must be developed in-house if specific or specialized processing requirements cannot be met by the system on which the data were acquired. Image format incompatibilities and data output capabilities also can become concerns related to the presentation of processed data in the form of pictures, slides, or figures. As a general consideration, image processing software is most efficient and useful if it contains the flexibility of user-defined parameters and multiple data formats.

VII. Applications

A. Platelet-Derived Growth Factor Isoform-Specific Induction of Nuclear Ca^{2+} Oscillations

Our laboratory has been examining the role of intracellular Ca^{2+} alterations in the signal transduction pathway of PDGF. Using LSCM with fluo-3/AM-loaded BALB/c 3T3 cells, we have been able to identify PDGF-induced changes in nuclear free calcium ($[Ca^{2+}]_n$) (Diliberto *et al.*, 1991). Scanned images were obtained from a single optical slice passing through the plane of the cell nuclei with a temporal resolution of 5–15 sec and were analyzed as a function of time. In quiescent cells, resting $[Ca^{2+}]_n$ levels were lower than those of the surrounding cytoplasm (Color Plates 5, 6). On addition to living cells, both AA-PDGF and BB-PDGF isoforms produced a transient increase in $[Ca^{2+}]_n$ that surpassed cytoplasmic levels. We observed that BB-PDGF stimulation resulted in the generation of $[Ca^{2+}]_n$ oscillations that diminished over time (Fig. 3). The frequency of BB-PDGF-stimulated oscillations was modulated by extracellular Ca^{2+} and could not be mimicked by increasing intracellular inositol 1,4,5-trisphosphate (IP$_3$) levels with AlF_4^- in the absence of

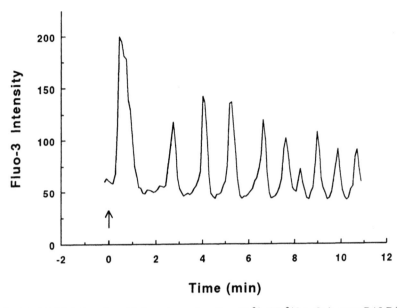

Fig. 3 BB-PDGF-induced oscillations in nuclear free Ca^{2+} ([Ca^{2+}]$_n$). Quiescent BALB/c 3T3 cells were loaded with fluo-3/AM and images similar to those depicted in Color Plate 5 were obtained by laser scanning confocal microscopy at 5-sec intervals. BB-PDGF (10 ng/ml) was added at the arrow. Represented on the *y* axis is the average fluorescent intensity of the mapped area of a single nucleus.

growth factor stimulation. Caffeine addition alone had no effect on [Ca^{2+}]$_n$ levels, but exposure of cells to caffeine after BB-PDGF stimulation augmented [Ca^{2+}]$_n$ oscillations, either by increasing the frequency or by reinitiating preexisting oscillations. In contrast, AA-PDGF stimulation resulted in the generation of one or two irregular transient [Ca^{2+}]$_n$ spikes. Caffeine pretreatment followed by AA-PDGF stimulation resulted in [Ca^{2+}]$_n$ oscillations very similar to those produced by BB-PDGF alone. Additionally, the AA-PDGF and BB-PDGF isoforms appeared to modulate distinct pools of nuclear Ca^{2+}, since BB-PDGF was still capable of inducing [Ca^{2+}]$_n$ oscillations subsequent to prior induction of oscillations by AA-PDGF with caffeine. Currently we are testing the hypothesis that these PDGF isoform-specific changes in nuclear free Ca^{2+} may be a mechanism by which isoform-specific cellular signaling pathways of biological function are manifested.

B. Electrical Stimulation of Nerve

Hernandez-Cruz and colleagues (1990) first described the use of confocal microscopic Ca^{2+} imaging to visualize cytoplasmic and nuclear Ca^{2+} transients in isolated voltage-clamped sympathetic neurons of the bullfrog. LSCM, with fluo-3 as the Ca^{2+} indicator, was used in either the line-scanning or the frame-scanning mode to monitor intracellular Ca^{2+} alterations with temporal resolution of 5 msec/line or 1 sec/image. Brief activation of voltage-dependent Ca^{2+} channels by electrical stimulation was observed to result in a nonuniform intracellular Ca^{2+} increase that spread from the periphery to the center of the cell in approximately 300 msec in a wave-like fashion, which then was followed by equilibration of Ca^{2+} levels across the cell and a return to resting values in 6–8 sec. The spatial characteristics of the response, in addition to its dependency on extracellular Ca^{2+}, were interpreted to suggest a mechanism of localized Ca^{2+} entry and spatial buffering on electrical stimulation. In addition, both spatial and temporal heterogeneity were observed in the relationship between cytoplasmic and nuclear Ca^{2+} levels in response to either electrical stimulation or caffeine exposure. The larger slower nuclear Ca^{2+} transients found to be induced by both conditions provide evidence for the involvement of an intracellular Ca^{2+} release process originating from sites near or within the nucleus. The relationship of electrically stimulated extracellular Ca^{2+} entry to nuclear Ca^{2+} increases hypothetically functions through a calcium-induced calcium-release amplification mechanism.

C. Calcium Waves in *Xenopus* Oocytes

Intracellular Ca^{2+} waves have been visualized using confocal microscopy in *Xenopus laevis* oocytes (Lechleiter *et al.*, 1991; Lechleiter and Clapham, 1992). LSCM and fluo-3 were used to obtain Ca^{2+} images through single optical sections with a temporal resolution of 1 sec. The spatial and temporal patterns of receptor-activated intracellular Ca^{2+} alterations were indicative of the presence of regenerative circular and spiral waves of Ca^{2+} release. These studies also eloquently demonstrated the utility of the combination of confocal Ca^{2+} imaging and caged compound release techniques to investigate the role of second messenger compounds (i.e., GTP and IP_3) in intracellular Ca^{2+} signaling mechanisms. Visible (fluo-3) and UV (caged GTPγS and caged IP_3) wavelength laser scanning were employed to image Ca^{2+} and to release caged compounds photolytically, respectively. The experimental results were found to be consistent with a model for intracellular Ca^{2+} release in which an IP_3-mediated release mechanism controls Ca^{2+} wave propagation with modulation by the diffusion and concentration of cytoplasmic Ca^{2+}. The involvement of G protein activation was found to be proximal to the IP_3-mediated Ca^{2+} release mechanism.

D. Sea Urchin Egg Fertilization and Development

Confocal microscopy has been used by Stricker and colleagues (1992) to demonstrate cytosolic and nuclear Ca^{2+} alterations that occur in living sea urchin eggs after fertilization. In these studies, either fluo-3 or Calcium Green (both loaded by microinjection) were used to obtain laser-scanned two-dimensional Ca^{2+} images of optical sections through the cell. Time-lapse analyses of these data revealed the initiation of Ca^{2+} waves that propagated throughout the interior of the cell within 20–60 sec of sperm addition at a rate of 3–10 μm/sec. A nuclear/cytoplasmic Ca^{2+} gradient was observed in prefertilization eggs (nuclear Ca^{2+} lower than cytoplasmic) that was reversed transiently on fertilization.

Note that these studies found normal development of the fertilized eggs subjected to confocal Ca^{2+} imaging measurements to occur only in those specimens loaded with Calcium Green and not in those loaded with fluo-3. Eggs subjected to LSCM alone or those loaded with fluo-3 but not imaged by laser scanning exhibited normal cleavage, pointing to the possibility of phototoxic effects specific to the use of fluo-3 in the imaging technique. More extensive evaluation of the potential toxicities of the various Ca^{2+} indicators is required in future confocal microscopic investigations as the field advances to attempt correlation of Ca^{2+}-imaging data with subsequent biological function at the single-cell level.

References

Buurman, E. P., Draaijer, A., Gerritsen, H. C., van Veen, J. J. F., Houpt, P. M., and Levine, Y. K. (1992). Fluorescence lifetime imaging using a confocal laser scanning microscope. *Scanning* **14**, 155–159.

Diliberto, P. A., and Herman, B. (1993). Quantitative estimation of PDGF-induced nuclear free calcium oscillations in single cells performed by confocal microscopy with Fluo-3 and Fura-Red. *Biophys. J.* **64**, A130.

Diliberto, P. A., Periasamy, A., and Herman, B. (1991). Distinct oscillations in cytosolic and nuclear free calcium in single intact living cells demonstrated by confocal microscopy. *In* "Proceedings of the 49th Annual Meeting of the Electron Microscopy Society of America" (G. W. Bailey and E. L. Hall, eds.), pp. 228–229. San Francisco: San Francisco Press.

Grynkiewicz, G., Poenie, M., and Tsien, R. Y. (1985). A new generation of Ca^{2+} indicators with greatly improved fluorescence properties. *J. Biol. Chem.* **260**, 3440–3450.

Haugland, R. (1992). "Molecular Probes Handbook of Fluorescent Probes and Research Chemicals," 1992–1994 edition. Junction City, Oregon: Molecular Probes.

Hernandez-Cruz, A., Sala, F., and Adams, P. R. (1990). Subcellular calcium transients visualized by confocal microscopy in a voltage-clamped vertebrate neuron. *Science* **247**, 858–862.

Kao, J. P., Harootunian, A. T., and Tsien, R. Y. (1989). Photochemically generated cytosolic calcium pulses and their detection by Fluo-3. *J. Biol. Chem.* **264**, 8179–84.

Lakowicz, J. R., Johnson, M. L., Lederer, W. J., Szmacinski, H., Nowaczyk, K., Malak, H., and Berndt, K. (1992a). Fluorescence lifetime sensing generates cellular images. *Laser Focus World* May, 1992, pp. 1–8.

Lakowicz, J. R., Szmacinski, H., and Johnson, M. L. (1992b). Calcium imaging using fluorescence lifetimes and long-wavelength calcium probes. *J. Fluorescence* **2**, 47–62.

Lechleiter, J. D., and Clapham, D. E. (1992). Molecular mechanisms of intracellular calcium excitability in *X. laevis* oocytes. *Cell* **69**, 283–294.

Lechleiter, J., Girard, S., Peralta, E., and Clapham, D. (1991). Spiral calcium wave propagation and annihilation in *Xenopus laevis* oocytes. *Science* **252**, 123–126.

Minta, A., Kao, J. P. Y., and Tsien, R. T. (1989). Fluorescent indicators for cytosolic calcium based on rhodamine and fluorescein chromophores. *J. Biol. Chem.* **264,** 8171–8178.

Rijkers, G. T., Justement, L. B., Griffioen, A. W., and Cambier, J. C. (1990). Improved method for measuring intracellular Ca^{++} with Fluo-3. *Cytometry* **11,** 923–927.

Roe, M. W., Lemasters, J. J., and Herman, B. (1990). Assessment of Fura-2 for measurements of cytosolic free calcium. *Cell Calcium* **11,** 63–73.

Stricker, S. A., Centonze, V. E., Paddock, S. W., and Schatten, G. (1992). Confocal microscopy of fertilization-induced calcium dynamics in sea urchin eggs. *Dev. Biol.* **149,** 370–380.

Wang, X. F., Periasamy, A., and Herman, B. (1992). Fluorescence lifetime imaging microscopy (FLIM): Instrumentation and applications. *Crit. Rev. Anal. Chem.* **23,** 369–395.

CHAPTER 11

Simultaneous Near Ultraviolet and Visible Excitation Confocal Microscopy of Calcium Transients in *Xenopus* Oocytes

Steven Girard and David E. Clapham

Department of Pharmacology
Mayo Foundation
Rochester, Minnesota 55905

I. Introduction

Changes in cytosolic Ca^{2+} were studied first with sensitive but indirect bioassays such as smooth muscle preparations that contract in response to drugs or hormones. More direct assays of intracellular Ca^{2+} are now available, including Ca^{2+}-regulated photoproteins, metallochromic dyes, Ca^{2+}-sensitive microelectrodes, and fluorescent Ca^{2+} dyes. Electrophysiological recording of

METHODS IN CELL BIOLOGY, VOL. 40

263

calcium-activated potassium or chloride currents also may be used to follow cytosolic Ca^{2+}.

The first direct measurements of cytosolic free Ca^{2+} were made in giant muscle cells with the photoprotein aequorin (Ridgeway and Ashley, 1967). Aequorin is a 21-kDa protein from the photocytes in the periphery of the umbrella of the jellyfish *Aequorea forskalea* (Blinks and Moore, 1985). This highly charged anion initially was microinjected into large cells, but since then it has been used regularly in cells as small as adrenal chromaffin cells (~13 μm) (Cobbold *et al.*, 1987). Liposome fusion, hypoosmotic shock, ethylene glycol bis(β-aminoethylether)-N,N,N',N'-tetraacetic acid (EGTA) or ATP permeabilization, scrape loading, or spin loading may permeabilize cells temporarily to allow uptake of aequorin (Cobbold and Rink, 1987). Once in the cytosolic compartment, native aequorin does not enter intracellular organelles (Blinks, 1986).

The Ca^{2+}-controlled luminescent reaction is the oxidation of coelenterazine, a luminophore tightly bound to aequorin. Oxygen bound to aequorin is required, but neither ATP nor molecular oxygen is consumed. The luminescence is proportional to $[Ca^{2+}]^{2.5}$; molecular cloning has revealed three Ca^{2+} binding sites on aequorin (Inoue *et al.*, 1985). Genetic engineering of aequorins has targeted a recombinant aequorin to the mitochondria, suggesting that a similar approach might be used to measure the Ca^{2+} inside the endoplasmic reticulum (ER) (Rizzuto *et al.*, 1992).

During the 1970s, bis-azo absorbance dyes and Ca^{2+}-sensitive microelectrodes were developed. Because metallochromic dyes change color on binding Ca^{2+}, absorption may be used to monitor the concentration of Ca^{2+} (Blinks, 1986). Experimentally, absorption is monitored at two wavelengths—one a strong function of Ca^{2+} concentration, the second independent of Ca^{2+} concentration. The latter measurement is proportional to the concentration of the dye, and both are used to calculate the Ca^{2+} concentration.

Ca^{2+}-sensitive electrodes record the electrical potential established between solutions of different Ca^{2+} concentration when separated by an ion-selective membrane. Hydrophobic charged ion exchangers or neutral carriers in the tip of the pipette function as the ion-selective membrane (Blinks, 1986). The most popular ligand is a neutral carrier, ETH 1001 (N,N'-di(11-ethoxycarbonyl)undecyl-N,N'-4,5-tetramethyl-3,6-dioxaoctane-1,8-diamide] (Oehme *et al.*, 1976). ETH 1001 solubilizes Ca^{2+} in media of low dielectric constant with tetraphenylborate (TPB) to balance the charge of the complex. Because of their high electrical resistance, Ca^{2+}-sensitive microelectrodes respond slowly to changes in Ca^{2+}; this feature is their principle disadvantage (Blinks, 1986).

Indicators labeled with ^{19}F have NMR signals that may be used to follow Ca^{2+} concentration (Smith *et al.*, 1983). For example, difluoro derivatives of 1,2-bis(o-aminophenoxy)ethane-N,N,N',N'-tetraacetic acid (BAPTA) with pCa of 6.2 can resolve changes in Ca^{2+} concentration without interference from physiological changes in pH or Mg^{2+}. Because large volumes of solution and long incubation times are necessary to obtain meaningful spectra, NMR cannot be used to follow intracellular Ca^{2+} transients (Blinks, 1986).

The development of fluorescent indicators for Ca^{2+} in the 1980s by Tsien and colleagues revolutionized the study of intracellular Ca^{2+} (Tsien, 1980). In an attempt to find a fluorescent molecule the emissions of which were altered specifically only by Ca^{2+}, quin-2, an analog of the Ca^{2+} chelator EGTA, was developed. The acetoxymethyl (AM) ester of quin-2 was synthesized as a membrane-permeant form of the indicator that was trapped in the cytosol after cleavage by cellular esterases. However, quin-2 has a small quantum yield and must be used in concentrations near 0.5 mM to achieve signals above autofluorescence. Further, quin-2 is extraordinarily photolabile, bleaching after absorbing only 3 or 4 photons.

The second generation of fluorescent Ca^{2+} indicators, indo-1 and fura-2, may be used to measure the concentration of free cytosolic Ca^{2+} ratiometrically (Grynkiewicz et al., 1985). The maximum of the fura-2 excitation spectrum shifts from 380 to 340 nm on binding Ca^{2+}. The maximum of the indo-1 emission spectrum shifts from 490 to 405 nm with increasing Ca^{2+}, at 351-nm excitation. Both dyes are ~30-fold brighter than their predecessor quin-2, permitting lower concentrations and less buffering. Ratiometric methods provide estimates of Ca^{2+} that are independent of dye concentration, instrument settings, and pathlength, provided the fluorescence contribution of a molecular species is linear in concentration.

Fluo-3 is excited with visible wavelengths (488 nm) and undergoes a 40- to 100-fold increase in fluorescence on binding Ca^{2+} (Eberhard and Erne, 1989). Visible excitation avoids autofluorescence, which contaminates signals from UV-excited indicators such as fura-2 and indo-1. However, indicators excited with visible wavelengths do not undergo a shift in their excitation or emission spectra, so the ratio method cannot be used. To quantitate the signals observed with fluo-3 or Calcium Green,™ the maximal fluorescence (F_{max}) and the minimal fluorescence (F_{min}) must be obtained. With an estimation of the affinity for Ca^{2+} under physiological conditions (K_d), the fluorescence signals (F) may be converted into Ca^{2+} concentration by Eq. 1 (Grynkiewicz et al., 1985):

$$Ca^{2+} = K_d \frac{F - F_{min}}{F_{max} - F} \tag{1}$$

II. Confocal Microscopy

A. Background

Over a century ago, Ernst Abbe demonstrated how the diffraction of light by a specimen and the objective lens influences resolution (Abbe, 1984). He derived the minimum distance (d) that the images of two points in the specimen can approach each other in the specimen plane and still be resolved

$$d = \frac{1.22 \; \lambda_0}{(NA_{obj} + NA_{cond})} \tag{2}$$

where NA_{obj} and NA_{cond} are the numerical apertures of the objective and condenser lenses and λ_0 is the wavelength of light in a vacuum (Abbe, 1984). Using the Rayleigh criterion, d is taken as the radius of the Airy disk or the radius of the first minimum of a unit diffraction image. The numerical aperture is given by

$$NA = \eta \sin \alpha \tag{3}$$

where η is the refractive index of the medium between the specimen and the objective and α is the half angle of the cone of light accepted by the objective lens or emerging from the condenser lens. The lateral resolution is also a function of the size of the field (di Tolardo, 1955) and may be improved by a limiting factor of $\sqrt{2}$ as the field decreases.

Axial (z axis) resolution may be defined as the radius of the first minimum of an "infinitely" small point object, but the precise energy distribution above and below focus cannot be deduced by geometric ray tracing alone (Inoue, 1990). For practical purposes, the axial resolution turns out to be half the lateral resolution. The depth of field is the depth of the image that appears to be in focus, measured along the z axis and translated to a distance in the specimen space. In conventional fluorescence imaging, the signal from each point spreads as a cone of light above and below the plane of focus, blurring the focused image of the specimen and reducing resolution. Because of this blurring, the depth of field in fluorescence imaging is very much greater than the axial resolution, unlike bright field microscopy in which the two are approximately equal (Inoue, 1990).

B. History

A confocal microscope was invented in 1951 by Marvin Minsky, a postdoctoral fellow at Harvard University studying neural networks in living brain (Minsky, 1988). In 1957, Minsky patented the concept of confocal imaging, the illumination and detection of a single diffraction-limited spot in a specimen (Fig. 1A). In the transmission configuration, the condenser is replaced with a second identical objective lens. The field of illumination is reduced to a diffraction-limited spot with a pinhole aperture in front of the illumination source. The objective lens is positioned to focus light from the same spot onto a second (confocal) pinhole in front of a detector. In the epi-illumination mode (Fig. 1B), a single objective is used to illuminate the specimen and to receive the returning fluorescence. Again, pinhole apertures are placed confocally in front of the illumination source and the detector to limit the field of excitation in the specimen. A dichroic mirror is critical to the epifluorescence mode because it reflects short excitation wavelengths and passes longer wavelength emissions. To create an image, the specimen must be scanned and the temporal output of the detector must be related to a position in the object. Either the stage is moved in the x,y plane to scan the object (stage scanner) or the beam is deflected across the specimen in a raster pattern (point scanner). The stage scanner uses paraxial illumination and avoids optical aberrations, but is limited by the mechanical scanning of the stage (Inoue, 1990).

The principle advantage of the confocal method is the rejection of out-of-focus fluorescence by the pinhole aperture in front of the detector (Fig. 1).

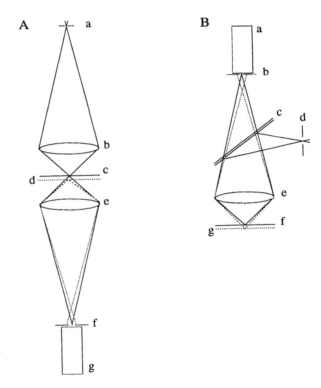

Fig. 1 Schematic of two configurations of confocal imaging. (A) Transmission configuration. Excitation energy is projected through a pinhole (a) and focused by the condenser lens (b) to a diffraction-limited spot in the specimen. A fraction of the resulting fluorescence is captured by the objective lens (e) and focused to a pinhole (f) in front of the detector (g). The condenser lens and objective lens must have identical optical characteristics (NA) and be aligned to the same diffraction-limited volume (voxel). Only the fluorescence from this voxel in the plane of focus (c) is captured at the detector. Signal from out-of-focus planes (d) is rejected by pinhole aperture (f). (B) Epi-illuminated configuration. The excitation is projected through a pinhole (d), reflected off a dichroic mirror (c), and focused by the objective to a diffraction-limited voxel in the specimen. A portion of the resulting fluorescence is captured by the objective (e) and follows the same path as the excitation beam (solid line). The dichroic reflector (c) passes the fluorescence because it is of longer wavelength than the excitation. Fluorescence from out-of-focus planes (g; dotted lines) is blocked by a pinhole aperture (b) and is not recorded at the detector (a).

This method allows noninvasive optical sectioning of living tissues, particularly for specimens greater than 10 μm thick. Three-dimensional reconstructions can be made without deconvolution of the images; dynamic events that previously were not observed can be studied with this method (Inoue, 1990). Further, confocal imaging avoids the artifacts introduced by the preparation of samples for electron microscopy. Because the lateral resolution of a microscope is a function of the field of view, the confocal configuration takes advantage of a diminishingly small illumination (at any instant), improving the lateral resolution by a factor of 1.4 (Inoue, 1990). Registration of images with dual labels is done automatically because a single diffraction-limited point is illuminated at any given instant in time.

Confocal microscopes are ideally suited for laser illumination. Although Maiman's initial manuscript describing the laser was rejected in 1960 by the editors of *Physical Review Letters* (Hecht, 1987), this invention revolutionized many branches of science, engineering, and medicine. Lasers were adapted to provide illumination for scanning confocal microscopes, overcoming limitations of conventional lamps that were too weak for many practical applications. In 1970, Egger at Yale University developed a laser-illuminated confocal microscope that scanned the preparation by oscillating the objective lens; this microscope was patented by Egger and Davidovits in 1972 (Egger, 1989). Then, an electrical engineering group at Oxford developed a stage scanner and advanced the fundamental theory of confocal imaging (Sheppard and Choudhury, 1977). In 1987, White, Amos, and Fordham in Cambridge developed the Lasersharp Bio-Rad MRC500 confocal microscope, providing a convincing demonstration of the power of confocal microscopy. We use an adapted version of the Bio-Rad MRC600 (Bio-Rad, Richmond, California) confocal scanner. The following section presents a full description of our experimental apparatus.

C. Bio-Rad Lasersharp MRC 600 Confocal Scanner

We mounted the Bio-Rad MRC600 scanner, a 25-mW Ion Laser Technology argon visible laser (488- and 514-nm lines) (Salt Lake City, Utah), a Coherent Innova-90 UV argon laser (Palo Alto, California), and an inverted Zeiss IM35 microscope (Batavia, Illinois) on an optical bench from Newport (Fountain Valley, California) (Fig. 2). The scanner was controlled with a Nimbus VX PC (Research Machines Ltd., Oxford, England) using SOM software from Bio-Rad. Digital images (8-bit) were stored on read–write optical disks and studied off line with ANALYZE software (Mayo Foundation, Rochester, Minnesota) using a Silicon Graphics Personal Iris computer (Mountain View, California).

A dichroic mirror in the first removable filter block (Fig. 2g) reflects the excitation laser beam(s) (solid line) into the scanning assembly, which consists of two rotating galvanometer mirrors and two spherical mirrors (Fig. 3). The scanning assembly deflects the beam in a raster-like pattern across the speci-

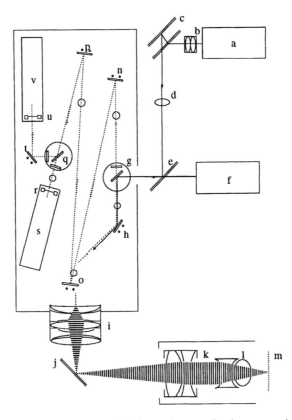

Fig. 2 Schematic of the Bio-Rad MRC600 Lasersharp confocal scanner adapted for simultaneous visible and UV excitation. The 351-nm line from a Coherent Innova 90-4 argon laser (a) is directed through a 5× beam expander (b) and a UV focusing lens (c) to correct for longitudinal chromatic aberrations (Fluharty *et al.*, 1991). A dichroic mirror placed at a 45° angle (e) reflects the 351-nm beam and passes the 488-nm line from an Ion Laser Technology 5425 argon laser (f) into the scanner. A dichroic mirror in the scanner's first filter block (g) deflects the beam to a steering mirror (h). The beam then is scanned by a galvanometer mirror (not shown; see Fig. 3). After scanning, the beam is focused by a 6× eyepiece (i), reflected off a 45° mirror (j) and into the epifluorescence port of a Zeiss IM35 inverted microscope. The tube lens (k) and an objective with its telon lens (l) bring the excitation beam to a diffraction-limited volume of focus (m) within the specimen. A fraction (~5%) of the stimulated fluorescence returns along this same path, is descanned by the galvanometer mirrors, and long-pass filtered at the first filter block (g). Three mirrors (n–p) increase the optical path-length to ~1.7 m. Because out-of-focus fluorescence will be either slightly converging or diverging, this increased path-length allows optical sectioning with millimeter pinhole diameters. A dichroic filter in the second filter block (q) may be placed at a 45° angle to reflect shorter wavelength fluorescence. These signals may be bandpass filtered with appropriate filters in the second filter block (q). Pinhole apertures (r,u) with diameters adjustable from 0.5 to 8 mm block out-of-focus fluorescence from the photomultiplier tubes (Thorn EMI, model 9828B) (s,v).

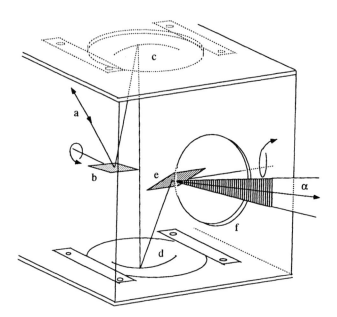

Fig. 3 Schematic of the Bio-Rad Lasersharp MRC600 confocal scanning head. The parallel incoming beam (a) is steered (mirror h on Fig. 2) to the first galvanometer mirror (b), which slowly rotates (~1 Hz) about its axis (± 6°). This deflection causes the beam to traverse arcs on the spherical mirrors (as shown) (c,d). The spherical mirrors direct the beam to the second galvanometer mirror (e), which quickly rotates (~500 Hz) about its axis (± 6°). In this manner, the angle at which the parallel excitation beam enters the eyepiece (not shown; f) is varied and the diffraction-limited focal point is scanned across the specimen in a raster pattern.

men; the incoming beam (solid line) deflects off the first galvanometer mirror which slowly rotates ± 6° (~0.7 Hz) and then is directed by two spherical aluminum mirrors onto the center of a second galvanometer mirror, which rapidly deflects the beam ±6° (500 Hz). Thus, the beam is deflected rapidly in the vertical direction and slowly in the horizontal direction as it exits the scanner through the eyepiece lens (Fig. 2f). The eyepiece, tube, telon, and objective lenses focus the excitation beam on a diffraction-limited spot in the focal plane of the specimen (Fig. 2m), stimulating fluorescence. A fraction of this fluorescence returns along the excitation beam path and passes through the same dichroic mirror that reflected the shorter wavelength excitation (Fig. 2g). After passing back through the optical train, emissions from the focal point are collimated perfectly while fluorescence from out-of-focus planes either converges or diverges. Three flat mirrors (Fig. 2n,o,p) increase the pathlength to the detector and take advantage of this filtering of the fluorescence. Out-of-focus fluorescence is blocked from detectors by variable diameter pinholes (Fig. 2r,u).

Because the refraction of a lens depends on the wavelength of light, corrections were necessary to focus the UV beam to the same focal plane in the sample as the visible laser beam (Bliton *et al.*, 1993). Bliton and co-workers used a 5× beam expander (Fig. 2b) and a 25-mm lens (Fig. 2d) to refocus the UV beam properly. Lateral chromatic aberrations in optical elements also cause similar problems, but we have developed a six-lens eyepiece to correct for lateral chromatic aberrations, permitting simultaneous full-field UV and visible excitation (Bliton *et al.*, 1993).

D. Alignment of UV and Visible Excitation

First we aligned the scanner for visible excitation. Briefly, the galvanometer mirrors were parked and a stationary 488-nm beam projected through the center of the eyepiece. Minor adjustments of M1 (Fig. 2h) were made to center the beam, if necessary. Next, we aligned the beam in the optical axis of the microscope by adjusting the 45° folding mirror just after the eyepiece (Fig. 2j); the beam was centered in a ground glass target, which replaced an objective. Then we positioned the microscope the proper distance from the eyepiece, after removing the turret. We moved the microscope along the optical bench until the 488-nm beam projecting through the tube lens was collimated (not converging or diverging). This adjustment insured that the visible beam was entering the telon and objective lenses properly. We then replaced the turret and secured the microscope to the optical bench.

Next we aligned the steering mirrors of the confocal scanner (Fig. 2n,o,p). With the blue excitation, high sensitivity (BHS) cube in position (510-nm dichroic mirror), we scanned a thick fluorescent specimen (30 μl 100 μM Calcium Green in saline solution). After placing a target in the site port just behind steering mirror M2 (Fig. 2n), we adjusted M1 to center the beam in the optical path. The other two flat mirrors (Fig. 2o,p) were adjusted similarly.

With the scanning box properly aligned for visible excitation, the 5x UV beam expander (Fig. 2b) and focusing mirror (Fig. 2c) were positioned to correct for longitudinal chromatic aberrations of the optics. First, the table mounts of the expander were adjusted so axial movement did not deflect the beam out of alignment. After removing the UV focusing mirror (Fig. 2d), we used folding mirrors (Fig. 2c) to center the UV beam on the dichroic mirror near the entry port of the scanner (Fig. 2e). The focusing lens was positioned in the center of the optical path without laterally deflecting the beam. Then, we scanned a fluorescent target (business card) with both visible and UV excitation. After adjusting for maximal signal with only visible excitation, we scanned using UV excitation, adjusting the dichroic mirror (Fig. 2e) and the beam expander (Fig. 2b) to produce the brightest possible image. Finally, the alignment was tested with a field of 0.1-μm diameter beads, fluorescent with UV and visible excitation.

III. Calibration of Instrumentation and Fluorescence Signals

We used a stage micrometer from Nikken (Tokyo, Japan) to determine the size of the pixels in images taken with our adapted Bio-Rad MRC600 scanner. The pixels had to be measured in both horizontal (x) and vertical (y) directions (Fig. 4). This calibration should be repeated if any element in the optical train is changed.

To estimate the thickness of the confocal section, we used the tilted reflector method (Cogswell and Sheppard, 1990) with the reflective filter block. The reflective dichroic mirror passes 50% of incident light, reflecting the rest. Briefly, we placed a mirror above the objective at an angle 5–10° from horizontal. Because the mirror was not perpendicular to the optical axis, the reflected confocal image was a linear streak and did not fill the entire field. We recorded the absolute z (axial) position of the objective with a calibrated Heidenhain length gauge. As the objective was focused, the reflection shifted horizontally in the image. We measured this shift to estimate the thickness of the optical section. The width of the linear reflective image was 247 μm at half-maximal intensity (not shown). In many of our experiments using *Xenopus* oocytes, low (10×) power objectives are used to observe a large field (~700 μm square). The principles described in the next section are the same for higher (60×, 100×) power objectives. For a discussion of UV/visible

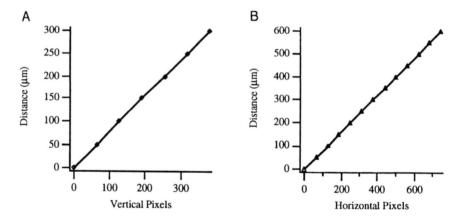

Fig. 4 Calibration of pixel dimensions. Transmitted images of a stage micrometer (Nikken, Tokyo) were used to calibrate the size of pixels. Using the UV/vis eyepiece and an Olympus 10× objective, the pixel dimensions were 0.79 and 0.80 μm in the horizontal and vertical directions, respectively. The images were acquired using zoom 3 of the SOM routine on a Nimbus personal computer with a box size of 1 and a horizontal box shape. The data show that the Bio-Rad MRC600 Lasersharp confocal scanner results in square pixels over the entire field of view; no significant deviation from the linear least squares best fit over the field occurs.

microscopy under high power, see Bliton *et al.* (1993). For the Olympus DPlanApo 10× UV (0.4 NA) objective used in our experiments, the reflection horizontally shifted 3.7 ± 0.2 μm in the image (measured in the specimen plane) per micron change in the focus of the objective (Fig. 5). Therefore, with this configuration, the confocal section was ~65 μm thick.

Indo-1 is particularly suited to confocal microscopy because its emissions, monitored at two different wavelengths, may be used to quantitate Ca^{2+} concentration. With 351-nm excitation, indo-1 undergoes a large shift in the peak emission wavelength from 485 to 405 nm on binding Ca^{2+}, whereas the signal monitored at 453 nm does not change with Ca^{2+} binding. Unfortunately, reduced pyridine nucleotides absorb similar wavelengths, so autofluorescence may be a problem with biological specimens.

The UV configuration of the MRC600 confocal scanner was used to perform an *in vitro* calibration of indo-1. We used a 380-nm long pass dichroic mirror in the first filter block to reflect the UV excitation to the specimen. Returning emissions passed through the first dichroic mirror and were long pass filtered at 390 nm to block stray excitation from reaching the detectors. The emissions were split with a 435-nm long pass dichroic mirror in the second filter block. Fluorescence of less than 435 nm reflected off this dichroic filter and were band pass filtered at 405 nm (35-nm full-width). Emission wavelengths longer than 435 nm passed through the dichroic mirror and were band pass filtered at 453 nm (10-nm full-width).

The emission spectrum of indo-1 changes dramatically with excitation wave-

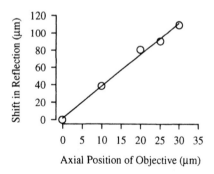

Fig. 5 Horizontal shift in the reflection of a tilted mirror as a function of the axial position (focus) of an Olympus DPlanApo 10× UV objective. Reflective confocal images were acquired with the objective in five different axial positions covering a range of 30 μm. The axial position was measured with a Heidenhain length gauge coupled to the mount of the objective. Horizontal position of the reflection was measured with ANALYZE software. Slope of the least squares best fit line through the data was 3.7 ± 0.2. The width of a reflective image was 247 μm at half-maximal intensity. The confocal aperture was set to its maximal opening (7 mm).

length (Fig. 6). At first, our attempts to calibrate the signal were unsuccessful because the Innova-90 argon laser provided both 351- and 365-nm lines. After installing a prism on the Innova-90 to select only the 351-nm line, a calibration of indo-1 was performed in mock intracellular solutions. We found that indo-1 has a single binding site (Hill coefficient ~1.1) for Ca^{2+} with a dissociation constant (K_d) of 225 nM at 22°C (Fig. 7), close to the previously reported K_d of 250 nM at 37°C (Grynkiewicz et al., 1985).

IV. Calcium Signaling in Oocytes

In 1971, Gurdon and colleagues demonstrated that oocytes of the South African clawed frog, *Xenopus laevis*, synthesize proteins encoded by exogenous mRNA (Gurdon et al., 1971). A decade later, Sumikawa et al. (1981) injected mRNA for nicotinic acetylcholine receptors and first demonstrated the expression of a foreign receptor or ion channel on the plasma membrane of *Xenopus* oocytes. Since then, the *Xenopus* oocyte has become a popular system for exogenous expression of receptors, channels, and enzymes (Soreq, 1985). Indeed, acetylcholine, serotonin, glutamate, substance K, neurotensin, thyrotropin-releasing hormone, α-thrombin, gonadotropin-releasing hormone, f-Met-peptide, dopamine, and angiotensin II receptors have been expressed in these cell (Dascal, 1987). A native Ca^{2+}-activated Cl^- current ($I_{Cl,Ca}$) provides a convenient assay to follow the expression of exogenous receptors coupled to

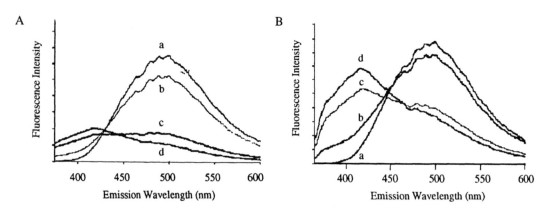

Fig. 6 Emission of indo-1 at 365- and 351-nm excitation. The emission spectra of 1 μM indo-1(potassium salt) in a mock intracellular solution with various free Ca^{2+} concentrations was measured using a Perkin-Elmer MPF 66 fluorescence spectrophotometer. The intracellular solution contained 126 mM KCl, 10 mM NaCl, 20 mM HEPES, 1 mM MgCl$_2$; pH 7.1 at 22°C. The estimated free Ca^{2+} concentration was: (a) less than 10 nM, (b) 100 nM, (c) 1 μM, and (d) 1 mM (Takahashi et al., 1987). (A) 365-nm excitation. (B) 351-nm excitation.

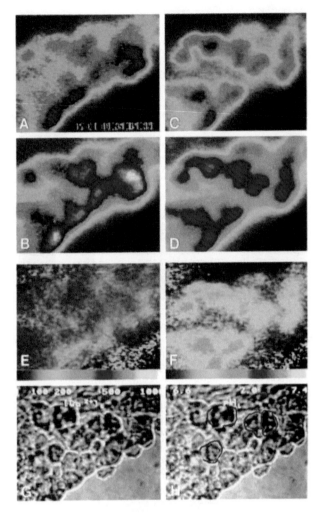

Color Plate 1 (Chapter 8) Simultaneous imaging of calcium and pH in primary pituitary melanotrope cells. (A–D) Raw data. Shown is an eight-frame average of the four simultaneous images formed from cells double labeled with indo-1 and SNARF-1, excited at 350 nm and 540 nm and imaged at (A) 405 nm, (B) 475 nm, (C) 575 nm, and (D) 640 nm. The false color pallete used corresponds to the 255-color look-up table displayed at the bottom of E and F, with gray level 0 set to black and 255 to white. The data are recorded in monochrome on video tape for further off-line analysis. After background subtraction and shading error correction, image A is divided by B to yield the Ca^{2+} map (E), and C is divided by D to yield the pH map (F). The equivalent calcium values for the ratios are calculated as described in Chapter 8, Section III, A and are displayed as the false color maps (E) for calcium and (F) for pH. (G) Phase contrast images are taken at the beginning and at the end of the experiment to help identify the cells. (H) The same image as G shown with the ROIs for the analysis of four cells. The software can handle up to 255 such irregular regions of interest (ROIs), which are defined using the phase contrast images for the off-line analysis. However, to check the progress of experiments, two to four ROIs are drawn and applied to the images in real time. The software can display the integrated gray levels and/or the uncorrected ratio for up to 15 ROIs. Images E–H were photographed from the RGB screen simultaneously using the color pallete discussed in the legend to Color Plate 2.

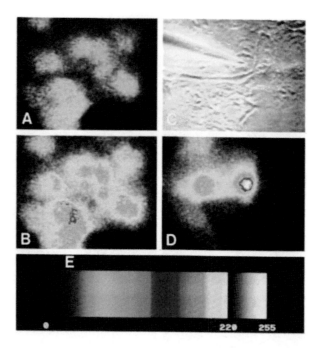

Color Plate 2 (Chapter 8) Simultaneous fluorescence and phase contrast imaging of indo-1-loaded Madin–Darby canine kidney cells being injected with 6-carboxy-*x*-rhodamine. Color Plate (A, B) The 405-nm and 475-nm wavelengths of indo-1 emission, as for Color Plate 1 A, B. The phase contrast image in C was formed at 575 nm, as discussed in Chapter 8. (D) Carboxyrhodamine excited at 540 nm and imaged at 640 nm. Three of the indo-1-labeled cells already have been injected. Twelve cells were injected over the course of about 1 min. Several cells were left uninjected to serve as controls for the injection process. Additional details are in Chapter 8, Section II, C, 6. (E) A 256-color palette look-up table that assigns various colors between 0 and 219 and a gray scale from 220–256. We find this type of palette very useful for simultaneous display of fluorescence and bright field or phase contrast images. The fluorescence image is recontoured to lie between gray levels 0 and 219; the monochrome image is recontoured to lie between gray levels 220 and 255. This manipulation can be done in real time, by carefully adjusting the offsets and gains of the four cameras and using the pallete as the input look-up table. This pallete was applied to images A–D and to images E–H of Color Plate 1.

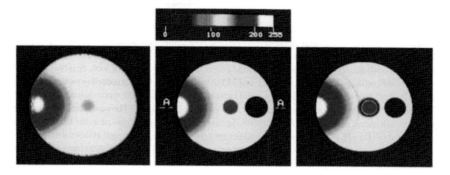

Color Plate 3 (Chapter 9) Ratio images of the test cell. A computer-generated test cell was constructed and simulated observed images were calculated and then processed (see Chapter 9). The 340/380 ratios of the "observed" images (*left*), actual in-focus data (*middle*), and processed images (*right*) are shown. A color scale is included to show how colors relate to gray-scale values. Intensity profiles across the centers of these images are shown in Chapter 9, Fig. 2. Before calculating the ratio shown in the right-hand image, the individual 340 and 380 images were processed with the no-neighbors technique using the following values for the constants: $c = 0.494$, $\alpha = 1.0$, and the 1-μm out-of-focus point spread function. The diameter of the test cell was 50 μm.

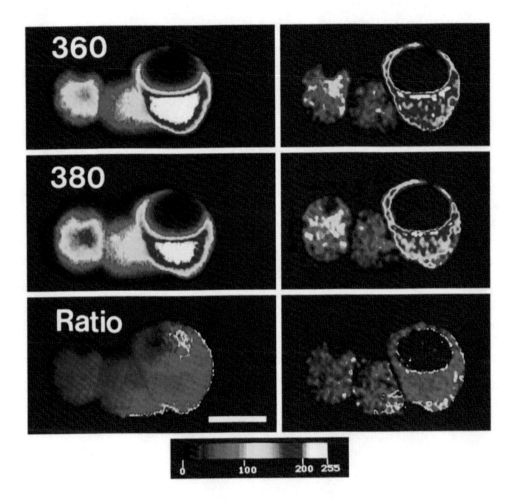

Color Plate 4   (Chapter 9) Fura-2 images of a cluster of three PC-12 cells. Undifferentiated PC-12 cells were loaded with fura-2/AM and plated onto poly-L-lysine-coated coverslips. The cells were observed with an Olympus 40 ×/1.3 NA oil immersion objective on a Nikon Diaphot inverted microscope. Images were collected using excitation light with a wavelength of 360 nm or 380 nm, using a Hamamatsu C2400-08 SIT camera. (*Left*) False-color representations of the observed images using the two excitation wavelengths and the 360/380 ratio of the two. (*Right*) The 360 and 380 images after processing with the no-neighbors technique, using empirically derived point spread functions (PSFs) and the following values for the constants: $c = 0.475$, $\alpha = 5.0$, and the 1-μm out-of-focus PSF. (*Bottom right*) The 360/380 ratio of the processed images. No masking operations have been applied to either ratio image. Bar, 10 μm. The color scale shows how colors relate to gray-scale intensities.

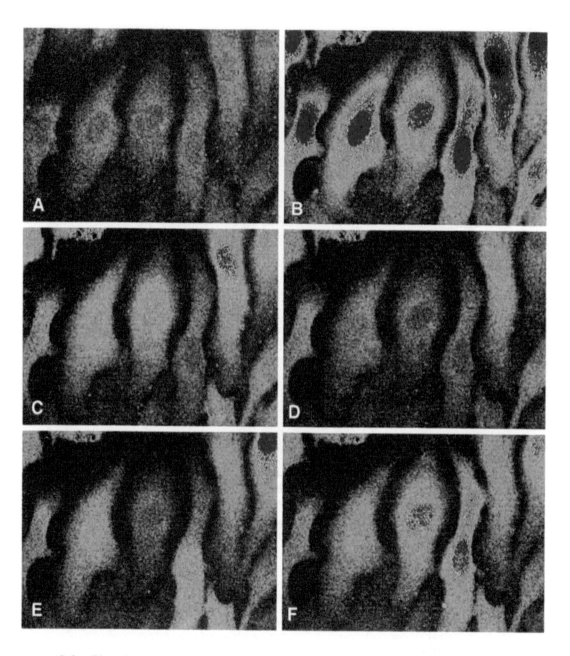

Color Plate 5 (Chapter 10) Confocal fluo-3 imaging of platelet-derived growth factor (PDGF)-induced oscillations in nuclear free Ca^{2+} ($[Ca^{2+}]_n$). Quiescent BALB/c 3T3 cells were loaded with fluo-3/AM (10 μM, 30 min). Images from an optical section ~0.7 μm thick, through the center of the nuclei, were obtained by laser scanning confocal microscopy (Bio-Rad MRC600) at 15-sec intervals. (A) Before growth factor addition, (B–F) 45, 105, 135, 180, and 195 sec after BB-PDGF (10 ng/ml) addition, respectively. On the pseudocolor scale, dark blue represents low pixel intensity (low Ca^{2+}), whereas red represents high pixel intensity (high Ca^{2+}).

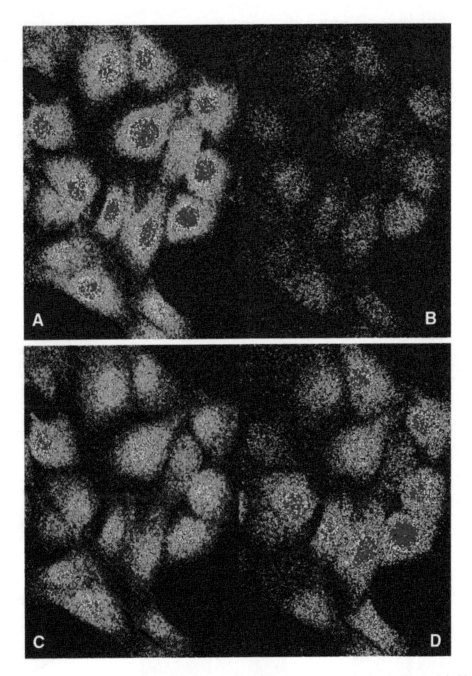

Color Plate 6   (Chapter 10) Nuclear free Ca^{2+} ($[Ca^{2+}]_n$) imaging in resting and stimulated cells coloaded with Fura-Red™/AM and fluo-3/AM. Quiescent BALB/c 3T3 cells were loaded with Fura-Red/AM ($10\,\mu M$, 45 min) and fluo-3/AM ($4\,\mu M$, 30). Images were obtained at 37°C from an optical section through the center of the nuclei by laser scanning confocal microscopy (Bio-Rad MRC600) with dual-channel detection. (A, C) Fura-Red images. (B, D) Corresponding fluo-3 images. (A, B) Resting cells before growth factor addition. (C, D) Same cells after addition of BB-PDGF (10 ng/ml). On the pseudocolor scale, dark purple/blue represents low fluorescence intensity, whereas red represents high fluorescence intensity.

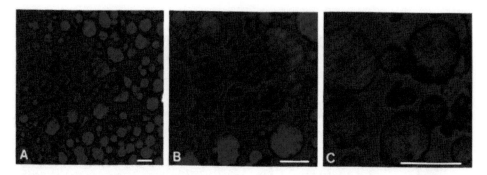

Color Plate 7 (Chapter 11) *Xenopus* oocyte stained with DiI, a hydrophobic carbocyanine membrane probe. Albino oocytes were stained with 10 μg/ml 1,1′-dioctadecyl-3,3,3′,3′-tetramethylindocarbocyanine perchlorate (DiI, Molecular Probes, Eugene, Oregon) in Barth's solution for 45 min at 22° C. The stock DiI solution was made by dissolving 2.5 mg in 1 ml ethanol, stored at −20° C. DiI is a member of a family of cyanine dyes used to investigate membrane dynamics and structure and becomes inserted into membranes or hydrophobic structures. Confocal images were acquired using the BHS filter set (long pass emissions at 515 nm) with 488-nm excitation. A 60 ×/1.4 NA Nikon objective was used at three different zooms (A, B, C). Bar, 5 μm.

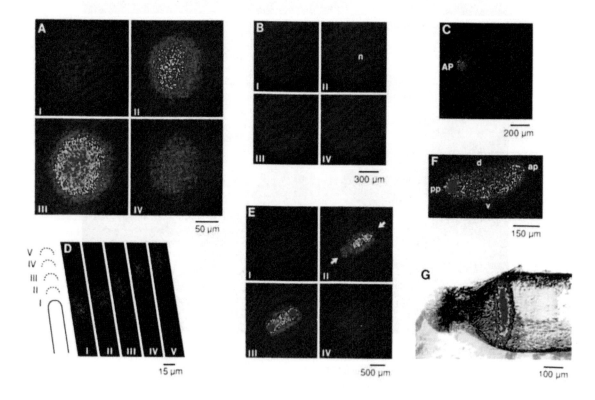

Color Plate 8 (Chapter 13) Seven examples of preliminary results when new recombinant or semi-synthetic recombinant aequorins have been used to image intracellular calcium in living cells. (A) A representative example of the calcium transient that accompanies nuclear envelope breakdown (NEBD) during the first cell cycle in an egg of the sea urchin *Lytechinus variegatus*. Each panel represents successive 3-min periods of accumulated light beginning 35 min after fertilization. Color scale: blue, 1–2 photons; turquoise, 3–4 photons; green, 5–6 photons; yellow, 7–8 photons; pink, 9–10 photons; red, 11–254 photons. (I) The resting level of calcium within the egg. (II, III) A pulse of elevated calcium accompanies NEBD, fills the whole egg, and lasts up to 6 min. (IV) The level then falls, eventually returning to a resting level similar to that indicated in I. Eggs were injected after fertilization with *h*-recombinant aequorin using the method described in Chapter 13, Section V, A, 1 (A. L. Miller, C. L. Browne, E. Karplus, R. Palazzo, and L. F. Jaffe, unpublished results). (B) An example of the calcium pulse that accompanies NEBD in the ctenophore *Beröe ovata*. We used intermittent video microscopy to track the migration of the female pronucleus until it fused with a male pronucleus. Each panel represents successive 5-min periods of accumulated light beginning in I, when it was clear that a zygote nucleus was present. A 10-min pulse of elevated calcium accompanied NEBD. However, unlike the small sea urchin egg, this calcium was restricted to the nucleus, rather than filling the whole cell. After *in vitro* fertilization, eggs were injected with *h*-recombinant aequorin using the method described in Chapter 13, Section V, A, 1 (A. L. Miller, C. Sardet, and L. F. Jaffe, unpublished results). (C) A representative example of the calcium signal that accompanies the emission of the first polar body from the egg of the leech *Theromyzon rude*. Intermittent video microscopy was used once again to track the supraequatorial ring of contraction that moves toward the animal pole (AP), ultimately resulting in the extrusion of the first polar body. The panel represents 10 min of accumulated light, beginning just as the ring of contraction reaches the pole, approximately 2 hr after the eggs were removed from the ovisacs of gravid females. The signal accompanies the appearance of the first polar body and eventually dies down to the resting level indicated over the rest of the egg. Shortly after removal from the ovisac, this egg was injected with recombinant aequorin using the method described in Chapter 13, Section V, A, 2 (A. L. Miller, Gonzalez-Plaza, and L. F. Jaffe, unpublished results). (D) An example of the elevated calcium zone at the tip of an elongating pollen tube of *Lilium longiflorium*. (I–V) Successive 3-min periods of accumulated light. The pollen tube is elongating at approximately 0.1 μm/sec during this period, suggesting that the elevated calcium zone is restricted to a thin peripheral zone right at the very tip. The pollen tube was injected with recombinant aequorin using an adaption of the method described in Chapter 13, Section V, A, 3 (B. Hepler, A. L. Miller, O. Pierson, and L. F. Jaffe, unpublished results). (E) a representative example of the calcium wave accompanying cytokinesis during the first cell division of a *Xenopus laevis* egg. Once again, photon collection was interrupted periodically to record intermittent video images of the morphological events. (I–IV) Successive 5-min periods of accumulated light. The egg is viewed from the vegetal pole. The sequence begins in I, just before the leading edges of the furrow have reached the equator and are, therefore, not in view. This view represents the resting level of calcium in the vegetal hemisphere prior to the appearance of the furrow. (II) Appearance of an elevated zone of calcium at the vegetal pole, which precedes the arrival there of the leading edges of the furrow. The arrows indicate the appearance of the leading edges as they pass over the equator. (III) In this particular case, the leading edge at the top right of the panel appears first, followed later by the other at the bottom left. (IV) On completion of furrowing, the level of calcium in the vegetal hemisphere of the egg falls back to a resting level similar to that illustrated in I. This egg was injected with recombinant aequorin using the method described in Chapter 13, Section V, A, 2 (A. L. Miller and L. F. Jaffe, unpublished results). (F) Representative example of elevated calcium zones of a *Drosophila* embryo developing through stages 3 to 4 (pp, posterior pole; ap, anterior pole; d, dorsal side; v, ventral side). An obvious zone of high calcium exists at the posterior pole, as do lesser zones at the anterior pole and in the mid-vegetal region. The panel represents 40 min of accumulated light, beginning about 90 min after the eggs were collected. This embryo was injected with recombinant aequorin using an adaptation of the method described in Chapter 13, Section V, A, 2 (A. L. Miller, D. Bearer, and L. F. Jaffe, unpublished results). (G) A representative example of the restricted spread of elevated calcium following transection of the giant axon of the squid *Loligo pealei*. This axon was loaded with recombinant aequorin prior to transection using the method described in Chapter 13, Section V, A, 2. This panel represents 1 min of accumulated aequorin-generated light collected 15 min after transection, and superimposed on a bright field video image taken just before. The medium bathing the axon contains $10 \, mM \, Ca^{2+}$, whereas the axoplasm of the interior contains only 50–$100 \, nM \, Ca^{2+}$. Aequorin has been discharged in the region from the pseudocolor band to the cut end. As a result, no signal is registered from this region. The axon does, however, possess an as yet undescribed mechanism that restricts the deleterious spread of elevated calcium down the axon from the cut end, thus maintaining the calcium at the resting level in the rest of the axon (Krause, Fishman, Shipley, A. L. Miller, and Bittner, unpublished results).

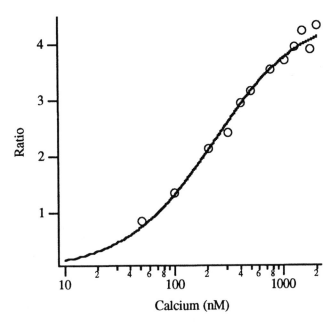

Fig. 7 Calibration of indo-1 in a mock intracellular solution. The affinity of indo-1 (1 μM of the potassium salt; Molecular Probes) was measured in a mock intracellular solution having 126 mM KCl, 10 mM NaCl, 20 mM HEPES, 1 mM MgCl$_2$; pH 7.1 at 22°C. The free Ca^{2+} concentrations was varied between 10 and 2000 nM by altering the ratio of CaK$_2$EGTA (Puriss grade, Fluka) and H$_2$K$_2$EGTA (Puriss grade, Fluka). The free Ca^{2+} concentrations were calculated from published affinity constants (Fabiato and Fabiato, 1979).

The 6.5× UV/vis eyepiece lens was adapted to the Bio-Rad MRC600 confocal scanning box (Fluharty *et al.*, 1991), permitting full-field confocal imaging with UV excitation. The 351-nm excitation from an argon laser reached the specimen with an energy less than 0.1 mW. A dichroic mirror placed in the second filter block (Fig. 4q) reflected emission wavelengths less than 435 nm and passed emissions of longer wavelength. A 405± 17-nm band pass filter selected the indo-1 emissions that increase with increasing Ca^{2+} concentration prior to detection by a PMT (channel 2). A 453 ±2-nm band pass filter selected the isostilbic emission wavelengths of indo-1, which are independent of Ca^{2+} concentration, before they were detected at a second PMT (channel 1). The data were background subtracted and the ratio of fluorescence (405 nm/453 nm) is given on the ordinate at free Ca^{2+} concentrations from 50 nM to 2000. The ratio was fit by the Hill equation (ratio $= r_{max}/[1 + (K_d/[Ca^{2+}]^h)]$ where r_{max} is the maximum ratio, K_d is the calcium dissociation constant for indo-1, and h is the Hill coefficient. The measurements show that indo-1 has a single binding site ($h = 1.08 \pm 0.8$) with a dissociation constant of 227 $\pm$ 17 nM for Ca^{2+} under these conditions.

the turnover of phosphatidylinositol and the release of inositol 1,4,5-trisphosphate (IP$_3$) (Dascal, 1987). This section discusses the confocal methods we used to investigate phosphatidylinositol (PI)-induced Ca^{2+} transients in *Xenopus* oocytes.

Adult female albino frogs were purchased from Xenopus I (Ann Arbor, Michigan), Carolina Biological Supply (Durham, North Carolina), and Nasco

(Fort Atkinson, Wisconsin); their oocytes lack pigment granules that block fluorescent signals. The animals were housed in plexiglass aquaria (12 × 12 × 24 in) at 19°C and fed toad brittle (Nasco) twice a week. Approximately 2 hr after feeding, water was exchanged from the tank—a 55-gallon aerated tank treated with 25 ml each NovAqua and AmQuel from Novalek (Hayward, California). NovAqua and AmQuel remove ammonia, chloramines, zinc, copper, and chlorine which are deleterious to frogs.

Before surgically harvesting oocytes, we anesthetized the frogs in an ice bath using the standard protocol for amphibians (Dascal, 1987). After testing for a response to noxious stimuli, we made a 1-cm paramedian incision in the inferior abdominal wall with a #10 carbon steel surgical blade (Bark-Parker, Franklin Lakes, New Jersey) and surgically removed a cluster of ~200 oocytes. We closed the abdominal wall and the peritoneum using three 6-0 Ethilon monofilament nylon sutures (Ethicon, Somerville, New Jersey). The skin incision was closed similarly. The animals were warmed slowly to room temperature (22°C) over ~2 hr and were returned to their aquaria the following day. Allowing 2–3 wk for the wounds to heal, each donor frog could be used repeatedly.

Oocytes were kept at 19°C in L-15 supplemented medium from Gibco (Grand Island, New York). Because of their large size, Dumont stage V–VI oocytes were distinguished easily from less-developed oocytes (Dumont, 1972). We manually removed the thecal and vascular epithelial layers surrounding oocytes (Dascal, 1987). Briefly, #55 Dumont forceps from Fine Science Tools (Belmont, California) were bevelled to ~30°; one forceps stabilized a group of ~10 oocytes while the second forceps removed the epithelium from oocytes in the group. In this manner, ~150 oocytes could be "defolliculated" manually in ~1 hr. The enveloping layer of small follicle cells is not removed, but collagenase treatment may be used to free the oocytes from their follicle cells completely (Dascal, 1987). The oocyte has few G protein-linked receptors (Kusano et al., 1977). Conversely, the surrounding follicular cells contain adenosine (Lotan et al., 1982), β-adrenergic (Kusano et al., 1982), angiotensin II (Fluharty et al., 1991), chorionic gonadotropin, follicle-stimulating hormone (FSH), and luteinizing hormone (LH) (Woodward and Miledi, 1987) receptors.

For some experiments, we injected the cRNA for muscarinic receptors into oocytes and studied receptor-activated Ca^{2+} transients with confocal imaging and a two-electrode voltage clamp. First, we subcloned the m3 acetylcholine receptor (AChR) cDNA into pBluescript SK^+, which allowed us either to transcribe cRNA in vitro with T7 polymerase or to use a vaccinia expression system. We digested 30 μl m3 construct (~30 μg) from E. Peralta (Harvard University) and 10 μl pBluescript II SK^+ (~10 μg) with EcoRI and BamHI. The m3 AChR cDNA insert (1932 bp) and the cut pBluescript SK plasmid (2943 bp) were purified. Then, we used bacteriophage T4 DNA ligase to insert the m3 cDNA clone directionally into the Bluescript vector XL1 blue Escherichia coli were transformed; then a recombinant plasmid was isolated and amplified using a large-scale plasmid preparation. We transcribed m3AChR

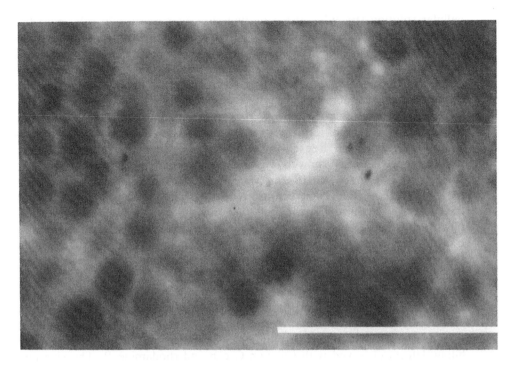

Fig. 8 Confocal image of a *Xenopus* oocyte stained with rhodamine B. Manually defolliculated albino oocytes were stained for 20 min at 22°C with 10 µg/ml of the hexyl ester of rhodamine B stain. Rhodamine B stains in the manner of the "short chain" carbocyanine $DiOC_6$ that was developed by Chen and co-workers to visualize the endoplasmic reticulum (Lee and Chen, 1988). The confocal image was taken with the BHS filter set (long pass at 515 nm) using a 60×/1.4 NA Nikon oil immersion objective. Images taken from an optical section slightly deeper in the oocyte showed larger filling defects (up to 10 µm). Bar, 5 µm.

cRNA *in vitro* using T7 polymerase and stored the RNA in TE buffer (1 µg/µl) at −80°C.

We pulled microelectrodes from 3.5-inch filamented glass capillaries using a model 700c vertical pipette puller from David Kopf Instruments (Tujanga, California). Under 100× magnification, the electrodes were broken manually to ~12 µm (inside diameter) to create injection pipettes. We baked cRNA injection pipettes at 450°C for 4 hr to destroy contaminating RNases and always wore gloves when handling cRNA solutions and pipettes. A Nanoject™ 47-nl automatic injector from Drummond Scientific (Broomall, Pennsylvania) was used for all intracellular injections. To minimize Ca^{2+} entry into oocytes, all injections were performed in a zero-Ca^{2+} solution containing 96mM NaCl, 5 mM $MgCl_2$; 2 mM KCl; 5 mM HEPES; 0.1 mM EGTA.

Ca^{2+} waves in voltage-clamped *Xenopus* oocytes are correlated temporally with membrane current oscillations (Lechleiter *et al.*, 1991a). In normal extracellular solution (Barth's), these oscillations reversed at -20 mV, near the Cl^- reversal potential in the oocyte (Barish, 1983). The current–voltage relationship was similar to the native Ca^{2+}-activated Cl^- current ($I_{Cl,Ca}$) described previously (Miledi, 1982; Barish, 1983; Miledi and Parker, 1984; Takahashi *et al.*, 1987). This result suggested that the fluorescent signals originate quite near the plasma membrane.

To investigate our ability to image the large oocyte, we labeled oocytes with rhodamine B, a lipophilic membrane probe thought to enter the ER preferentially (Terasaki and Reese, 1992). A representative oocyte stained with rhodamine B (Fig. 8) showed prominent spherical filling defects with diameters ranging from 1 to 10 μm. When observed with higher magnification ($60\times$ 1.4 NA objective), oocytes injected with 25 μM 10 kDa Calcium Green–dextran also had these filling defects (not shown), suggesting that large structures or organelles displace the ER and the cytoplasm. This result is in contrast to that seen in other smaller cells such as sea urchin eggs, in which rhodamine B stains in a reticular pattern throughout the cytosol (Terasaki and Reese, 1992). Next we stained oocytes with 1,1-dioctaecyl-3,3,3′,3′-tetramethylindocarbocyanine perchlorate (DiI) (Honig and Hume, 1986), a lipid-soluble label, in an attempt to characterize these filling defects further. After a 35–45 min incubation at room temperature, DiI stained spherical structures with diameters of 1–10 μm (Color Plate 7), consistent with the size of the filling defects observed with Calcium Green and rhodamine B. The size and lipophilic nature of these structures suggests that they are the lipophilic platelet granules previously preserved with electron microscopy (Dumont, 1972). More importantly, these opaque structures block fluorescent signals. To quantitate this effect, we measured the full-field Calcium Green signal with the $100\times$ (1.3 NA) objective in different axial (z) positions ($\sim\mu$m depth of confocal section with the settings used). The average Calcium Green signal was a symmetrical function of the axial position (focus) of the objective (Fig. 9), having a width of ~20 μm. Therefore, the fluorescence signals observed with confocal microscopy originate from within 20 μm of the plasma membrane; we cannot image deeper into these rather opaque cells.

One conclusion from our work was that $I_{Cl,Ca}$ oscillations in *Xenopus* oocytes result from waves of elevated intracellular Ca^{2+} propagating in the periphery of these cells. We observed repetitive (regenerative) Ca^{2+} waves following submaximal stimulation (0.1–1 μM ACh) of m3 AChR-expressing oocytes, but not with saturating stimuli (50 μM ACh) (Leichleiter *et al.*, 1991a, b; Lechleiter and Clapham, 1992). These data suggest that the waves occur within a certain concentration range of IP$_3$, assuming that saturating stimulation of PI-coupled receptors generates more IP$_3$ than submaximal stimulation.

We wished to clarify the response to saturating ACh in m3 AChR-expressing *Xenopus* oocytes. Specifically, we wanted to know whether the response was a single propagating wave, initially described with a velocity of 10–30 μm/sec (Lechleiter *et al.*, 1991a). One might speculate that this appar-

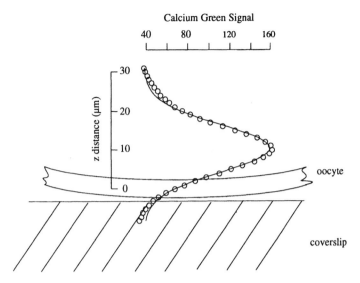

Fig. 9 An axial profile demonstrates that fluorescent signals originate from the most peripheral 30 μm of *Xenopus* oocytes. The full-field Calcium Green™ fluorescence as a function of the axial (z) position of an Olympus 100× 1.3 NA objective. The confocal aperture was set to 1 mm to minimize the thickness of the optical section. A confocal image was taken every 1 μm over a 40-μm range of focus. The mean fluorescence signal is plotted as a function of the axial position of the objective and fitted to a Gaussian curve where

$$\text{signal} = k_0 + k_1 \exp -[(z - k_2)/k_3]^2$$

with $k_0 = 27.8 \pm 1.2$; $k_1 = 125 \pm 1$; $k_2 = 21.7 \pm 0.1$; $k_3 = 8.6 \pm 0.2$. The cell margin and coverslip are shown for reference (not to scale).

ent m3 AChR-induced wave was caused by the limited access of agonist in the restricted space between the cell and the cover slip; recall that we use an inverted confocal system. Therefore, we attached a polyethylene woven mesh (750-μm opening) to the cover slip to elevate oocytes off the cover slip slightly. Indeed, with the mesh in place, the kinetics of m3-induced intracellular Ca^{2+} transients were accelerated ~6-fold (Fig. 10; Table I). The entire confocal field released Ca^{2+} within several seconds ($n = 5$). Clearly, saturating stimulation of m3 AChR-expressing oocytes causes a rapid generalized increase in Ca^{2+}, not a solitary propagating wave. Propagating waves that traverse the confocal field many times clearly were observed under the conditions previously described (Leichleiter *et al.*, 1991b; Lechleiter and Clapham, 1992).

We observed focal release sites just after the application of agonist (Fig. 10) with the mesh in place, suggesting that the cytosol localized Ca^{2+} stores with differing sensitivity to IP_3 (Parker and Ivorra, 1991).

A

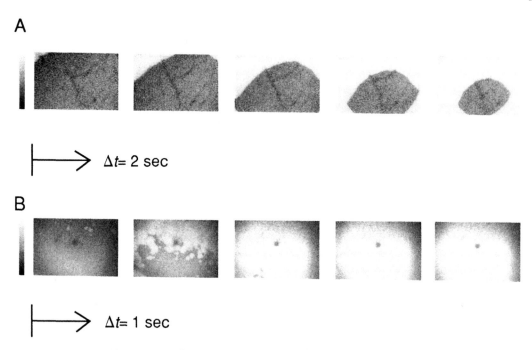

Δt = 2 sec

B

Δt = 1 sec

Fig. 10 Pattern of m3AChR-induced Ca^{2+} transients influenced by mesh. Saturating ACh (50 μM) was applied to Calcium Green™-loaded (12 μM) oocytes expressing human m3ACh receptors. Confocal images were taken at 1 Hz. (A) An oocyte was observed without a mesh, as before. The images are displayed with time increasing to the right (2-sec intervals), showing what appears to be a propagating wave of elevated intracellular Ca^{2+}. (B) A mesh was used to elevate oocytes slightly ($\sim\mu$m) off the bottom of the dish. Images are shown in 1-sec intervals and demonstrate that saturating stimulation of m3AChRs causes a rapid generalized increase in Ca^{2+}. Localized areas of Ca^{2+} release (hot spots) were observed only with the mesh in place, and probably indicate a heterogeneous distribution in the sensitivity of Ca^{2+} stores to InsP$_3$ (Parker and Ivorra, 1991). Note the difference in time scales. Propagating waves do occur under the conditions, as previously described (Lechleiter *et al.*, 1991b; Lechleiter and Clapham, 1992).

V. Mobile Buffers

Ca^{2+} indicators may change the mobility of intracellular Ca^{2+}, thereby altering the transients they were designed to follow (Meyer, 1991). In the extreme sense, one could argue that the addition of mobile buffers artifactually might create Ca^{2+} waves. To address this issue, we studied IP$_3$S$_3$-induced Ca^{2+} waves using indicators with a 70-fold range of molecular weights. The steady-state Ca^{2+} wave velocity was independent of affinity and mobility of the indicator (Table I). Further, the wavelength of excitation between 351 and 488 nm had no significant effect on the wave velocity. When used in low micromolar concentrations, these indicators faithfully report Ca^{2+} transients without significantly buffering Ca^{2+}. Therefore, wave velocity and frequency are not affected by the addition of low concentrations of mobile Ca^{2+} buffers because the same phenomena are observed using dextran-immobilized indicators.

Table I
Steady–State Ca^{2+} Wave Velocity Is Independent of Indicator Affinity and Molecular Mass

Indicator[a]	Size (kDa)	K_d at 22°C (nM)	Concentration (μM)	Wave velocity (μm/s ± SEM)	n	Oocytes
Fluo-3[b]	0.85	316	50	25 ± 2	214	—
Calcium Green™	1.1	189	13	22.2 ± 0.7	97	10
Calcium Green™ dextran	10	251	20	22.7 ± 1.3	78	5
Calcium Green™ dextran	70	356	25	22.9 ± 1.4	45	5
Indo-1	0.84	210	50	23.9 ± 1.0	11	1

[a] The affinity of indicators was measured by Molecular Probes in 100 mM KCl, buffered to pH 7.2 with 5 mM MOPS. The final concentrations were estimated assuming a 1-μl volume of distribution. Wave velocity was the steady-state velocity observed after injection of 1–5 μM IP_3S_3, in the absence of extracellular Ca^{2+}.

[b] The fluo-3 data are from the literature (DeLisle and Welsh, 1992). The number of oocytes and the extracellular Ca^{2+} concentration were not specified.

We observed Ca^{2+} waves, using indo-1 as the indicator, in oocytes stimulated with 5% horse serum. Preliminary data suggest that the resting Ca^{2+} is ~100 nM and the wave amplitude is 370 ± 80 nM (mean ± SD, n = 11) (Fig. 11). Miledi and co-workers have identified a 67-kDa protein in human serum that activates oscillatory $I_{Cl,Ca}$ currents in *Xenopus* oocytes through the PI cascade (Tigy *et al.*, 1991). Additional data suggests that a lipid bound to serum albumin (lysophosphatic acid) may have triggered these Ca^{2+} waves (Tigyi and Miledi, 1992).

VI. Conclusions

Adaptations of the Bio-Rad MRC600 allow us to utilize fluorescent indicators simultaneously with excitation from near UV through visible wavelengths (Bliton *et al.*, 1993). We have used Calcium Green with 488-nm excitation to record the Ca^{2+} transients following the photolytic cleavage of caged IP_3, caged GTPγS, and caged calcium with ~360-nm excitation in a peripheral confocal section of *Xenopus* oocytes (Lechleiter and Clapham, 1992). Based on our experiments (Bliton *et al.*, 1993; Lechleiter *et al.*, 1991a; Lee and Chen, 1988; Girard *et al.*, 1992; Lechleiter and Clapham, 1992), the current model for calcium waves in *Xenopus* oocytes is founded on a single pool of endoplasmic reticular IP_3 receptors that release calcium on stimulation by IP_3. Calcium acts as a coagonist to increase the sensitivity of the receptor to IP_3 at low concentrations of cytoplasmic calcium, but decreases the sensitivity of the receptor to IP_3 at higher (physiological) concentrations of calcium. The regenerative wave behavior relies on diffusion-linked excitable stores; a full model incorporates the complex behavior of the IP_3 receptor, the Ca^{2+}-ATPase pumps, and the calcium store using a set of partial differential equations. We currently are combining this model for regenerative wave behavior with new data on calcium entry from the extracellular space.

In this chapter, we have presented some of the details of our imaging experi-

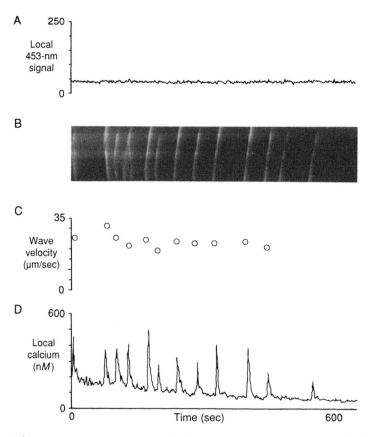

Fig. 11 Ca^{2+} waves observed using indo-1. Oocytes were injected with ~50 μM indo-1 and observed with confocal microscopy as before. When the extracellular solution was supplemented with 5% horse serum, Ca^{2+} waves were observed (3 of 5 oocytes from one donor frog). (A) The 453-nm (isostilbic) emission as a function of time demonstrates a stable background, no photobleaching. (B) Volume projection of 600 images with time increasing to the right. (C) Wave velocity. (D) Local calcium concentration (nM) within a 100 μm^2 box in the center of the image plane.

ments that cannot be given in the abbreviated "Methods" section of primary publications. We have shown the calibrations necessary for quantitative imaging, including a procedure to estimate the thickness of confocal sections. Preliminary data indicate that the amplitude of the calcium waves exceeds 300 nM with a base resting value between 50 and 100 nM. Experiments shown here demonstrate that propagating, regenerative waves in *Xenopus* oocytes are not altered by the addition of mobile calcium indicators at low concentrations. Our refinements in UV/visible confocal microscopy, combined with the increasing number of caged second messengers and specific fluorescent labels, should allow for new experimental methodologies in elucidating the mechanisms of intracellular signaling.

References

Abbe, E. (1984). Note on the proper definition of the amplifying power of a lens or a lens-system. *J. R. Microsc. Soc.* **4**, 348–351.

Barish, M. E. (1983). A transient calcium-dependent chloride current in the immature *Xenopus* oocyte. *J. Physiol.* **342**, 309–325.

Blinks, J. R. (1986). Intracellular [Ca^{2+}] measurements. *In* "The Heart and Cardiovascular System" (H. A. Fozzard, E. Haber, R. B. Jennings, A. M. Katz, and H. E. Morgan, eds.). pp. 671–701. New York: Raven Press.

Blinks, J. R., and Moore, E. D. W. (1985). Practical aspects of the use of photoproteins as biological calcium indicators. *In* "Optical Methods in Cell Physiology" (P. deWeer and B. Salszman, eds.) pp. 229–238. New York: Wiley.

Bliton, C., Lechleiter, J., and Clapham, D. E. (1993). Optical modifications enabling simultaneous confocal imaging with dyes excited by ultraviolet- and visible-wavelength light. *J. Microsc.* **169**, 15–26.

Cobbold, P. H., and Rink, T. J. (1987). Fluorescence and bioluminescence measurement of cytoplasmic free calcium. *Biochem. J.* **248**, 313–328.

Cobbold, P. H., Cheek, T. R., Cuthbertson, K. S., and Burgoyne, R. D. (1987). Calcium transients in single adrenal chrommafin cells detected with aequorin. *FEBS Lett.* **211**, 44–48.

Cogswell, C. J., and Sheppard, C. J. R. (1990). *In* "Confocal Microscopy (T. Wilson, ed.). Academic Press, London.

Dascal, N. (1987). The use of *Xenopus* oocytes for the study of ion channels. *CRC Crit. Rev. Biochem.* **22**, 317–387.

DeLisle, S., and Welsh, M. J. (1992). Inositol trisphosphate is required for the propagation of calcium waves in *Xenopus* oocytes. *J. Biol. Chem.* **267**, 7963–7966.

di Tolardo, F. G. (1955). Resolving power and information. *J. Opt. Soc. Am.* **45**, 497–501.

Dumont, J. N. (1972). Oogenesis in *Xenopus laevis*. Stages of oocyte development in laboratory maintained animals. *J. Morphol.* **136**, 153–164.

Eberhard, M., and Erne, P. (1989). Kinetics of calcium binding to fluo-3 determined by stopped-flow fluorescence. *Biochem. Biophys. Res. Commun.* **163**, 309–314.

Egger, M. D. (1989). The development of confocal microscopy. *Trends Neurosci.* **12**, 11.

Fabiato, A., and Fabiato, F. J. (1979). Calculator programs for computing the composition of the solutions containing multiple metals and ligands used for experiments in skinned muscle cells. *J. Physiol. (Paris)* **74**, 463–505.

Fluharty, S. J., Reagan, L. P., and White, M. M. (1991). Endogenous and expressed angiotensin II receptors on *Xenopus* oocytes. *J. Neurochem.* **56**, 1307–1311.

Girard, S., Luckhoff, A., Lechleiter, J., Sneyd, J., and Clapham, D. (1992). Two-dimensional model of calcium waves reproduces the patterns observed in *Xenopus* oocytes. *Biophys. J.* **61**, 509–517.

Grynkiewicz, G., Poenie, M., and Tsien, R. Y. (1985). A new generation of Ca^{2+} indicators with greatly improved fluorescence properties. *J. Biol. Chem.* **260**, 3440–3450.

Gurdon, J. B., Land, C. D., Woodland, H. R., and Marbaix, G. (1971). Use of frog eggs and oocytes for the study of messenger RNA and its translation in living cells. *Nature (London)* **233**, 177–182.

Hecht, E. (1987). "Optics," 2d Ed. Reading, Massachusetts: Addison-Wesley.

Honig, M. G., and Hume, R. I. (1986). Fluorescent carbocyanine dyes allow living neurons of identified origin to be studied in long-term cultures. *J. Cell Biol.* **103**, 171–187.

Inoue, S. (1990). Foundations of confocal scanned imaging in light microscopy. *In* "Handbook of Biological Confocal Microscopy (J. B. Pawley, ed.) pp. 1–14. New York: Plenum Press.

Inoue, S., Noguchi, M., Sakaki, Y., Miyata, T., Iwanaga, S., Miyata, T., and Tsuji, F. I. (1985). Cloning and sequence analysis of cDNA for the luminescent protein aequorin. *Proc. Natl. Acad. Sci. U.S.A.* **82**, 3154–3158.

Kusano, K., Miledi, R., and Stinnakre, J. (1977). Acetylcholine receptors in the oocyte membrane. *Nature (London)* **270**, 739–741.

Kusano, K., Miledi, R., and Stinnakre, J. (1982). Cholinergic and catecholaminergic receptors in the *Xenopus* oocyte membrane. *J. Physiol. (London)* **328,** 143–170.

Lechleiter, J. D., and Clapham, D. E. (1992). Molecular mechanisms of intracellular calcium excitability in *X. laevis* oocytes. *Cell* **69,** 283–294.

Lechleiter, J., Girard, S., Clapham, D., and Peralta, E. (1991a). Subcellular patterns of calcium release determined by G protein-specific residues of muscarinic receptors. *Nature (London)* **350,** 505–508.

Lechleiter, J., Girard, S., Peralta, E., and Clapham, D. (1991b). Spiral calcium wave propagation and annihilation in *Xenopus laevis* oocytes. *Science* **252,** 123–126.

Lee, C., and Chen, L. B. (1988). Dynamic behavior of endoplasmic reticulum in living cells. *Cell* **54,** 37–46.

Lotan, I., Dascal, N., Cohen, S., and Lass, Y. (1982). Adenosine-induced slow ionic currents in the *Xenopus* oocyte. *Nature (London)* **298,** 572–574.

Meyer, T. (1991). Cell signalling by second messenger waves. *Cell* **64,** 675–678.

Miledi, R. (1982). A calcium-dependent transient outward current in *Xenopus laevis* oocytes. *Proc. R. Soc. London B.* **215,** 491–497.

Miledi, R., and Parker, I. (1984). Chloride current induced by injection of calcium into *Xenopus* oocytes. *J. Physiol.* **357,** 173–183.

Minsky, M. (1988). Memoir on inventing the confocal scanning microscope. *Scanning* **10,** 128–138.

Oehme, M., Kessler, M., and Simon, W. (1976). Neutral carrier Ca^{2+} microelectrode. *Chimia* **30,** 204–206.

Parker, I., and Ivorra, I. (1991). Localized all-or-none calcium liberation by inositol trisphosphate. *Science* **250,** 977–979.

Ridgeway, E. B., and Ashley, C. C. (1967). Calcium transients in single muscle fibers. *Biochem. Biophys. Res. Commun.* **29,** 229–234.

Rizzuto, R., Simpson, A. W., Brini, M., and Pozzan, T. (1992). Rapid changes of mitochondrial Ca^{2+} revealed by specifically targeted recombinant aequorin. *Nature (London)* **358,** 325–327.

Sheppard, C. J. R., and Choudhury, A. (1977). Image formation in the scanning microscopy. *Optica* **24,** 1051.

Smith, G. A., Hesketh, R. T., Metcalfe, J. C., Feeney, J., and Morris, P. G. (1983). Intracellular calcium measurements by ^{19}F NMR of fluorine-labeled chelators. *Proc. Natl. Acad. Sci. U.S.A.* **80,** 7178–7182.

Soreq, H. (1985). The biosynthesis of biologically active proteins in mRNA microinjected *Xenopus* oocytes. *CRC Crit. Rev. Biochem.* **18,** 199–238.

Sumikawa, K., Houghton, M., Emtage, J. S., Richards, B. M., and Barnard, E. A. (1981). Active multisubunit ACh receptor assembled by translation of heterologous mRNA in *Xenopus* oocytes. *Nature (London)* **292,** 862–864.

Takahashi, T., Neher, E., and Sakmann, B. (1987). Rat brain serotonin receptors in *Xenopus* oocytes are coupled by intracellular calcium to endogenous channels. *Proc. Natl. Acad. Sci. U.S.A.* **84,** 5063–5067.

Terasaki, M., and Reese, T. S. (1992). Characterization of endoplasmic reticulum by colocalization of BiP and dicarbocyanine dyes. *J. Cell Sci.* **101,** 315–322.

Tigy, G., Henschen, A., and Miledi, R. (1991). A factor that activates oscillatory chloride currents in *Xenopus* oocytes copurifies with a subfraction of serum albumin. *J. Biol. Chem.* **266,** 20602–20609.

Tigyi, G., and Miledi, R. (1992). Lysophosphatidates bound to serum album activate membrane currents in *Xenopus* oocytes and neurite retraction in PC12 pheochromocytoma cells. *J. Biol. Chem.* **267,** 21360–21367.

Tsien, R. Y. (1980). New calcium indicators and buffers with high selectivity against magnesium and protons: Design, synthesis and properties of prototype structures. *Biochemistry* **19,** 2396–2404.

Woodward, R. M., and Miledi, R. (1987). Hormonal activation of ionic currents in follicle-enclosed *Xenopus* oocytes. *Proc. Natl. Acad. Sci. U.S.A.* **84,** 4135–4139.

PART IV

Use of Aequorin for $[Ca^{2+}]_i$ Imaging

CHAPTER 12

Inexpensive Techniques for Measuring [Ca^{2+}]$_i$ Changes Using a Photomultiplier Tube

Kenneth R. Robinson, Thomas J. Keating, and R. John Cork

Department of Biological Sciences
Purdue University
West Lafayette, Indiana 47907

I. Introduction

The measurement of Ca^{2+} in living cells has been revolutionized by two developments: the invention of calcium-sensitive fluorescent probes and the availability of powerful, low-cost computers that can be used for video image analysis. These developments have made it possible to create two- (and even three-) dimensional maps of the Ca^{2+} concentration in cells, so spatial gradients and temporal changes can be visualized. The biological literature and the covers of scientific journals are replete with the beautiful and often spectacular false-color pictures of Ca^{2+} distributions and waves. Although the cost of the equipment has fallen, a calcium-imaging system is still an expensive proposition for an individual investigator. The least expensive low light level video camera and the appropriate image-processing hardware and software cannot be assembled for less than $25,000; complete systems from commercial suppliers cost much more. A second limitation to imaging Ca^{2+} distribution is the time required for such experiments. At video rates, collecting one image requires ~30 msec and averaging a number of frames is usually necessary. If a ratiometric method is used, images must be collected at two wavelengths. The practical result is that collecting complete images more rapidly than 1/sec is difficult, placing a significant constraint on the visualization of calcium transients that may be associated with rapid cellular events.

An investigator wishing to measure Ca^{2+} but lacking the appropriate equipment will need to consider at least two aspects of the problem before deciding how to proceed. The first is cost; as pointed out earlier, the capital outlay for video image processing is substantial. In many cases, a map of the Ca^{2+} distribution in the cell may not be necessary; simply knowing whether changes in Ca^{2+} occur in response to a signal or in conjunction with a cellular event may be sufficient. Under such circumstances, a lower cost alternative to image processing may be adequate and may be possible for investigators with limited budgets. The data from this simpler arrangement can be used to decide whether the purchase of imaging equipment is justified. The situation may arise, as it has in this laboratory, in which an imaging setup exists but is used heavily and is not available for all the experiments for which Ca^{2+} measurements are required. A low-cost second system for Ca^{2+} measurements is quite useful. The second consideration is the time course of the expected Ca^{2+} changes. If the changes are likely to occur on a time scale of much less than 1 sec, imaging systems may be inadequate and alternatives should be considered.

In the remainder of this chapter, we describe a photometric system that is based on a photomultiplier tube. This system has more than adequate sensitivity for fluorescence measurements, a response time in the submillisecond range, and a cost of less than $1,500, not including, of course, the cost of a fluorescence microscope. We assume that the investigator has access to a machinist who can make the required adaptor. We also describe the use of photon counting for aequorin measurements of Ca^{2+}, although the cost of this system is substantially higher.

II. Photomultipliers

Heinrich Hertz first observed in 1887 that charged particles, later shown to be electrons, are emitted from the surfaces of metals when they are irradiated with light. Einstein subsequently extended Planck's quantum theory to explain this phenomenon, known as the photoelectric effect, by proposing that light is quantized into packets, each carrying energy of $h\nu = hc/\lambda$, where h is Planck's constant (6.63×10^{-34} J · s), ν is the frequency of the light, c is the speed of light, and λ is the wavelength. This energy is imparted to a single electron in the metal surface and, if the energy is sufficient to overcome the binding energy, the electron is emitted with a kinetic energy equal to the difference between hν and the binding energy.

The photoelectric effect is the underlying phenomenon of the photomultiplier tube (PMT). (Much of the information in the following discussion was extracted from material provided by two of the major manufacturers of PMTs: Thorn EMI Electron Tubes Ltd. and the Hamamatsu Corporation. Addresses and phone numbers of these companies and other suppliers are presented in the Appendix.) When photons strike the active surface of the PMT, the photocathode, electrons are released into the tube. These primary electrons are accelerated through an electrostatic field and are focused onto a small area of a metal oxide plate, called a dynode, that is maintained at a positive potential with respect to the photocathode. Since they have gained energy in the electrostatic field that is maintained between the photocathode and the first dynode, the primary photoelectrons release a shower of secondary electrons that are accelerated toward the second dynode. This event is the "multiplier" step of a photomultiplier; the total current multiplication depends on the number of dynodes and the accelerating voltage. Typical PMTs have ~10 dynodes, have a total anode-to-cathode voltage of 1 kV, and produce a current gain of 10^6. The anode collects the electrons that are released by the last dynode, completing the circuit for the amplified photocurrent. Two fundamentally different ways of measuring the amplified current are available. One is to measure and record the photocurrent directly as an analog signal and the second is to resolve the current that results from a single photon into a pulse and to count the pulses. These methods will be discussed later.

Modern PMTs use semiconductors as their photocathode material. Modification of the energy band structures has resulted in reduced electron affinity, which permits the escape of excited electrons and improves quantum efficiency. A common photocathode material is K_2CsSb, called the bialkali photocathode. This material is maximally sensitive at about 400 nm, at which its quantum efficiency is about 30%, that is, the number of emitted electrons will be 30% of the number of impinging photons. This sensitivity also can be expressed as the number of amperes of current per watt of light; 30% quantum efficiency corresponds to about 90 mA/W. Note that this value is the cathode radiant sensitivity and is the current produced before any multiplication. The anode radiant sensitivity is the cathode radiant sensitivity times the current gain. The peak sensitivity of bialkali photocathodes is at ~400 nm; however, the sensitivity of this material falls off dramatically at longer wavelengths. At 600 nm, bialkali photocathodes have quantum efficiencies of only about 2.5%, and are useless at longer wavelengths. Other photocathode materials have reasonable sensitivity throughout the visible spectrum, although the required compromise is higher noise and greater cost. The two widely used designs of

PMTs are the side-on configuration and the head-on configuration. In a side-on tube, light enters from the side of the glass envelope whereas in the end-on configuration, light enters the end of the cylindrical glass envelope.

In the analog mode, the amplified photocurrent is converted to a voltage and is recorded. The current-to-voltage amplifiers used with PMTs often have feedback resistors of 1 MΩ, producing an output of 1 V per 1 μA of photocurrent. All PMTs produce a certain amount of current in the absence of light (the dark current), which shows up as an output voltage. Depending on the particular PMT, the anode dark current may be between 0.03 and 10 nA. If the conversion gain of the amplifier is 1 MΩ, a dark current of 0.03 nA will produce an output voltage of 30 μV. This output voltage can be easily offset; however, the fluctuation in the output voltage—the noise—will remain. This noise is the ultimate limitation on the lower level of light detection, although in actual practice other sources of noise are likely to dominate. The noise of a particular PMT is expressed often as an equivalent noise input (ENI), which is the amount of light power that will produce a signal as large as the noise. In general, a signal several-fold larger than the ENI is necessary for reliable measurements. Manufacturers often will report the ENI for a PMT; if not, it can be estimated from other data by the equation

$$\text{ENI} = (2q_e \cdot I_a \cdot G \cdot \Delta f)^{\frac{1}{2}}/S$$

where q_e is the charge on an electron (1.6×10^{-19} coul), I_a is the anode dark current, G is the current amplification by the PMT, Δf is the bandwidth of the system in hertz, and S is the anode radiant sensitivity (in A/W) at a particular wavelength, usually the wavelength of maximum response. Often the ENI is quoted for a bandwidth of 1 Hz. This number can be used to compare PMTs and can be measured for a particular PMT to determine whether the device is operating as well as it should. The units of the ENI are $W/\sqrt{\text{Hz}}$; typical ENIs at a bandwidth of 1 Hz are in the range of 0.1–0.01×10^{-15} W.

The PMT output can be recorded in a number of ways after it has been amplified and converted to a voltage, including displaying the signal on a chart recorder. The major disadvantage of this method is the loss of temporal resolution; the fastest response of a chart recorder is ~0.1 sec and most are slower. Alternatives include using a digital storage oscilloscope, or digitizing the output and using a computer for storage and analysis. A system we have put together in our laboratory uses the PC-LPM-16 multifunction I/O board from National Instruments. This board, which costs $400, can run in any MS-DOS-based computer that has an available 8-bit (XT) slot. The board can perform analog data acquisition at a rate of up to 50 MHz and comes with a library of functions that can be called from programs written in Microsoft C, QuickBASIC, or Turbo PASCAL.

The connector on the PC-LPM-16 is a 50-pin male ribbon-cable type; the appropriate cable can be purchased from National Instruments for $150. This cable includes a connector block that allows one to screw wires into each of the 50 positions. However, since only two connections are needed for analog data acquisition (signal and ground), a female connector that accepts single wires is sufficient. Such a connector can be purchased from Newark Electronics for about $8 (3M Scotchflex $0.100'' \times 0.100''$ socket connector with polarizing slots, stock #90F8811).

The board samples the analog input with a maximum resolution of 1.2 mV so, if the signal is only on the order of tens of millivolts, amplifying the signal before sampling it would be useful. This amplification can be achieved by a simple operational amplifier circuit, the noninverting amplifier (Horowitz and Hill, 1989).

The other mode in which PMTs may be used is photon counting. This method is by far the most sensitive method of light detection and should be considered if aequorin measurements are contemplated. The underlying basis for photon counting is that the current pulse at the anode that results from a single photon striking the cathode lasts only a few tens of nanoseconds. Therefore, if the rate at which photons are absorbed by the cathode is low enough ($<10^6$/sec), the individual current pulses can be detected and counted; little chance exists that two photons will coincide. When photons of a given wavelength strike the photocathode of a PMT, the current pulses that are produced are not of a uniform size, but have a distribution. The first step (after amplification) in photon counting is to send the pulses through a discriminator, which does not pass pulses that are larger or smaller than some experimenter-determined levels. By appropriately setting the upper and lower levels of the discriminator, most of the pulses resulting from the photons of interest can be passed while many of the pulses resulting from other sources (noise) can be rejected. In this way, the signal-to-noise ratio can be maximized. Photon counters can have dark counts of a few per second at room temperature; these values can be lower if the tube is cooled. After discrimination, the pulses are shaped and counted. Equipment for amplifying, discriminating, shaping, and counting pulses is substantially more expensive than the simple amplifier needed for analog recording so, comparatively, the cost of the PMT itself is minor. The advantage is that the elimination of some of the extraneous pulses gives an improvement in the signal-to-noise ratio that is not possible with analog recording. If the analog signal from a PMT that has an ENI of 0.03 fW $\sqrt{\text{Hz}}$ (a very good PMT) is filtered at 1 kHz, the noise is equivalent to 1 fW of light. At 400 nm, this value represents 1800 photons/sec. As mentioned earlier, photon counters can have dark counts <10 photons/sec.

III. Specific System for Cellular Fluorescence Measurements

The Hamamatsu Corporation has introduced a new series of PMTs in which the phototube is integrated into a housing that also contains the high voltage power supply and the current-to-voltage amplifier. The only required external circuitry is a ±15 V power supply and a 25-turn, 10-kΩ potentiometer. The three different versions of this instrument are the HC120 series, which uses a ½-inch side-on PMT; the HC124 series, which uses a 1-⅛-inch end-on PMT; and the HC125 series, which uses a 1-inch end-on PMT. Each of these models is available with a choice of three or four PMTs. Complete units range in price from $400 to $750 (1992 prices).

The major new technology that makes these compact units possible is the adaptation of the Cockroft–Walton voltage multiplier, which originally was designed for generating high voltages for particle accelerators. The system consists of a network of capacitors and rectifiers and replaces the conventional resistive voltage divider used to supply the voltage to each dynode. Because its power requirements are much less than those of a resistor network, this unit generates less heat and can be incorporated into a housing with the PMT. The low power usage also simplifies the requirements for the external power supply. According to Hamamatsu, the units can be battery powered.

The choice of which of the three models and which PMT to buy depends on the specific application. The HC120 series is the most compact, but its rectangular shape makes it more difficult to adapt to the cylindrical camera and video ports of microscopes. The photocathodes of these PMTs are small, generally 4 mm × 13 mm, imposing limitations on where the PMT can be mounted in the light path. On the other hand, these PMTs have lower noise than those with larger photocathodes. The HC124 and HC125 series units are cylindrical and are thus easier to mount. The larger photocathodes (21-mm diameter for the HC125 and 25-mm diameter for the HC124) allow more flexibility in positioning and facilitate interposing filters and pinholes in the light path. A valuable reference in the design of photometers and the adaptation of them to microscopes is *Microscope Photometry* (Piller, 1977).

In this laboratory, we use an HC120-05 assembly with an R4457 PMT. Shown in Fig. 1 is a full-scale diagram of the unit and a schematic diagram of the circuit. The assembly weighs about 100 g. The R4457 PMT was chosen because of its increased sensitivity in the red part of the visible spectrum and its low noise. Shown in Fig. 2 is the typical spectral response of this PMT. The measured ENI of our particular tube is 0.03 fW. Spectral response is an important consideration, since long wavelength calcium indicators are now available. The cost of the unit, including the PMT, was $720 in 1992. Shown in Fig. 3 are photographs of the unit and the adaptor that is used to couple the assembly to the video port of a Nikon Diaphot microscope. Note the shutter that is used to protect the PMT from bright light; some such arrangement must be included in any design.

Care must be taken to insure that the projected image from the microscope is not larger than the photocathode, which in this case is only 4 mm wide. A Nikon 5× projection lens (CF PL5×) works well. A white card can be held near the port to measure the diameter of the projected image at various distances from the microscope; this information then can be used in designing the

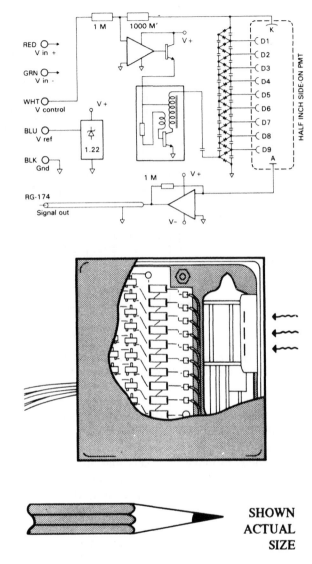

Fig. 1 (*Left*) A full-size diagram of the Hamamatsu 120 series photomultiplier assembly. The $\frac{1}{2}$-in. side-on PMT and the window to admit light can be seen. (*Right*) A schematic diagram of the assembly. This information is reprinted from the Hamamatsu catalog, with permission.

adaptor. With the 5× projection lens, placement of the photocathode 2 cm from the distal surface of the lens is a workable configuration. The shutter and slots for filters can be placed in this 2-cm separation.

In our system, we do not use filters in the PMT housing. The emission filter is mounted in a standard Nikon filter cube that sits underneath the objective

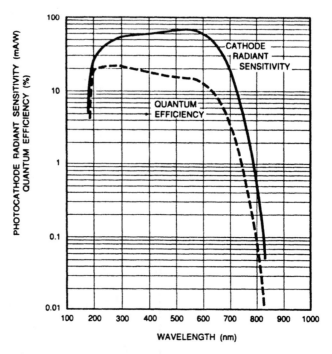

Fig. 2 The typical spectral response of the R4457 PMT. This tube has good responses at wavelengths from 200 nm to 700 nm, making it useful for all optical probes for calcium.

lens. Filter cubes with dichroic mirrors and excitation and emission filters are available from Nikon. Perhaps better alternatives are filters from the Omega Optical Company; this company has a cooperative arrangement with Molecular Probes to make filter sets that are matched to the excitation and emission spectra of the various calcium-sensitive fluorescent dyes. If only relative changes in Ca^{2+} are to be measured, then only a single set of filters is needed; this set can be mounted in the filter cube. If one is planning to use a ratiometric method, however, provision must be made for changing either the excitation filters or the emission filters, depending on which calcium indicator is to be used. We have designed several sliders that can be mounted on our Nikon Diaphot microscope to allow the excitation or emission filters to be changed manually. For the excitation filters, we have modified the end of the nosepiece between the mercury lamp and the dichroic mirror. Our machine shop built a screw-in holder that incorporates a slider holding up to 3 filters. Having the sliding lamp shutter between the lamp and the filters is important since they will burn if constantly exposed to the high intensity light.

If the emission filters need to be changed, for example if indo-1 is being used, then a similar sliding filter holder can be designed to fit into the side port of the microscope. For emission filters, having the whole assembly in a light

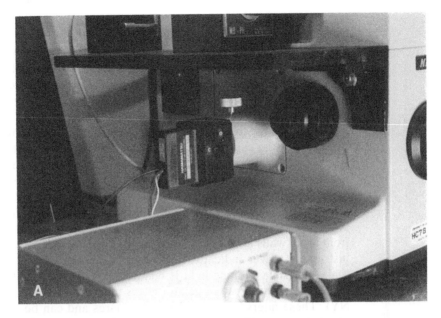

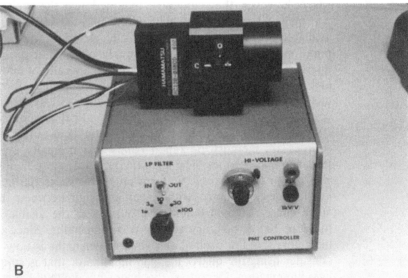

Fig. 3 (A) Photograph of the PMT assembly in place on the video port of a Nikon Diaphot microscope. (B) PMT assembly, adaptor, and control unit. The adaptor was machined by Delrin. The shutter is controlled by the knob extending from the adaptor, and consists of a rotatable cylinder with a hole drilled through it. This arrangement permits the adaptor to be made as one piece. If a slot for filters were included, it would be located where the shutter is. The control unit includes a selectable filter as well as the low voltage power supply and the potentiometer to control the high voltage.

tight box is more important so that stray light does not degrade the measurements.

Such filter changers can be built by a competent machine shop for $100 and $200. If more money is available, Nikon sells a filter wheel and shutter assembly that can be mounted on the excitation side or in the emission light path. This wheel, which costs about $7000, can be controlled by a computer or with a push-button remote control.

One advantage to using low noise PMTs to measure fluorescence is that the excitation beam can be attenuated with neutral density filters because the detector is so sensitive. This attenuation reduces or eliminates photobleaching of the fluorescent dye and minimizes damage to cells. However, practical limits exist to how much the excitation beam can be attenuated, limits that are not imposed by the sensitivity of the PMT. At very low signal levels, avoiding stray light is difficult. For example, the light from an arc lamp can be reflected from room walls and thus can find its way into the microscope objective. Even instrument power indicator lights can cause problems on the most sensitive scale. If visible exciting light is being used, a convenient type of neutral density filters is the Wratten gelatin filters made by Eastman Kodak (Rochester, NY). These filters are sold in 3-inch squares and can be cut to fit almost any space; however, they cannot be mounted near the excitation lamp since they will melt. Glass neutral-density filters that are usable in the UV are readily available from a variety of sources such as Oriel Corporation and Corion Corporation. These companies also supply other optical components, including interference filters. Their catalogs contain a wealth of information about optics and optical systems, a valuable adjunct for any laboratory using optical methods. These suppliers make the investigator less dependent on microscope companies that routinely overcharge for optical components.

Another consideration is the excitation light source. The most versatile sources are mercury or xenon lamps that normally are supplied with fluorescence microscopes. The spectral output of these lamps is shown in Fig. 4. The xenon lamp gives a uniform output whereas the output of the mercury lamp is discontinuous, with lower emission between the peaks. Note that 50-W quartz halogen bulbs have outputs in the visible spectrum that equal those of 100-W mercury and 75-W xenon bulbs. Since a variety of calcium indicators that are excited at visible wavelengths is now available, one could consider using the quartz halogen illuminator that is supplied for transmitted light illumination. Of course, this light source would have to be attached to the epi-illumination setup, but that could be done with a simple adaptor. Other considerations include the size of the filament compared with the arc of a mercury or xenon lamp and the associated difficulty in collimating the beam (the arc is so small that it behaves as a point source), but the experience with our PMT system suggests that measuring the fluorescence from a dye-loaded cell when the beam of a 100-W mercury lamp is attenuated by a factor of 10^3 is possible, allowing for inefficiency in the illumination system; one simply can

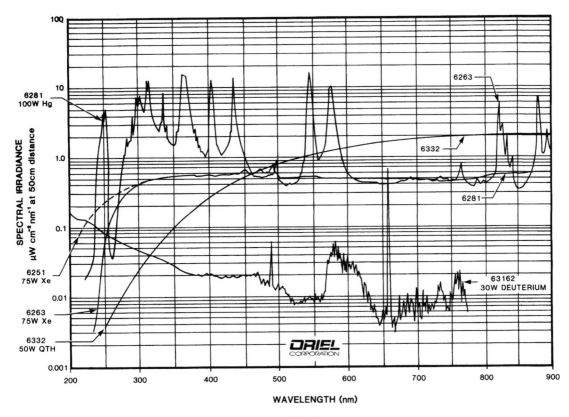

Fig. 4 The spectral radiance of common fluorescence excitation light sources. The substantial output of the 50-W quartz-tungsten-halogen lamp above 450 nm makes it suitable for exciting the longer wavelength calcium indicators. These data are reprinted from the catalog of the Oriel Corporation, with permission.

remove neutral-density filters. Using a quartz halogen lamp instead of an arc lamp has a number of advantages, of which cost is a major one. Quartz tungsten halogen bulbs cost about \$10 whereas arc lamps cost \$100. The radio frequency interference produced when an arc lamp is turned on requires that sensitive amplifiers and computers in the vicinity be turned off or be shielded rigorously (Inoué, 1986); this problem does not arise with incandescent lamps.

An example of calcium-dependent signals recorded from a single naive [i.e., not treated with nerve growth factor (NGF)] PC-12 cell is shown in Fig. 5. Cells were loaded with fluo-3 by incubating them in the cell-permeant acetoxymethyl (AM) form of the dye. A single cell was isolated optically by stopping down the diaphragm of the mercury light source so only one cell was illuminated. A 40× phase contrast objective (0.85 NA) was used; even with this relatively low numerical aperture, adequate signal from the unstimulated cell

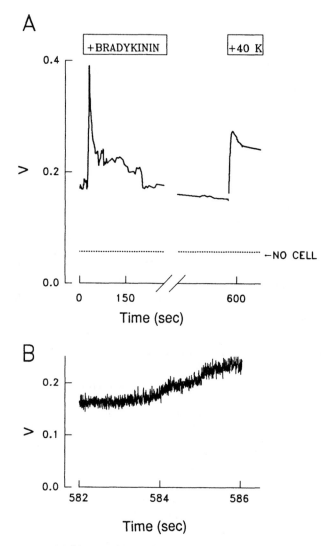

Fig. 5 Measurement of Ca²⁺ changes in a PC-12 cell using the equipment described in the text. Cells were loaded with fluo-3 by incubating them in the AM ester for 15 min at 37°C. The cells had not been exposed to nerve growth factor. At the times indicated in A, bradykinin (100 n *M*) and 40 m*M* K⁺ were present; the increases in Ca²⁺ are apparent. The output of the PMT assembly was amplified and filtered at 200 Hz, and digitized by a Labmaster interface (Scientific Solutions, Solon, OH) at 1 msec/point and 1024 points per trace. The curve in A shows the average voltage per trace. In the expanded time scale of B, traces taken at the indicated times are shown. Brady-kinin and K⁺ were added as small aliquots and allowed to diffuse to the cell. A Nikon 40×/0.85 NA dry objective was used. The excitation light source was a 100-W mercury arc lamp attenuated 10× with a neutral density filter.

was recorded when the exciting light was attenuated by a factor of 10. The responses of the cell to bradykinin and 40 mM K^+ were detected easily. In the expanded time scale of Fig. 5B, the response of the cell to depolarization by K^+ can be resolved readily. These data were collected by S. Scott and J. Strong (Purdue University) on their Nikon Diaphot microscope using the PMT system that we have assembled. One advantage of this system is its simplicity and small size, making it possible to transfer the system from one laboratory to another with a minimum of trouble.

IV. Aequorin Measurements

Despite the obvious utility of fluorescent calcium probes, in many ways the photoprotein aequorin remains the gold standard (see Chapter 13). This molecule is luminescent, that is, the photoprotein emits light when it reacts chemically with Ca^{2+}. Since no exciting light is involved, autofluorescence is not a factor; few cells are autoluminescent. The disadvantages to using aequorin include the difficulty in introducing it into cells (in general, the molecule must be microinjected) and the low light output. The latter problem is relevant to this chapter. Because of the weak signal, the expense of imaging discussed earlier is increased. Extremely sensitive low-noise detectors such as cooled charge-coupled devices (CCDs) or imaging photon detectors costing $30,000 or more must be used (e.g. Fluck *et al.,* 1991). As for fluorescence measurements, photomultipliers can be used to measure aequorin signals for far less money, with the sacrifice of spatial information.

In general, aequorin measurements of Ca^{2+} require photon counting, but this requirement is not absolute. For example, the first cellular aequorin measurements (Ridgway and Ashley, 1967) used analog recording. However, detecting resting levels of Ca^{2+} without photon counting is difficult. We have assembled a lensless system for aequorin-based Ca^{2+} measurements that utilizes a photon counting PMT at ambient temperature (Cork *et al.,* 1987a), based on a design suggested by Lionel Jaffe. The system uses a Hamamatsu R464 PMT; our particular tube was selected especially by the company for low noise. This tube is an end-on, bialkali photocathode type, the peak sensitivity of which closely matches the wavelength of the aequorin emission. Although the diameter of the tube is 2 inches, the effective size of the photocathode is 5 mm × 5 mm. The advantage of this tube is that it has a dark counting rate of only a few per second (ours is about 2 counts per second) at 20°C. The fact that the tube does not require cooling allows the aequorin-loaded cells to be brought in close proximity to the photocathode, as indicated in Fig. 6. This design permits a high efficiency of capture of the emitted photons. Resting signals from aequorin-loaded *Xenopus* eggs and oocytes are typically in the range of hundreds of counts per second, two orders of magnitude above the noise. Since

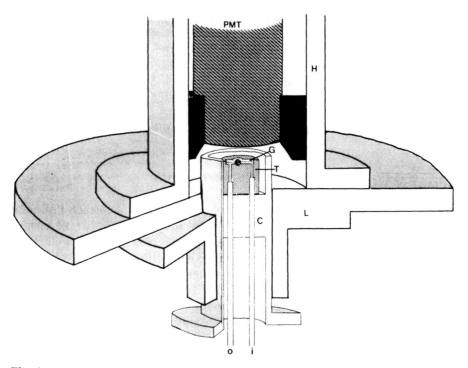

Fig. 6 A sectional diagram of a PMT holder and cell chamber used for aequorin measurements on *Xenopus* oocytes. The cell rests in a depression in a channel (G) cut in a Teflon block (T), which fits into the upper part of an aluminum holder (C). Stainless steel inlet (i) and outlet (o) tubes allow perfusion of the cell. The holder then slides into a baseplate/stand (L), which is bolted to the bottom of the PMT housing. (Reproduced from Cork *et al.*, 1987a, with permission.)

the light emission of aequorin increases approximately with the square of Ca^{2+} concentration, calcium transients produce clearly detectable signals.

The cost of this system is substantially more than that of the analog system already described. The current price of the R464 PMT is $850, plus a small surcharge for selecting for low noise. In addition to the PMT, a housing that shields against magnetic fields is needed at a cost of about $350. A high voltage power supply is also necessary. The electronics required for photon counting are described earlier in this chapter. Our equipment was made by Pacific Precision Instruments (Concord, CA), but unfortunately, that company no longer makes photon counters. Complete systems can be purchased from Hamamatsu and other suppliers, but they are expensive. A complete system, including PMT, housing, socket, high voltage supply, amplifier, discriminator, pulse shaper, and counter with analog and digital outputs costs about $8,000. Of course, photon counting equipment also can be adapted for fluorescence measurements. We have used our photon counter for measuring fura-2 fluorescence from single cells by coupling it to a microscope (Cork *et al.*, 1987b;

Strautman *et al.*, 1990). Although this setup provides an extremely sensitive photometric system, ours is mechanically awkward. Newer photon counting devices, such as those sold by the Hamamatsu Corporation, could be used for both aequorin and fluorescence measurements more readily than our older system.

V. Future Directions

One emerging area of study is that of photodiodes. Advanced Photonix has developed a large area (0.5-inch diameter) avalanche photodiode that is claimed to be "functionally equivalent to PMTs." These photodiodes are expensive (about $2,000) and require a high voltage power supply, but they are rugged, compact, and have excellent response throughout the visible spectrum (although they lose sensitivity in the blue range). The company claims a typical ENI of 30 fW/$\sqrt{\text{Hz}}$, which does approach that of some PMTs but it is still three orders of magnitude larger than that of the Hamamatsu R4457 PMT. Another version of an avalanche photodiode is made by EG&G Optoelectronics. This corporation sells a photon counting module that has a typical dark count of <25/sec; however, these devices have active areas of only about 150 μm diameter and they cost nearly $4,000. The module includes a high voltage power supply and outputs 2 V, 200-nsec pulses. The difficulty is in focusing the light on the tiny active area. Clearly the photodiodes have a way to go before they are truly competitive with PMTs for low light levels, but rapid progress can be expected and anyone building a photometer system should become informed about new developments. Because of their ruggedness, compact size, and spectral response, photodiodes could become a preferred method for measuring fluorescence. The detector could even be located internally, within the microscope.

VI. Summary

Photomultiplier tubes remain among the most sensitive methods for detecting light. Their cost is one to two orders of magnitude less than that of other comparably sensitive detectors. Advances in the associated electronics have lowered the cost and reduced the size of the instruments. If an investigator is willing to go to the primary suppliers and has access to a machine shop, systems sensitive enough for fluorescence measurements on single cells (or portions of a single cell) can be assembled for remarkably little money. In this chapter, we have emphasized a design using particular suppliers because we have used this system in our laboratory. The reader should not assume that we have explored all possibilities; further, progress in these areas tends to be rapid and new developments may open up additional opportunities.

Appendix I: List of Suppliers

Source	Comments
Advanced Photonix 1240 Avenida Acaso Camarillo, California 93012 (805) 987-0146 FAX: (805) 484-9935	Maker of avalanche photodiodes.
Corion Corporation 73 Jeffrey Avenue Holliston, Massachusetts 01746 (508) 429-5065	Supplier of interference and other filters at reasonable prices.
EG&G Optoelectronics 22001 Dumberry Road Vandreuil, Quebec J7V 8P7 Canada (514) 424-3300 (514) 424-3411	Maker of avalanche photodiodes and general electro-optical equipment.
Hamamatsu Corporation 360 Foothill Road P.O. Box 6910 Bridgewater, New Jersey 08807-0910 (800) 524-0504 FAX: (908) 231-1539	Major supplier of PMTs, as well as other light detecting equipment. Catalog is a useful source of general information about PMTs.
Molecular Probes P.O. Box 22010 4849 Pitchford Avenue Eugene, Oregon 97402-9144 (503) 465-8300 FAX: (503) 344-6504	Major source of fluorescent probes. Catalog has a wealth of information about their properties and extensive references concerning their use.
National Instruments 6504 Bridge Point Parkway Austin, Texas 78730-5039 (800) 433-3488 (512) 794-0100 FAX: (512) 794-5794	Also sells I/O boards for EISA bus, Mac NuBus, Mac LE, Mac SE.
Newark Electronics 4801 North Ravenswood Avenue Chicago, Illinois 60640-4496 (312) 784-5100 FAX: (312) 784-5100, Ext. 3107	General electronics parts supplier.

Source	Comments
Omega Optical, Inc. 3 Grove Street P.O. Box 573 Brattleboro, Vermont 05302-0573 (802) 254-2690 FAX: (802) 254-3937	Supplier of specialized filters for fluorescence microscopy, some specially designed to match the characteristics of the calcium indicators sold by Molecular Probes.
Oriel Corporation 250 Long Beach Boulevard P.O. Box 872 Stratford, Connecticut 06497 (203) 337-8282 FAX: (203) 378-2457	Suppliers of a wide range of optical equipment including filters, lenses, polarizers, mirrors, light sources, detectors, and positioners. The three volume catalog has useful tutorials.
Thorn EMI Electron Tubes 100 Forge Way, Unit F Rockaway, New Jersey 07866 (201) 586-9594 FAX: (201) 586-9771	Major supplier of PMTs and photon counting equipment. Catalog has a nice description of PMT operating principles.

Acknowledgment

The development of the instrumentation described in this chapter was funded by grants from the National Science Foundation.

References

Cork, R. J., Cicirelli, M. F., and Robinson, K. R. (1987a). A rise in cytosolic calcium is not necessary for maturation of *Xenopus laevis* oocytes. *Dev. Biol.* **121,** 41–47.

Cork, R. J., Reinach, P., Moses, J., and Robinson, K. R. (1987b). Calcium does not act as a second messenger for adrenergic and cholinergic agonists in corneal epithelial cells. *Curr. Eye Res.* **6,** 1309–1317.

Fluck, R. A., Miller, A. L., and Jaffe, L. F. (1991). Show calcium waves accompany cytokinesis in Medaka fish eggs. *J. Cell Biol.* **115,** 1259–1265.

Horowitz, P., and Hill, W. (1989). "The Art of Electronics." Cambridge: Cambridge University Press.

Inoué, S. (1986). "Video Microscopy." New York: Plenum Press.

Piller, H. (1977). "Microscope Photometry." Berlin: Springer-Verlag.

Ridgway, E. B., and Ashley, C. C. (1967). Calcium transients in single muscle fibers. *Biochem. Biophys. Res. Commun.* **29,** 229–234.

Strautman, A. F., Cork, R. J., and Robinson, K. R. (1990). The distribution of free calcium in transected spinal axons and its modulation by applied electrical fields. *J. Neurosci.* **10,** 3564–3575.

CHAPTER 13

Imaging $[Ca^{2+}]_i$ with Aequorin Using a Photon Imaging Detector

Andrew L. Miller, Eric Karplus, and Lionel F. Jaffe

Calcium Imaging Laboratory
Marine Biological Laboratory
Woods Hole, Massachusetts 02543

METHODS IN CELL BIOLOGY, VOL. 40
Copyright © 1994 by Academic Press, Inc. All rights of reproduction in any form reserved.

I. Introduction

We dedicate this chapter to John Blinks and Osamu Shimomura, two light-houses of luminescence whose contributions in this field have guided us and many others through the sometimes confusing waters of measuring intracellular calcium.

Advances in the aequorin method have opened up exciting new possibilities in calcium physiology, including the synthesis of a large family of semisynthetic and recombinant hypersensitive aequorins (Shimomura, 1991). These advances also include the first successful transformations of eukaryotic cells to induce the expression of apoaequorin (aequorin's polypeptide component) in yeast cells (Nakajima-Shimada *et al.*, 1991), *Nicotiana* seedlings (Knight *et al.*, 1991), mammalian cell line mitochondria (Rizzuto *et al.*, 1992), mammalian cell lines (Sheu *et al.*, 1993; Button and Brownstein, 1993), and *Dictyostelium* cells (Cubitt *et al.*, unpublished observations). Moreover, some remarkable new applications of the aequorin method have been made in cell biology (Fluck *et al.*, 1991b,1992, cleavage furrow and polar calcium in fish eggs, respectively; Browne *et al.*, 1992, nuclear envelope breakdown in response to calcium) as well as in organismic biology (Knight *et al.*, 1992, wind-induced calcium release in plants).

Several reviews on the aequorin method are available (Campbell, 1988; Blinks, 1989; Cobbold and Lee, 1991; Shimomura, 1991; Yoshimoto and Hiramoto, 1991). This chapter updates these reviews with an emphasis on imaging, not merely measuring total light output. We also present practical details about when the aequorin method is or is not the method of choice.

The flickering patterns of free calcium in living systems have been studied successfully with absorption dyes, microelectrodes, and various fluorescent dyes as well as with the chemiluminescent photoproteins that include aequorin. Currently, however, most imaging work is done with fura-type fluorescent dyes or with luminescent aequorins. We, therefore, discuss the advantages and limitations of the aequorins by comparing them with the fura-type dyes. As a result of various advances, the aequorin method clearly is often the method of choice and should, therefore, be given serious consideration by investigators.

II. Rationale for Choosing between Aequorin Luminescence and Fluorescence Ratio Imaging of Ca^{2+}-Sensitive Fluorophores

A. Advantages of Aequorin

Aequorin is usually the best molecule to use to image calcium, *provided it gives enough light,* because aequorin is nondisturbing, has a very wide dynamic range, involves ultralow background signals, shows inherent contrast enhancement, and has very long lifetimes in the cytosol. Moreover, investigators are learning to direct aequorin into specific noncytosolic compartments as well as into particular cell types by genetic means. These advantages are detailed here.

1. Aequorin Is Nondisturbing

"Empirically, . . . [its] presence has not interfered detectably with calcium-mediated functions in any of the [mature] cells into which it has been injected so far" (Blinks, 1982, p. 30). For example, aequorin-loaded muscle cells and neurons contract and conduct normally for hours. Even photoreceptor cells can be aequorin loaded and studied effectively without apparent disturbance (McNaughton *et al.*, 1986; Payne and Fein, 1987).

Moreover, eggs can be aequorin loaded and studied effectively without disturbing normal development. Although natural (or heterogeneous) aequorin can interfere seriously with the development of eggs, with the introduction of recombinant (and purer?) aequorin, this problem has somehow vanished (Shimomura *et al.*, 1990). In our own laboratory, we now find that sea urchin, ascidian, fish, and frog eggs all can be injected with recombinant aequorin (at the usual concentrations) and develop into normal swimming larvae (Miller *et al.*, 1990; Speksnijder *et al.*, 1990; Fluck *et al.*, 1991a; Browne *et al.*, 1992; A. L. Miller *et al.*, unpublished observations).

On the other hand, the use of fluorescent dyes to study calcium patterns continuously seems to be incompatible with extended development. We know of no reports of such use in which the studied cells underwent more than a few cell divisions. Fluorescent methods presumably disturb development via the needed exciting light as well as the calcium buffering action of the dyes. The use of a luminescent indicator such as aequorin involves no exciting light; moreover, loading cells with aequorin introduces negligible calcium buffering, because aequorin is a very weak buffer at physiological pCa values. Moreover, aequorin is not loaded into cells at a concentration of more than 10 μM, which is far below the level at which the prolonged presence of even optimally effective calcium buffers should affect cell physiology (Speksnijder *et al.*, 1989).

2. Aequorin Has a Very Wide Dynamic Range

The wide dynamic range of aequorin facilitates measuring changes and differences of free cytosolic calcium over a thousand-fold range, from 0.1 to 100 μM in terrestrial or fresh water cells and 0.3 to 300 μM in marine cells. This range is indicated by experiments done *in vitro* (Blinks, 1989) and *in vivo*. Measurements of natural cytosolic calcium levels with the aid of natural aequorin have been reported from about 0.1 μM (Snowdowne and Borle, 1984; Woods *et al.*, 1987; Speksnijder *et al.*, 1989; Iaizzo *et al.*, 1990) up to 20–

30 μM (Ridgway *et al.*, 1977; Speksnijder *et al.*, 1989) and even 200–300 μM (Llinas *et al.*, 1992). This thousand-fold range contrasts with various fura-type fluorescent probes that have an effective dynamic range of only 10- to 30-fold (and generally saturate at 1–2 μM). Although *in vitro*, fura-2 yields measurements of free calcium over a more than 200-fold range (Williams and Fay, 1990a), whatever the cause, we know of no measurements of calcium *in vivo* made with fura-type dyes that show convincing data over a range of more than 10- to 30-fold (Williams and Fay, 1990b). Perhaps the minute changes in ratio that are available at the extremes of the ranges of these dyes generally are confounded by changes in autofluorescence.

The wide dynamic range of aequorin is of considerable importance since many or most natural calcium shifts or gradients are so large they are measurable only with such a reporter. Indeed, despite their obvious advantages in certain applications, the widespread use of the fura-type dyes has contributed to a widespread tendency to underestimate seriously peak cytosolic Ca^{2+} levels.

3. Aequorin Measurements Involve Ultralow Background Signal

Autoluminescence (in contrast to autofluorescence) is practically absent from most natural systems. This advantage has proven particularly important in the study of certain eggs. *Ciona* eggs, for example, show strong and rapidly changing patterns of autofluorescence in the blue range (Deno, 1987). When autoluminescence is absent, the only background signals to be overcome are purely instrumental. With our imaging photon detector system, this background can be reduced easily to well under 1 photon/sec.

4. Aequorin Shows Inherent Contrast Enhancement

Above a few hundred nanomolar calcium, aequorin light is proportional to the 2.5–3.0 power of calcium, which provides a large amount of inherent contrast enhancement. This subtle advantage has helped us visualize the calcium waves that accompany the forming cleavage furrows in dividing medaka fish eggs (Fluck *et al.*, 1991a) and in dividing *Xenopus* eggs (Miller *et al.*, 1990; Color Plate 8E). Such cleavage waves had not been reported before. This characteristic also may have helped visualize subsurface microdomains of calcium within the squid giant synapse, microdomains that apparently contain free calcium of a few hundred micromolar (Llinas *et al.*, 1992). More generally, such contrast enhancement should help visualize highly localized regions of relatively high Ca^{2+} within living cells. Suspicion is growing that such regions are both important and widespread (Jaffe, 1990; Kao *et al.*, 1990).

5. Aequorin Is Very Long Lasting inside the Cytosol of Living Cells

Blinks (1982) published Lee's observation that aequorin-injected frog muscle fibers generate useful light signals for more than 3 days. Iaizzo *et al.* (1990, p. 520) reported that "the potentially reactive concentration of aequorin in [rat] hepatocytes was [only] 5 times lower following 48 hr of incubation versus 24 hr." Fluck *et al.* (1991b) injected medaka eggs with aequorin and clearly observed calcium waves associated with blastoderm contraction waves 24 hr later. We have observed a similar persistence of effective aequorin inside developing sea urchin embryos up to the pluteus stage (A. L. Miller *et al.*, unpublished results).

On the other hand, unconjugated fura-type dyes are reported to move into organelles (and confuse observations) within 1 min to 1 hr after loading (Read *et al.*, 1992). The new dextran-conjugated dyes last far longer if they are not irradiated (Miller *et al.*, 1992). Assertions that these conjugated dyes quickly move out of the cytosol and into organelles (Read *et al.*, 1992) seem to be unreliable since these authors failed to determine the distribution of the cytosol in their images. However, because of photobleaching and photodynamic damage to the cells, even these improved dyes may not remain useful for substantially longer periods when used to continuously monitor calcium patterns.

6. Aequorin Is Being Genetically Engineered

The pioneering paper Rizzuto *et al.* (1992) reports the genetic introduction of apoaequorin into the mitochondria of a line of bovine endothelial cells, the subsequent reconstitution of aequorin within these organelles by exposure of the cells to coelenterazine, and the subsequent (and unprecedented) demonstration that agonist-induced rises in cytosolic Ca^{2+} evoke rapid increases in mitochondrial Ca^{2+}. These researchers also report their plans to extend such genetic targeting techniques to the endoplasmic reticulum and to the nucleus. This technique is described in detail in Chapter 14. Moreover, in still unpublished work, A. B. Cubitt *et al.* are using such genetic techniques to introduce apoaequorin into specific groups of differentiating cells within developing slugs of *Dictyostelium*. Since currently no method is available to localize fura-type dyes similarly, the aequorin method offers the only way to image free calcium within specific organelles and/or specific populations of differentiating cells.

B. Limitations of the Aequorin Method

To the extent that the system to be studied is too small, the free calcium levels too low, the spatiotemporal resolution needed too high, or the available methods of introducing aequorin too inefficient, the aequorin method simply may not provide enough light. One situation in which aequorin is likely to generate too dim a light is any multicellular system in which the genetic introduction of apoaequorin has not yet been accomplished. In such cases, to use the aequorin method one must inject aequorin into the egg and wait for the multicellular stage desired or resort to mass loading by mechanical means or by electroporation. To date, these techniques are extremely inefficient, that is, the cytosolic aequorin concentrations that can be attained in these ways currently are orders of magnitude lower than those that are attainable by microinjection or by genetic means (see Section V). In such cases, the reader still may find it best to use fluorescent fura-type dyes since they can be introduced easily *en masse* by permeation in the ester form. Some striking examples of such cases, in which fluorescent dyes work well and the aequorin method is not yet applicable, lie in studies of calcium waves through airway epithelia (Sanderson *et al.*, 1990), through hippocampal brain slices (Dani *et al.*, 1992), and through groups of cultured astrocytes (Finkbeiner, 1992).

Good examples of intracellular phenomena in which fluorescent dyes have worked well, and the aequorins available at the time would not have, are the calcium waves seen in cultured heart myocytes by Takamatsu and Wier (1990) and in cultured hepatocytes by Rooney *et al.* (1990). The then-available aequorins would have given too dim a light to attain the spatiotemporal resolution needed to see such fast moving waves in such small cells. However, these and other fluorescent measurements of absolute levels of $[Ca^{2+}]_i$ which exceed 1 μM may have underestimated the true values grossly (L. F. Jaffe, unpublished observations). Note also that these particular cells were not developing.

C. Situations in Which the Aequorin Method Is the Method of Choice

The balance of advantages is very different in developing systems. High calcium in cleavage furrows has been seen only using aequorin-injected eggs (Miller *et al.*, 1990; Fluck *et al.*, 1991b; see Color Plate 8E); steady zones of high calcium in developing eggs likewise have been seen only by using aequorin (Fluck *et al.*, 1992). Moreover, the remarkable postfertilization waves that cross hamster eggs (Miyazaki *et al.*, 1986) and ascidian eggs (Speksnijder *et al.*, 1990) have been seen only with the aequorin method. All the advantages of aequorin contributed to these successes. The nondisturbing nature and long cytosolic lifetime of this molecule are particularly helpful in developing systems; the wide dynamic range helped reveal calcium zones in the 3–10 μM range; the independence of autofluorescence is especially important in large egg cells; the inherent contrast enhancement was particularly helpful in revealing the very shallow (but intense) zones of calcium found in cleavage furrows. Moreover, these initial successes were attained using natural aequorin or recombinant aequorin reconstituted with the natural luminophore coelenterazine and, thus, were of near natural sensitivity. As the new semisynthetic, hypersensitive aequorins are applied to these problems, far more detailed knowledge will be gained and the advantage of the aequorin method will increase. Color Plate 8A–G shows examples of preliminary experiments we have done on a variety of developing systems using both unmodified and sensitized recombinant aequorins.

Another developmental calcium phenomenon that has been confirmed, refined, and expanded through application of the aequorin method is the large calcium pulse that accompanies nuclear envelope breakdown (NEBD) during the first cell cycle in sea urchin eggs (Browne *et al.*, 1992; illustrated in Color Plate 8A). The use of fluorescent dyes had indicated previously that an NEBD calcium pulse was associated with dispermic or ammonia-activated eggs (Poenie *et al.*, 1985; Steinhardt and Alderton, 1988). Using hypersensitive *h*-aequorin, an NEBD calcium pulse has been seen regularly in normally developing monospermic eggs (Browne *et al.*, 1992). Figure 1 indicates the advantage of using hypersensitive *h*-aequorin rather than unmodified recombinant aequorin to investigate this phenomenon. In addition, effective undischarged *h*-aequorin continued to reveal changing levels of free calcium for up to 17 hr as the embryo developed to the pluteus stage (A. L. Miller, C. L. Browne, E. Karplus, R. E. Palazzo, and L. F. Jaffe, unpublished results; see Fig. 2). On the other hand, another quite sophisticated application of fluorescent dyes to sea urchin eggs, while providing useful new details about the fertilization wave, failed to show an NEBD pulse at all (Stricker *et al.*, 1992).

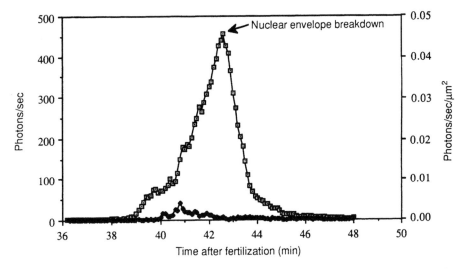

Fig. 1 Comparison of luminescence from recombinant (◆) and *h*-recombinant (□) aequorin-loaded eggs accompanying nuclear envelope breakdown during the first cell cycle in the eggs of the sea urchin *Lytechinus variegatus* (A. L. Miller, C. L. Browne, E. Karplus, R. Palazzo, and L. F. Jaffe, unpublished results).

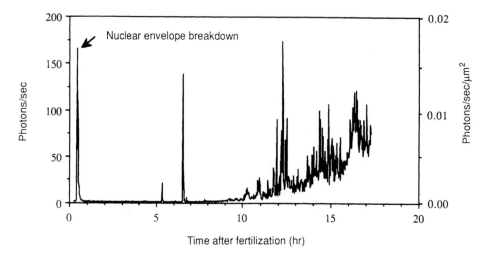

Fig. 2 A long-term luminescent record from an *h*-aequorin-loaded *Lytechinus* egg. This sensitized recombinant aequorin continued to reveal changing patterns of free calcium within the developing embryo for up to 17 hr, until the pluteus larva swam out of the field of view (A. L. Miller, C. L. Browne, E. Karplus, R. Palazzo, and L. F. Jaffe, unpublished results).

In general, then, aequorin is clearly the method of choice for imaging calcium in developing systems, but what of the nondeveloping ones that concern most investigators? The fundamental underlying advantages of aequorin are clear enough for most situations; however, the balance of advantages also depends on a number of practical considerations that are currently in transition, including the practical availability of the new hypersensitive aequorins for general use (see Section IV). These considerations also involve the availability of calibration curves for these new aequorins comparable to those now available for natural aequorin. At present, only limited information is available. The availability and variety of different cells or organisms that express apoaequorin is also limited, but will surely expand. One is also restricted in the number of manufacturers that build and supply the type of positional detector and processing hardware and software required for this kind of research. Practical details and guidance with respect to these matters are discussed in Section VI. Our next major step toward improving the aequorin imaging technique is the development of software needed to deconvolute the limited luminescent output from different planes of focus, to attain resolution in the *z* axis, so-called computational confocal microscopy. Readers who wish to be updated on these matters are encouraged to contact the authors directly.

III. Types of Aequorin

In this section, we briefly describe the aequorins that are currently available. Aequorin samples previously used over the last two decades were "natural" or "heterogeneous" aequorins. These molecules were extracted from collected *Aequorea* jellyfish and, as the heterogeneous designation suggests, were in fact a mixture of many isoaequorins. Ten types of isoaequorin have been separated from heterogeneous aequorin, namely, aequorins A–J (Shimomura, 1986; Shimomura *et al.*, 1990).

A variety of so-called "semisynthetic" aequorins has been prepared by replacing the coelenterazine moiety of these natural isoaequorins with synthetic analogs of coelenterazine. Some semisynthetic aequorins greatly increase their calcium sensitivity over that of natural aequorin whereas some decrease their sensitivity. Shimomura (1991) described 31 types of semisynthetic aequorin. This collection has expanded the detection range of calcium concentrations to which one can apply the aequorin method. One can, therefore, select a semi-synthetic aequorin to suit a particular experimental requirement. As an example one can cite the study of Llinas *et al.* (1992) on microdomains of calcium in the squid giant synapse, in which a less sensitive aequorin was used.

Some of these semisynthetic aequorins show a bimodal luminescence spectrum (see Table 1 in Shimomura, 1991). The ratio of their peak heights is dependent on the concentration of calcium in the range pCa 5–7. Such aequorins may allow the determination of intracellular calcium concentration directly from the ratio of the two peak intensities (Shimomura *et al.*, 1989).

Another significant improvement in the aequorin method was the successful production of aequorin using molecular cloning techniques (Inouye *et al.*, 1985; Prasher *et al.*, 1985). The cDNA for apoaequorin is cloned and expressed in *Escherichia coli*, extracted, purified, and then reconstituted into active aequorin on addition of coelenterazine to give "recombinant aequorin", or with synthetic analogs of coelenterazine to give "recombinant semisynthetic aequorin." The properties of these molecules are described by Shimomura *et al.* (1988,1989,1990).

We seldom use any of the natural isoaequorins or semisynthetic isoaequorins. We have found many of them to be toxic to cells after several hours. Almost all our imaging work uses recombinant aequorin or one of the semisynthetic recombinant aequorins. These molecules do not seem to be toxic in any way. We (and others) routinely find that eggs injected with these aequorins develop into normal adult forms (Speksnijder *et al.*, 1990; Fluck *et al.*, 1991b,1992; A. L. Miller and L. F. Jaffe, unpublished results).

To check and correct for nonuniformities in the cytosolic distribution of aequorin, one can perform a kind of ratio imaging by injecting cells with a combination of unmodified recombinant aequorin and recombinant aequorin labeled with fluorescein (Shimomura, 1991). When waves or pulses of calcium are of interest, they are very unlikely to originate from uneven distributions of aequorin. However, distribution is of particular significance when imaging standing gradients of calcium distribution or localized domains of elevated calcium. We designed our imaging system to allow us to switch among luminescent, fluorescent, and bright field images (see Section VI). Luminescent ratios simply are divided by corresponding fluorescent ratios to correct for differences in aequorin distribution. We have used this technique in our study of elevated calcium domains at the poles of developing fish eggs (Fluck *et al.*, 1992) and of standing gradients of calcium between the oocyte and nurse-cell complex of the silk moth *Cecropia* (Woodruff *et al.*, 1991).

IV. Sources, Handling, and Storage of Aequorin

A. Sources

Semisynthetic aequorins (identifiable by an italicized prefix, e.g., *h*-aequorin) are produced by O. Shimomura at the Marine Biological Laboratory (Woods Hole, Massachusetts). These molecules are made from purified isoaequorins and coelenterazine analogs synthesized by Y. Kishi at Harvard University (Cambridge, Massachusetts). Details of the properties of these semisynthetic aequorins are described by Shimomura *et al.* (1989). Various forms of recombinant aequorin also are prepared by Shimomura, produced from apoaequorin prepared by S. Inouye (Chisso Corporation, Yokohama, Japan) and coelenterazines synthesized by Kishi. The properties of these recombinant and recombinant semisynthetic aequorins are described by Shimomura *et al.* (1990). How these aequorins are supplied can vary. A common method of supply is in 1-mg aliquots freeze-dried from solutions of 1–10 mg/ml in isotonic KCl buffer [plus 0.05 mM ethylene glycol bis(β-aminoethylether)-$N,N,N,'N'$-tetraacetic acid (EGTA) or ethylenedianine tetraacetic acid (EDTA) and 5 mM HEPES or MOPS set to pH ~7.2], or as a frozen solution of aequorin in buffer.

Unmodified recombinant aequorin also is produced by SeaLite Sciences (Atlanta, Georgia) and marketed under the name AquaLite™. This molecule is supplied in a lyophilized form. Highly purified forms of natural heterogeneous aequorin also are supplied by J. R. Blinks (P.O. box 3050, Friday Harbor, Washington) and by the Sigma Chemical Company (St. Louis, Missouri) as "Aequorin Type III."

B. Storage and Handling

The solutions of aequorin generally used for microinjection and other experimental purposes can be stored indefinitely at −70°C or below, without any serious loss of activity. These solutions also tolerate several freeze–thaw cycles. We do, however, recommend that the supply of aequorin solution be split into experimental aliquots (e.g., 10 μl) that are stored and used completely one after the other. This procedure protects against potential catastrophes that may discharge the entire stock solution. At $-20°C$, solutions of natural aequorin and recombinant aequorin can be stored for a few weeks, but the highly sensitive semisynthetic aequorins are inactivated to a certain degree (even in a few days) and are better stored at $-70°C$. These highly sensitive aequorins also tolerate several freeze–thaw cycles without significant loss of activity. How to store aequorin will depend on the strategy of experimentation. For details of storage methods, we refer readers to some of the excellent reviews on aequorin (Blinks, 1982,1989; Campbell, 1988; Cobbold and Lee, 1991; Shimomura, 1991; Yoshimoto and Hiramoto, 1991). We would recommend not storing aequorin in a freezer likely to contain dry ice.

We also stress the care required even when handling aequorin-bearing containers (especially in the case of the highly sensitive aequorins) when they are removed from $-70°$ storage. Condensation may bring a water film in contact with one's fingers, leading to calcium contamination of the container. We use EGTA-washed forceps and avoid all contact between fingers and the container.

V. Methods for the Introduction of Aequorins into Cells

A. Microinjection

Almost all microinjection techniques currently used can be adapted to introduce aequorin successfully into cells for temporal measurements. The real limitation to using the aequorin method lies in the ability to image (not just measure) the resulting luminescent output.

The maximum concentration of various aequorins in aqueous buffer solution is 7–8% w/v (Shimomura, 1991), although a concentration of 15% w/v has been reported for natural heterogeneous aequorin (Cobbold and Rink, 1987). However, because of the increased viscosity of aequorin solutions of high concentrations, the highest concentration of aequorins for practical microinjection use is 1–2% w/v. Therefore, to introduce sufficient amounts of aequorin into large cells and eggs, one cannot merely increase the concentration of the injectate. One must inject larger volumes. This problem should be borne in mind when choosing an injection method for a particular cell type.

Thus, the microinjection technique utilized generally depends on the cell type to be injected. In practical terms, the method also depends on the equipment available or affordable. We have microinjected aequorin successfully into a large variety of cell types: from large *Xenopus* (1200-μm diameter) to small *Spisula* (65-μm diameter) eggs; from cultured *Aplysia* bag cell neurons to the giant axons of squid; and from germinating lily pollen tubes to growing root hairs. This section, however, is not intended to be a review of microinjection techniques, but an attempt to describe how a variety of techniques can be adapted to injecting aequorin. For microinjection details, we refer the reader to an appropriate paper or review. The major concern throughout the microinjection process is preventing accidental discharge of the aequorin in the injectate.

We describe three general methods of aequorin microinjection that should address most experimental contingencies. One important factor to bear in mind is that, in most cases, only one or a few aequorin-loaded eggs or cells will be viewed and imaged at any one time. Experiments generally involve viewing these cells over a period of time to record some event (e.g., fertilization, cytokinesis, response to an added hormone). Therefore only one or a few cells at a time must be microinjected with aequorin, allowing one to concentrate on the "quality" of microinjection rather than the "quantity" of cells microinjected. If the investigator feels that the injected cell has been damaged in any way during the insertion or removal of the micropipette, or during the introduction of aequorin while the micropipette is in the cell, that cell should be disregarded and another injected. The first cell may recover and eventually develop or function normally. However, a fair chance exists that a combination of calcium entering the injection site from the external medium, combined with calcium released from internal stores as a result of a "brutal" injection, will create a localized high calcium environment resulting in the discharge of a significant portion of aequorin in the injection bolus. This injury-induced localized calcium elevation is particularly damaging since the aequorin itself is colocalized and concentrated in the injection bolus. One can minimize this problem by microinjecting aequorin into cells bathed in a calcium-free medium when possible, but this situation can be a "Catch-22" since cells recover from injection injury much more quickly when calcium is present in the bathing medium.

Aequorin also can be protected by the inclusion of magnesium ions in the injectate buffer solution. Magnesium ions compete with calcium for the binding sites on the aequorin complex and, thus, desensitize the aequorin to calcium ions (Blinks, 1982). This procedure is especially helpful when microinjecting the ultrasensitive aequorins. Kikuyama and Hiramoto (1992), in their successful study of intracellular calcium changes during maturation of starfish oocytes, used a solution of ultrasensitive *hcp*-aequorin (which emits light 700 times faster than heterogeneous aequorin at pCa~7) that included ~5 mM MgCl$_2$.

If the research being done requires imaging calcium in a large number of small cells (e.g., in a tissue), microinjecting aequorin is not the method of choice. One might be able to use some of the alternative techniques of aequorin introduction mentioned in a subsequent section (especially genetic engineering) or an alternative fluorescent marker that can be mass loaded in an ester form.

1. Modified Hiramoto/Kiehart Microinjection Method

The first microinjection method we describe is the simplest and least expensive when first beginning these procedures. This method is basically an adaptation of the method devised by Hiramoto (1962) and developed and described in detail by Kiehart (1982); thus, we refer to it as the Hiramoto/Kiehart method. This technique is particularly suited to injecting eggs 60–150 μm in diameter, although we have adapted it successfully to inject other smaller cells such as cultured neurons and epithelial cells. As a rule for all *egg* injections, we never inject more than 10% of the egg volume. With respect to somatic cells, we refer the reader to Cobbold and Lee (1991). For details about this microinjection technique, we refer the reader to Kiehart's excellent discussion (Kiehart, 1982).

To adapt this technique to microinject aequorin, we observe the general rule that every surface that comes in contact with the aequorin solution must be prewashed with 100 μM EGTA solution, especially when the surface is glass. To insure this prewashing, we strictly adhere to the following procedures.

1. We cut 4- to 5-mm lengths of ethanol-washed pyrex capillary tubing (0.8-mm OD, 0.6-mm ID; Drummond Scientific, Broomall, Pennsylvania). These sections are soaked in a 100 μM EGTA solution. One must be careful that no air bubbles are trapped within the capillaries to prevent the EGTA solution from coming in contact with the inner walls. The ethanol prewash helps minimize this occurrence. Capillaries are always handled using forceps to prevent calcium contamination from the fingers. We store the cut capillaries in EGTA solution. To remove this washing solution from the lumen of a capillary, the capillary is blown dry with compressed air.

2. Use of a vegetable oil is an essential part of the Hiramoto/Kiehart microinjection method. As a result, the oil comes in contact with the injectate during several phases of the microinjection process. All oils have some calcium contamination and, therefore, pose a problem when contact is made with the aequorin-containing injectate solution. To address this problem, we prewash a few milliliters Wesson™ vegetable oil with a 100 μM EGTA solution by vigorously shaking roughly equal volumes of the two solutions for several minutes in a plastic centrifuge tube. This tube is left to stand overnight, resulting in the "calcium-free" oil partitioning to the surface. The importance of this precaution of washing the oil will become apparent in later sections.

3. Also prewashed are several 35 × 10 mm plastic Petri dish lids (Falcon 1008), again with 100 μM EGTA solution. These lids are filled with washing solution, drained, and left to air dry.

4. To prepare for the removal of an experimental aliquot of aequorin from storage, we use two auto-pipettes (e.g., Gilson P200 and P20) and set them to 50 μl and 1 μl, respectively. A plastic tip is placed on the P200 and its tip is

washed thoroughly on the inside with a 100 μM EGTA solution by filling and emptying it several times. One must insure that the outside of the yellow tip is submerged sufficiently in the 100 μM EGTA washing solution to insure thorough washing of the outside surface as well. The tip then is transferred carefully to the P20, taking care that one's fingers only come in contact with the distal part of the tip, well away from the portion that comes into contact with the aequorin solution. This procedure results in a calcium-free tip that will not cause accidental discharge when inserted in the aequorin storage aliquot, nor will any aequorin be discharged in the microliter volume drawn into the tip lumen.

5. A storage aliquot of aequorin is removed from the $-70°C$ freezer and allowed to thaw. The aequorin container is held, and the cap removed, using forceps (again to prevent calcium contamination from the fingers); the washed micropipette tip is inserted carefully into the container without touching the inner walls. A 1-μl volume is loaded into the tip, which is carefully removed once again. The storage container is recapped with forceps and immediately returned to the $-70°C$ freezer.

6. The 1-μl drop is ejected from the tip onto the inside of a prewashed Petri dish lid (evaporation is reduced by immediately placing the base on the lid; also, once one is familiar with this operation, Steps 6, 7, and 8 take ~30 sec).

7. This tip is then discarded and replaced by a new one that had been prewashed as described in Step 4. This tip is used to place a 1-μl drop of EGTA-washed Wesson oil next to the aequorin drop. A fresh tip is used for the oil to prevent contamination of the oil by traces of aequorin left in the old tip. The latter event could lead to confusing signals from injected cells since a small droplet of oil is introduced into the cell during the microinjection procedure (see Step 12).

8. A length of prewashed capillary (from Step 1) is picked up with forceps and touched in rapid succession first into the aequorin drop, then into the oil drop. Capillary action draws the aequorin solution, then the oil, up into the capillary. Hesitation in transferring the capillary from the aequorin drop to the oil drop may result in air bubbles trapped between the two fluids. However, this is not a real problem but merely an optical nuisance. The opposite end of the capillary is quickly sealed with soft valap (a $1:1:1$ mixture of beeswax, vaseline, and lanolin that is melted and allowed to cool). This sealing halts the progression of the aequorin/oil column down the capillary and effectively prevents evaporation of the aequorin solution from this end of the capillary. Evaporation from the other end is prevented by the oil plug. We call this loaded capillary our "aequorin reservoir," which effectively serves a whole day of experiments. This procedure is summarized in Fig. 3.

9. Following exactly the same procedure, a second capillary is prepared. This capillary is loaded from a 1-μl droplet of 100 μM EGTA solution rather than one of aequorin. When attempting this procedure for the first time, we

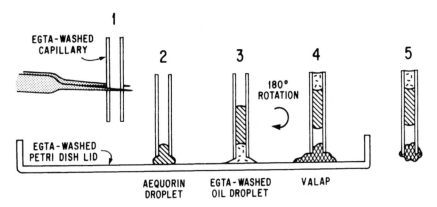

Fig. 3 Steps 1–5 illustrate preparation of the "aequorin reservoir" capillary. The EGTA "washing reservoir" capillary is prepared in exactly the same manner, substituting the aequorin droplet with 100 μM EGTA solution.

recommend loading the EGTA capillary first for practice. If a mistake is made, the EGTA capillary can be thrown away and no valuable aequorin is lost. We call this capillary the "washing reservoir."

10. We prewash the capillaries used to pull micropipettes in 100 μM EGTA solution. On its own, however, this procedure is not sufficient to prevent accidental discharge of aequorin in the micropipette since the pulling process itself exposes new calcium-contaminated glass in the crucial tip region. This condition explains the function of the "washing reservoir."

11. The "aequorin reservoir" and the "washing reservoir" are placed side by side on the support chamber holding the eggs or cells to be injected (see Kiehart, 1982). The microinjection pipette is inserted first through the oil plug into the EGTA solution in the washing capillary. The inner walls of the micropipette are then washed several times by loading and ejecting 100 μM EGTA solution. A slug of 100 μM EGTA (~20 pl) is left in the pipette to give the lumen a final wash as the oil and aequorin are drawn into the pipette.

12. The micropipette now can be inserted into the aequorin reservoir and an appropriate volume of EGTA-washed oil, aequorin, and oil plug loaded. Steps 11 and 12 are illustrated in Fig. 4. The oil plug in the tip serves a double purpose of protecting the aequorin in the pipette from the passage through the medium bathing the preparation (which often contains calcium) and, since it is also injected into the cell, serving as a handy marker for the cells that have been injected. This marker is especially helpful when a few dozen, or even a hundred, eggs are located in an injection chamber.

13. We also recommend not ejecting the injectate into the cell immediately after penetration. We wait several minutes, giving the cell time to begin to recover from the localized elevation of calcium generated by the injection

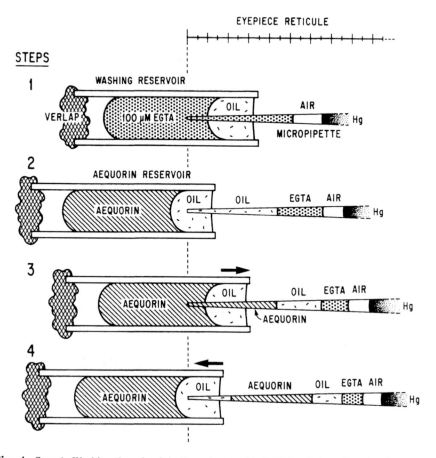

Fig. 4 Step 1. Washing the microinjection pipette with EGTA solution. Step 2. Filling with a calibration slug of EGTA-washed oil. Step 3. Filling with an equal volume of aequorin solution. Step 4. Capping off with an oil plug. The micropipette is aligned under the eyepiece reticule which, for the sake of clarity, is shown at the top of the figure. This procedure adapts a routine described by Kiehart (1982) to aequorin injection.

process (leaking in from the external medium as well as released from internal stores). Once again, the oil plug protects the aequorin in the micropipette during this period. We also do not remove the micropipette immediately after ejection of the aequorin solution, but again wait several minutes, giving the injectate time to diffuse away from the injection bolus before re-injuring the plasma membrane by withdrawing the micropipette. If a clump of adhering cytoplasm is dragged out when removing the micropipette (this will vary among cell types), we recommend silanizing the outside of the micropipettes before Step 11. This process can be done by baking the micropipettes with dimethylaminonitrimethylsilane vapor (Fluka Chemical, Ronkonkoma, New York) for 30 min at 175°C.

We find this aequorin-modified Hiramoto/Kiehart method to be reliable and quantitative, allowing one to quantify the aequorin concentration within the injected cell. The method is not fast but, since only a few cells need to be injected, we find it perfectly adequate for certain cell types. We store the aequorin reservoir at 2–3°C in the refrigerator between experiments.

2. High Pressure Microinjection in the Nanoliter Range

We use the high pressure microinjection method to inject large eggs or cells because it makes the reliable delivery of larger volumes more convenient. Examples of such cases are *Xenopus* eggs, *Drosophila* eggs, giant axons of squid, and root hairs. A typical *Xenopus* egg has a volume of ~900 nl. Sticking to our rule of thumb of not injecting more than 10% of the total egg volume, the volume of injectate required must be in the tens of nanoliters range.

To ease insertion and lessen the injury during injection (and the subsequent release of calcium), we bevel micropipette tips used during this technique. Several pipette bevelers are available on the market, or one can be designed and built specifically. We use a Narishige EG-4 beveller that we have modified in the following way. The micropipette holder is connected to a simple amplifier/headphone arrangement via a phonograph stylus. This arrangement provides an inexpensive and accurate means of determining when the un-bevelled pipette tip makes contact with the grinding surface of the beveller. Following such a procedure results in micropipettes of a consistent shape and size (outside diameter of 3–5 μm at the widest part of the bevel) that do minimal damage to experimental cells. For cells that are extremely difficult to penetrate (e.g., a root hair), we use a capillary with a filling fiber to pull our micropipettes. We then orient the micropipette in the holder of the beveller in such a manner that, when the bevelling process is complete, the filling fiber runs all the way down to the tip of the bevel. This procedure results in an extremely stiff tip that aids in the penetration of certain cell types.

Any commercially available or comparable laboratory-built high pressure injection unit suffices for this type of microinjection. We would, however, recommend buying or designing a unit that allows suction in addition to applying positive pressure (for example, we use a Medical Systems PLI-100), which facilitates tip-loading of the aequorin solution. Tip-loading is preferable since only the inner walls of a small region of the tip rather than the inside of the whole micropipette must be prewashed and it helps minimize the problem of clogging (discussed in Section V,A,3).

We tend to inject larger eggs and cells with an axial drive motion at an angle of ~15° from the horizontal. Cells can be held in a variety of ways, from a simple back support to a suction holding pipette. The procedure generally is carried out under a stereomicroscope at magnifications between 50 and 100×. The nature of this setup makes prewashing the tip region of the micropipette (both inside and outside), loading the pipette with aequorin solution, and injecting experimental material a simple matter (see Fig. 5). The general procedure follows.

1. A micropipette is pulled, beveled, silanized, etc., depending on the experimental protocol; then it is mounted in the pipette holder. Several 50-μl drops of filtered 100 μM EGTA solution are placed on the inside of a lid of a 35 × 10 mm Petri dish (Falcon 1008). We pass the EGTA solution through a

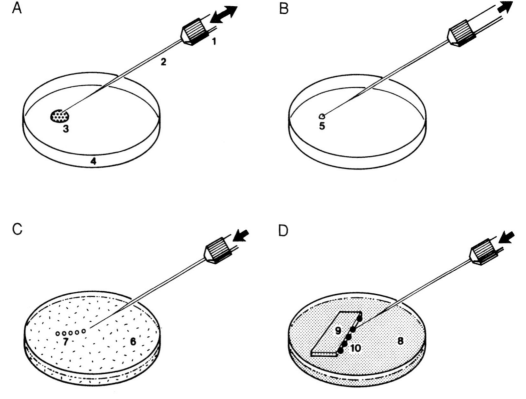

Fig. 5 Illustration of (A) washing, (B) loading, (C) calibrating, and (D) microinjecting with a simple high pressure microinjection setup. 1, Micropipette holder; 2, micropipette; 3, 50-μl droplet of a 0.2-μm millipore-filtered 100 μM EGTA washing solution; 4, EGTA-washed Petri dish lid; 5, 1-μl droplet of aequorin solution; 6, EGTA-washed Wesson vegetable oil; 7, calibration droplets, diameters measured with eyepiece reticule; 8, solution bathing the experimental material; 9, simple backstop cut from a glass microscope slide; and 10, experimental cells (e.g., *Xenopus* eggs, in this case).

0.2-μm millipore filter to remove any dust particles that may contribute to clogging the micropipette. These droplets then act as a tip-washing reservoir.

2. We suck up and extrude 100 μM EGTA solution several times, carefully marking the extent of washing on the micropipette shaft with a marker pen. During this prewash, we do an initial calibration of injection volume, varying the volume by adjusting the injection pressure and/or the timing of the injection pressure pulse, injecting droplets of EGTA solution under oil, and measuring their diameters.

3. Up to a 1-μl droplet of aequorin is removed and placed in a prewashed Petri dish lid in exactly the same manner described in Section V,A,1, Steps 5 and 6.

4. The prewashed tip of the calibrated micropipette is immersed in the droplet and the aequorin solution tip-loaded into the micropipette by suction.

5. The calibration of the micropipette is readjusted by ejecting droplets of aequorin solution under EGTA-washed Wesson oil (because the aequorin solution is more viscous than EGTA). The tip of the aequorin-loaded micropipette is kept in the EGTA-washed oil between injections, with sufficient "back-pressure" on the injectate to prevent oil from entering the pipette tip. This step prevents evaporation from the tip that would result in clogging.

6. As the micropipette passes through the experimental bathing medium, a close watch is kept on maintaining a slightly positive back-pressure to insure a slight efflux of aequorin solution from the pipette tip. This efflux insures that no medium contaminates the aequorin solution within the micropipette, which could result in accidental discharge. If injections are taking place in a medium that contains calcium, the back-pressure adjustments can be made by observing the aequorin luminescence around the tip of the micropipette in a reduced light environment to which the eyes are adapted. If the bathing medium is calcium free (with some EGTA), the efflux from the tip can be discerned as a change in the refractive index of transmitted light through the more viscous aequorin solution flowing from the tip.

7. We often carry out injections in calcium-free medium, then immediately transfer the injected cells back to the regular medium in which they develop, which usually contains calcium.

8. If the equipment available can apply only positive pressure, the micropipette can be back-filled with aequorin. When, under certain experimental conditions, we have been forced to do this, we insure that the entire lumen of the pulled fiber-filled micropipette is washed in EGTA beforehand. We prewash a 1-μl Hamilton syringe with 100 μM EGTA and use this syringe to place 1 μl aequorin as far down toward the tip of the injection pipette as possible.

9. Whether one tip-loads or back-fills, the aequorin-loaded micropipette can be used for many repeated injections. This method proves to be simple and reliable for microinjecting aequorin into large cells and eggs.

3. Injecting Very Small Cells Using Micropipettes with Very Fine Tips: Front-Loading/High Pressure Microinjection Method

A combination of methods is the choice for very small cells such as chromaffin cells (12 μm in diameter), hepatocytes (25–30 μm in diameter), and a wide variety of single flat mammalian cells with approximate diameters of 70 to 100 μm. Variations of this method are used in many laboratories in which it has been developed to microinject aequorin. These variations have been described in some detail by Cobbold and Lee (1991) and Blinks (1989). We once again point out several pertinent points with regard to aequorin, but refer the reader to these reviews for details.

Aequorin solutions can be ejected from micropipettes with tips small enough to have resistances of at least 3–5 MΩ in 3 M KCl (Brown and Blinks, 1974) and up to 10 MΩ in 3 M KCl (Cobbold and Lee, 1991). The biggest problem faced when injecting aequorin from such fine-tipped micropipettes is that of tip clogging. Clumps of aequorin protein tend to cause clogging, which increases with concentration. Thus, the most suitable solution for this type of injection may not be the 1–2% aequorin solutions used in the other microinjection methods. Further, Kohama *et al.* (1971) have reported that aequorin is particularly prone to aggregation in acidic solutions. Therefore, we recommend particular care when buffering injectate solutions to insure slightly alkaline pH levels. Dithiothreitol (DTT; 1 mM) also can be included in the aequorin injection buffer, since it improves the flow of aequorin into and out of fine micropipettes. Cobbold and Lee (1991) suggest that DTT acts to reduce the rate of oxidative denaturation of aequorin and thus prevent aggregation of the denatured protein.

Another source of clogging that is unrelated to aequorin itself is particulate matter that contaminates the glassware or solutions used in the process. One should, therefore, insure that glass capillaries and solutions are scrupulously clean.

In addition to the suggestions just made, the clogging problem can be approached in two basic ways. First, one can attempt to remove particulate matter by passing the aequorin solution through a fine millipore filter (Brown and Blinks, 1974) or by using a chilled ultracentrifuge (A. L. Miller *et al.*, unpublished results). A second, elegantly simple method of circumventing this problem is described by Cobbold and Lee (1991). These researchers use one-shot, tip-loaded pipettes. The premise is that, if a solution gets *in* through a fine tip, it can be ejected *out* through the same tip! If the pipette clogs during the loading process, it simply is discarded and another one is filled. The volumes of aequorin solution loaded through such fine tips by capillary action are minute, but sufficient for the small cell being injected. Thus, the amounts of aequorin lost through discarding clogged micropipettes are insignificant.

Once again, one must take great care not to discharge the aequorin accidentally during this whole process. All glassware, plastics, oil, and so on used in the process should be prewashed with 100 μM EGTA as described for the other microinjection techniques. A version of this particular method of aequorin microinjection, to which we refer as the "Cobbold Method", is described clearly by Cobbold and Lee (1991).

B. Transgenic Expression

The expression of cDNA for apoaequorin in a variety of cell types—yeast (Nakajima-Shimada *et al.*, 1991), tobacco plant cells (Knight *et al.*, 1991), mammalian cell lines (Sheu *et al.*, 1993; Button and Brownstein, 1993), *Dictyostelium* cells (A. B. Cubitt *et al.*, unpublished), and even mammalian cell line mitochondria (Rizzuto *et al.*, 1992)—followed by the subsequent reconstitution of aequorin by introducing the luminophore represents the most exciting advance in the aequorin method for imaging [Ca^{2+}]$_i$. This technique is discussed in detail in Chapter 14. Aequorin introduced into cells in this manner is nonperturbing. Its reconstitution with a variety of synthetic coelenterazines can be targeted to calcium sensitivities in the cytoplasm or in specific organelles. One major step forward conferred by this technique is the advancement of the aequorin method beyond the confines of imaging single, or several, large cells that had to be microinjected laboriously with sufficient quantities of the photoprotein to generate a detectable signal.

C. Other Methods

In general, one could summarize the "other methods" used to load photoproteins into cells as being, even at their best, very inefficient. The cytosolic aequorin concentrations that can be attained currently using these methods are orders of magnitude lower than those that are attainable by either microinjection or genetic means. With due respect to the investigators who, over the years, have struggled with them, we only recommend their use as a very last resort. The adoption of a genetic approach is suggested, when possible, as an alternative and much more worthwhile route.

If the reader is forced to consider other techniques or wishes to do so out of curiosity, they include liposome fusion (Dormer *et al.*, 1978), osmotic shock (Campbell and Dormer, 1978), ultrasound application (Yates and Campbell, 1979), and incubation with EGTA (Sutherland *et al.*, 1980).

VI. Techniques for Detecting Aequorin Luminescence for the Purpose of Imaging [Ca^{2+}]$_i$

The application of photon imaging technology to biological research is still in the early stages of development. A practical guide for the design of a photon imaging system for biological and other microscope based research applications is presented here. We have developed a prototype system to collect meaningful data from light levels in the range of 3–10,000 photons/sec with background levels on the order of 1 photon/sec over the image of an aequorin-loaded cell (Miller *et al.*, 1990; Speksnijder *et al.*, 1990; Fluck *et al.*, 1991a,b,1992; Browne *et al.*, 1992). The advantage of this type of photon imaging is that it involves such low levels of light that it is practical to store video frames and continuous records of luminescence from single experiments on readily available mass storage media for desktop computer systems. The disadvantage is that the amount of light is so low that meaningful spatial and temporal measurements can be obtained only using a position-sensitive photon counting device called an "imaging photon detector" (IPD), with noise levels on the order of 1 photon/sec over the entire image field. Peak signal levels during an experiment may reach 10,000 photons/sec in a restricted area of interest, but most signals are on the order of a few tens to a few hundreds of photons per second.

A. Microscope Platform

Imaging systems for biological research applications are designed around a microscope. The main consideration when selecting a microscope for this type of imaging is light loss in the optical pathway between the specimen and the photon detector. We based our systems around the Zeiss Axiovert 100 TV platform, which has an image port on the same axis as the objective (Carl Zeiss, Inc., Thornwood, New York). For our purposes, the infinity corrected optics of the Zeiss instrument transmit the maximum amount of light with the minimum distortion. Lens selection is also critical, since most light is lost in transmission through lenses. As a rule, we use lenses with the highest numerical apertures that transmit reasonably well at 470 nm (the emission spectrum of aequorin). The microscope also serves as a useful tool for manipulating and orienting the experimental material. In addition to supporting the IPD, the microscope also provides a platform for many other observational techniques including standard bright field, phase contrast, Differential Interference Contrast (DIC), epifluorescence, and video imaging. The data collected with an IPD are most useful when they can be combined with images from these more familiar microscopy techniques. The latter, however, involve levels of light that would saturate and could permanently damage an IPD. In practical terms, an imaging system requires either two image acquisition devices—an IPD and a video camera (with a means to switch between the two)—or a single detector in which the gain can be altered dramatically and rapidly. After surveying the technologies available, we opted to develop a system based on two detector types: one to generate a photon image, the other a video image of the experimental material. This system allows us to maximize the application of both devices. Figure 6 illustrates the basic outline of our Aequorin Imaging System.

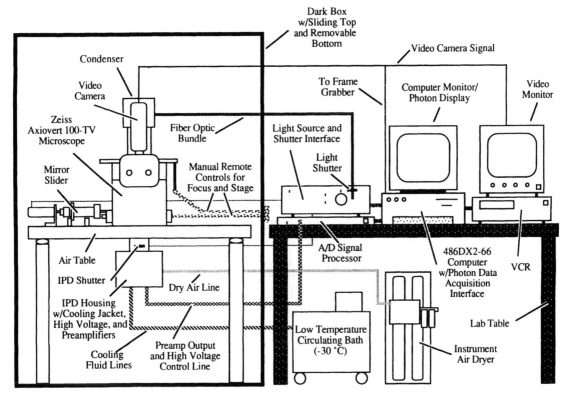

Fig. 6 Outline of the imaging system developed at the Marine Biological Laboratory, Woods Hole, for the purpose of imaging aequorin luminescence in combination with other microscopy techniques.

B. Imaging Photon Detectors

Photon imaging is achieved by acquiring and storing single photon event data from an image field. A photon event consists of the time and two-dimensional position at which a single photon strikes the surface of a suitable detector oriented in the image field. The best known detector technologies available for performing photon imaging are shown in Fig. 7. Both utilize a photocathode that converts single photons into single electrons and two (or three) stages of microchannel plates that collectively amplify each photoelectron (an electron generated by a photon) into a bundle of about one million electrons. The operation of a microchannel plate array is well described in the literature (Wiza *et al.*, 1977). Each bundle of electrons from the microchannel plates has a temporal and spatial distribution that is determined by the time and position at which the photon triggering the shower was converted into an

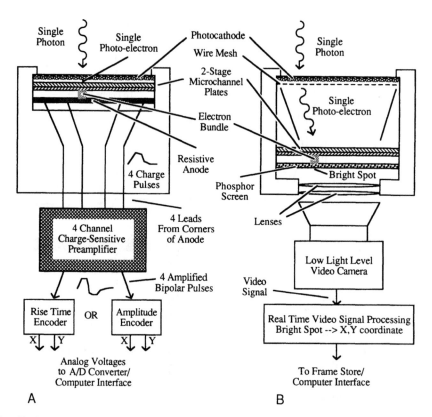

Fig. 7 Imaging photon detectors. (A) Resistive anode type with either rise-time or pulse-amplitude encoding. (B) Phosphor-video type with real-time video signal processor.

electron by the photocathode. The probability that a photon striking the photocathode of an IPD will result in a photon event being detected by the device is known as the detection efficiency of the detector. Detection efficiency is limited primarily by the quantum efficiency of the photocathode, which refers to the probability that a photon striking the photocathode with a given energy (i.e., light of a certain wavelength) will stimulate the release of a photoelectron inside the detector. In the blue–green spectrum, which includes aequorin luminescence at wavelengths of ~470 nm, the best bialkali photocathodes can achieve quantum efficiencies of about 20%.

The dynamic range of the photon imaging detector is limited by the charge throughout capabilities of the microchannel plates and the speed of processing this information. At least two different techniques have been applied successfully to convert each bundle of electrons coming from a microchannel plate array into recorded photon-event data. (1) The resistive anode technique uses a resistive anode in the shape of a flat plate to divide each bundle of electrons into four parts that go to four leads located at each corner of the plate. The signals coming out of these leads can be processed to determine the exact time and position at which a bundle of electrons hit the plate. (2) The phosphor-video technique uses a phosphor screen in front of a very sensitive video camera. When a bundle of electrons strikes the phosphor screen, a bright spot appears for a brief period of time. This bright spot is detected by a video camera and the electronic video signal coming out of the camera is processed in real time to determine the position of the bright spot in the field of view of the camera. The operation of resistive anode and phosphor screen detectors has been described in detail in other literature (Wiza *et al.*, 1977; Wick, 1987,1989). The imaging system we developed is based on a resistive anode detector because it directly digitizes the time and position of each detected photon. This method completely avoids the distortion and noise introduced by generating an intermediate (phosphor screen) image.

IPDs are extremely sensitive to light and can be damaged permanently by exposure to visible white light. We use an IPD with a primarily blue-sensitive bialkali photocathode. Any procedure that might expose the detector surface to visible light levels should be performed only with red light and with the high voltage across the microchannel plates turned off.

The noise of an IPD is expressed in terms of dark counts, events that are not triggered by photoelectrons but nevertheless result in a bundle of electrons propagating through the device. Dark counts can be reduced appreciably by cooling the photocathode and microchannel plate assembly. Practical improvements can be made to about −30°C. Below this temperature, cooling techniques become troublesome and expensive, and a meaningful reduction in noise level is not attainable because of dark counts resulting from the intrinsic properties of potassium-based microchannel plates and phenomena induced by cosmic radiation. We use a Neslab ULT-80 low temperature bath circulator, which maintains our detectors near −30°C (Neslab Instruments, Newington, New Hampshire).

Regardless of the cooling apparatus used, condensation forming on the detector surface inevitably becomes a problem at a few degrees below room temperature. Therefore a low humidity environment must be provided for the detector surface. A simple technique is to flow dry air or nitrogen over the detector surface. We use an air-drying system that reduces the dew point of our laboratory compressed air supply to around −100°F (Balston 75–20 Instrument Air Dryer; Balston, Haverhill, Massachusetts).

Suppliers for both resistive anode and phosphor-video detectors are Photek (Photo-Emissive Technology, St. Leonards-on-Sea, East Sussex, UK) and Hamamatsu Photonic Systems Corporation (Bridgewater, New Jersey). Science Wares (Falmouth, Massachusetts) supplies an imaging system based on the resistive anode type detector alone.

C. Data Acquisition and Processing

The signal available from each bundle of electrons produced by an IPD must be processed to yield a position coordinate and an event time for each photon detected. Two methods can be used with the resistive anode technique to determine the position at which a photon struck the detector. (1) The amplitude encoding method analyzes the relative number of electrons from each of the four leads draining charge from the anode. (2) The rise-time encoding method analyzes the relative times at which charge leaves the anode via each of the four leads. Both methods require the use of charge-sensitive preamplifiers immediately after the anode and before the processing unit. Rise-time encoding allows for a shorter processing cycle because waiting for all the electrons from the anode to enter the processing circuit before performing a calculation is not necessary. In both encoding schemes, two analog voltages are produced to represent the position at which the photon struck the detector—one for the horizontal position and the other for the vertical position. These analog position voltages can be converted easily into digital integer indices on a rectangular grid describing the detector surface. One method of collecting this information is by generating an ordered sequential list that will record the time and position of each photon as it is detected. Another is by forming a two-dimensional array representing a digitized video frame and recording the number of photons occurring in each region of the image over a given period of integration. The advantage of the sequential list method is that it can be used to replay the photons in the same manner as they were produced, to produce a two-dimensional array with an arbitrary period of integration. The two-dimensional array method requires that the period of integration be selected before the data have been acquired.

When using the resistive anode technique, each photon event must be processed in sequence. Ambiguities in photon event detection can arise when electron bundles from more than one photoelectron enter the processing circuitry during a single event processing cycle. This ambiguity can be eliminated by ignoring output from the processing circuit when the number of electrons entering the circuit exceeds a predetermined upper limit for the number of electrons in a bundle generated by a single photoelectron. The duration of the event processing cycle of the circuit imposes an upper limit on the number of photons per second that can be handled. Most circuits can manage at least 10^5 photons/sec easily. Above this light level, the detector can be damaged; a protection circuit is used to shut off the high voltage from the microchannel plates.

A photon imaging system involves the use of a computer to acquire, store, and analyze data. The discussion of data acquisition, handling, and storage that follows is structured around the choice of a 386 or 486 computer platform.

D. Relating a Photon Image to a Video Image

In our imaging system, in addition to a photon detector, a video camera is used intermittently to record a bright field image of the experimental material. Therefore, we must consider how video images can be acquired and stored in a format compatible with the photon imaging data. Video images can be stored in analog format on video tape, but much higher quality images can be stored in digital format using a computer; also, a digital format is more compatible with the photon image acquisition techniques described earlier. Each square in the digital format image is called a pixel; the number assigned to each pixel is called the pixel value. One of the most useful image digitizers for a personal computer is known as a frame grabber, an expansion card that plugs into a computer. This device accepts an electronic signal from a video camera as input, usually in the form of a composite video signal containing image intensity information as well as display synchronization information, and converts it into digital pixel values in a large, fast, two-dimensional memory array. The benefit of having the image transformed into a digital memory array is that the computer hosting the frame grabber can be used to perform operations to enhance and analyze the image stored in the memory array. The memory array also provides an intermediate buffer that allows the acquisition and storage of digitized video images on mass storage media.

The general procedure for acquiring a video image during a photon imaging experiment involves closing a shutter in front of the IPD and redirecting the optical path to a video camera (while activating the appropriate external illumination sources). After a few tenths of a second, the video camera should be generating a stabilized video signal corresponding to the image it sees; the frame grabber can digitize this signal in a single video frame scan period, which is approximately 1/30th of a second. Then the system can be returned to a photon counting state by reversing this procedure: deactivating the external illumination sources, redirecting the optical path, and opening the shutter in front of the photon detector. If properly automated, the entire process of acquiring a video image can be performed in a few seconds.

E. Data Storage

The storage requirements for photon data and video images must be considered. Photon data consist of position coordinates and time information. For each photon event, the position coordinates are stored most efficiently as a pair of single byte values (8-place binary numbers), one byte for each coordinate. This convention allows the position resolution of a 256 × 256 grid to be realized over a selected region of the detector surface. The physical dimensions represented by this grid depend on the magnification of the objective and the size of the area on the detector represented by the grid. The time information is best recorded with the position information in one of two ways. One method records the exact time at every nth photon event, where n is on the order of 100. Another method records an index of pointers to a sequential photon position list at regular time intervals on the order of 0.5 sec. The actual time of detection for each photon is approximated by assuming a constant detection rate between recorded times or pointers. Both these methods are suitable for slowly varying photon detection rates. The latter method can be made more accurate for quickly varying rates by recording the time interval between individual photon events in a separate list of values, using one byte per photon event.

Video images that have been digitized by a frame grabber can be stored directly in a computer file. Most reasonably priced frame grabbers will digitize an image into an array on the order of 512×512 single byte pixel values. This format allows approximately four uncompressed images to be stored per megabyte (10^6 bytes) of media. Compression algorithms can improve the storage density by a factor of 2 or more, depending on the content of the images. The video camera images that are stored during an experiment will be intermittent views of the subject being studied, because most of the experimental time is spent collecting photons. If a video image is stored once each minute, the storage requirements will be 15 megabytes per hour of experimental time.

Affordable mass storage media for personal computers include hard disks, magnetic tapes, and magneto-optical disks. These media are most commonly available in 100–250 MB capacity range. Magnetic tapes and magneto-optical drives have the added advantage of being removable media, a feature that is important if data are to be retained outside the computer. In practice, our storage requirements are within reach of the media described here. Our experience with the prototype system at the Marine Biological Laboratory (MBL) indicates that most aequorin luminescence experiments conceived of to date will produce on the order of 10^7 photons or less. When combined with a set of 60 digitized video images, the mass storage requirements for a typical experiment should be less than 50 MB.

F. Data Analysis Techniques

The general method for analyzing photon image data begins with the composition of a two-dimensional array that contains all the photons collected over a known time interval. Photon fluxes then can be calculated from this array as the number of photons per unit area per unit time. The memory array in a frame grabber is an excellent host for photon image data analysis because it allows the data to be visualized directly on a display monitor as a photon image. The number of photons in each element of the two-dimensional analysis array then becomes a pixel value in a digitized video image.

To generate a video output of a photon array corresponding to the digitized image stored in memory, frame grabbers use output look-up tables. A look-up table converts a pixel value into a display scale. We program our look-up tables with pseudocolor functions to highlight contrast with a color code. The rainbow spectrum provides an intuitive color code for identifying bright spots in a photon image.

A very basic analysis method for handling photon image data sets involves simply looking at a sequence of photon images with constant exposure times. This observation quickly will reveal the locations of interesting events in the data set. To focus in on certain parts of an experiment, one must have an idea of where the interesting events might be. Ideally, as in our system, a set of video images should be available to correlate with the photon images. With a sequential list data set, trying different exposure times to explore different aspects of the data set is relatively easy. Once the interesting events in a data set have been identified, one can perform quantitative analysis and reproduce hard copies of color images via a color video printer or a color printer.

Quantitative measurements can be obtained from photon images with the assistance of a host computer containing a frame grabber. The computer can be programmed to extract flux calculations or generate cross-sectional profiles from the same regions in a sequence of images; the results can be displayed in a customized format or stored in files that can be imported into readily available signal processing and data presentation software. Such data can be used to extract standing gradient or wave speed information, depending on the analysis algorithms. Analysis algorithms generally must be customized to provide the desired information from a particular data set.

G. Three-Dimensional Luminescent Imaging

We are developing the ability to obtain photon images from different focal planes during an experiment so we can use this information to analyze the three-dimensional distribution of photon generation associated with aequorin or other types of luminescence. Such methods will depend largely on our ability to obtain sufficient data from different focal planes in a reasonable period of time. We propose to apply a mathematical deconvolution function to the data obtained from different focal planes to improve the spatial resolution of the information in any given photon image. Our approach will be similar to the deconvolution methods used with higher light intensities. We will use larger and fewer pixels to provide sufficient photons per pixel to reduce the variability associated with finite numbers of photons.

VII. Troubleshooting Experimental Difficulties with Aequorin

In this section, we summarize some typical problems that can occur during an aequorin-based imaging experiment, as well as how to identify, isolate, and address such problems. We deal with the most common type of experiment in which aequorin is introduced into cells by microinjection. Experiments involving transgenic expression of aequorin provide their own peculiar set of problems. The experimental "problem" that immediately springs to mind is a failure to detect a signal where or when one is expected. For example, a previously successful experiment is repeated but no signal is detected, or a calcium event is not imaged after an indirect experimental method—such as calcium buffer injection—indicated that one possibly should have occurred.

A recommended step before every experiment is checking the activity of the aequorin sample after removing it from storage. Several methods ranging from the simple to elaborate, are available. Blinks *et al.* (1978) describe equipment and a method that remains as good as any other. This step removes the uncertainty that the aequorin sample has, for some reason, deteriorated during the freezing–storage–thawing cycle, and provides a quantitative measure of the luminescent activity of the sample used for that particular series of experiments.

The next question that arises is, "Was aequorin successfully introduced into the experimental cell during the microinjection process?" A common problem when injecting eggs is penetrating the fertilization membrane but not the plasma membrane. As a result, one injects aequorin into the perivitelline space where, if any calcium is present, the aequorin is discharged immediately. One can determine whether one has penetrated the plasma membrane of a cell successfully in several ways. We usually inject a specific recombinant aequorin selected for a particular experiment, mixed with ~15% fluorescein-labeled aequorin (also supplied by O. Shimomura, MBL). The latter serves the double purpose of acting as an indicator of a successful introduction and (as discussed previously) of correcting for aequorin distribution within the injected cell when analyzing subsequent luminescent signals (Woodruff *et al.*, 1991; Fluck *et al.*, 1992).

Several investigators determine when a cell plasma membrane has been penetrated by monitoring the potential sensed by the tip of the micropipette. To employ this technique with respect to aequorin injection, refer to Blinks (1989).

Even when a fluorescent signal or detection of a membrane potential clearly indicates that the micropipette tip—and, therefore, the injectate—entered the experimental cell, one could still fail to detect a signal. Aequorin may have been discharged accidentally during any of several stages of the microinjection protocol: loading the micropipette, passing through the bathing medium, or ejection into the cell. To check for this possibility and thus reject a class of negative results, we routinely lyse our experimental cells at a predetermined stage in the experiment. This process indicates whether any active aequorin is localized within the cell and, if so, how much is left. This output added to that collected during the experimental period gives a good indication of how much active aequorin was introduced in the first place. Such a "burn-out" step, as we call it, is simply done by perfusing a solution containing a detergent (such as Triton X) and calcium ions through the experimental chamber.

VIII. Conclusions

The aequorin method has always had some important advantages over fluorescent dye techniques. However, in the past it also had some limitations that included low levels of attainable luminescence, the need to use microinjection for efficient loading, and poor depth resolution. Greatly increased levels of luminescence now have been attained through the development of hypersensitive aequorins. Efficient mass loading of cells, irrespective of their size, is now attainable through the use of genetic techniques. Improved depth resolution should be attainable by developing software to correct for out-of-focus light (which we call computational confocal luminescent microscopy). These developing improvements make the aequorin method a valuable tool with which to study calcium physiology for cell, developmental, and neurobiologists.

Acknowledgments

This work was supported by grants awarded to L. F. Jaffe and A. L. Miller (NSF BIR 9211855) and to L. F. Jaffe (NSF DCB 9103569). We also thank Richard A. Fluck and Jane A. McLaughlin for critically reading the chapter, and Bob and Linda Golder of the MBL's Photo/Graphics Lab for assistance with the figures.

References

Blinks, J. R. (1982). The use of photoproteins as calcium indicators in cellular physiology. *In* "Techniques in Cellular Physiology—Part II," (P. F. Baker, ed.) pp. 1–38. Amsterdam: Elsevier/North Holland.

Blinks, J. R. (1989). Use of calcium-regulated photoproteins as intracellular Ca²⁺ indicators. *Meth. Enzymol.* **172**, 164–203.

Blinks, J. R., Mattingly, P. H., Jewell, B. R., van Leeuwen, M., Harrer, G. C., and Allen, D. G. (1978). Practical aspects of the use of aequorin as a calcium indicator: Assay, preparation, microinjection and interpretation of signals. *Meth. Enzymol.* **57**, 292–328.

Brown, J. E., and Blinks, J. R. (1974). Changes in intracellular free calcium concentration during illumination of invertebrate photoreceptors: Detection with aequorin. *J. Gen. Physiol.* **64**, 643–665.

Browne, C. L., Miller, A. L., Palazzo, R., and Jaffe, L. F. (1992). On the calcium pulse during nuclear envelope breakdown (NEBD) in sea urchin eggs. *Biol. Bull.* **183**, 370–371.

Button, D., and Brownstein, M. (1993). Aequorin-expressing mammalian cell lines used to report Ca^{2+} mobilization. *Cell Calcium* **14,** 663–671.

Campbell, A. K. (1988). "Chemiluminescence. Principles and Applications in Biology and Medicine." Weinheim, Germany: VCH.

Campbell, A. K., and Dormer, R. L. (1978). Inhibition by calcium ions of adenosine cyclic monophosphate formation in sealed pigeon erythrocyte "ghosts". A study using the photoprotein obelin. *Biochem. J.* **176,** 53–66.

Cobbold, P. H., and Lee, J. A. C. (1991). Aequorin measurements of cytoplasmic free calcium. *In* "Cellular Calcium: A Practical Approach" (J. G. McCormack and P. H. Cobbold, eds.), pp. 55–81. Oxford: IRL Press.

Cobbold, P. H., and Rink, T. J. (1987). Fluorescence and bioluminescence measurements of cytoplasmic free calcium. *Biochem. J.* **248,** 313–328.

Dani, J. W., Chernjavsky, A., and Smith, S. J. (1992). Neuronal activity triggers calcium waves in hippocampal astrocyte networks. *Neuron* **8,** 429–440.

Deno, T. (1987). Autonomous fluorescence of eggs of the ascidian *Ciona intestinalis*. *J. Exp. Zool.* **241,** 71–79.

Dormer, R. L., Hallett, M. B., and Campbell, A. K. (1978). The incorporation of the calcium-activated photoprotein obelin into isolated rat fat-cells by liposome-cell fusion. *Biochem. Soc. Trans.* **6,** 570–572.

Finkbeiner, S. (1992). Calcium waves in astrocytes—Filling in the gaps. *Neuron* **8,** 1101–1108.

Fluck, R. A., Miller, A. L., and Jaffe, L. F. (1991a). Calcium waves accompany contraction waves in the *Oryzias latipes* (medaka) blastoderm. *Biol. Bull.* **181,** 352.

Fluck, R. A., Miller, A. L., and Jaffe, L. F. (1991b). Slow calcium waves accompany cytokinesis in medaka fish eggs. *J. Cell Biol.* **115,** 1259–1265.

Fluck, R. A., Miller, A. L., and Jaffe, L. F. (1992). High calcium zones at the poles of developing medaka eggs. *Biol. Bull.* **183,** 70–77.

Hiramoto, Y. (1962). Microinjection of the live spermatozoa into sea urchin eggs. *Exp. Cell Res.* **27,** 416–426.

Iaizzo, P. A., Olsen, R. A., Seewald, M. J., Powis, G., Stier, A., and Van Dyke, R. A. (1990). Transient increases of intracellular Ca^{2+} induced by volatile anesthetics in rat hepatocytes. *Cell Calcium* **11,** 515–524.

Inouye, S., Noguchi, M., Sakaki, Y., Takagi, Y., Miyata, T., Iwanaga, S., Miyata, T., and Tsuji, F. I. (1985). Cloning and sequence analysis of cDNA for the luminescent protein aequorin. *Proc. Natl. Acad. Sci. U.S.A.* **82,** 3154–3158.

Jaffe, L. F. (1990). The roles of intermembrane calcium in polarizing and activating eggs. *In* "Mechanisms of Fertilization: Plants to Humans" (B. Dale, ed.), pp. 389–418. New York: Springer-Verlag.

Jaffe, L. F. (1993). Classes and mechanisms of calcium waves. *Cell Calcium* **14**(10), in press.

Kao, J. P. Y., Alderton, J. M., Tsien, R. Y., and Steinhardt, R. A. (1990). Active involvement of Ca^{2+} in mitotic progression of Swiss 3T3 fibroblasts. *J. Cell Biol.* **111,** 183–196.

Kiehart, D. P. (1982). Microinjection of echinoderm eggs: Apparatus and procedures. *In* "Methods in Cell Biology: The Cytoskeleton, Part B" (L. Wilson, ed.), Vol. 25, pp 13–31. New York: Academic Press.

Kikuyama, M., and Hiramoto, Y. (1992). Change in intracellular calcium ions upon maturation in starfish oocytes. *Dev. Growth Differ.* **33,** 633–638.

Knight, M. R., Campbell, A. K., Smith, S. M., and Trewavas, A. J. (1991). Transgenic plant aequorin reports the effects of touch and cold-shock and elicitors on cytoplasmic calcium. *Nature* (*London*) **352,** 524–526.

Knight, M. R., Campbell, A. K., and Trewavas, A. J. (1992). Wind-induced plant motion immediately increases cytosolic calcium. *Proc. Natl. Acad. Sci. U.S.A.* **89,** 4967–4971.

Kohama, Y., Shimomura, O., and Johnson, F. H. (1971). Molecular weight of the photoprotein aequorin. *Biochemistry* **10,** 4149–4152.

Llinas, R., Sugimori, M., and Silver, R. B. (1992). Microdomains of high calcium concentration in a presynaptic terminal. *Science* **256,** 677–679.

McNaughton, P. A., Cervetto, L., and Nunn, B. J. (1986). Measurement of the intracellular free calcium concentration in salamander rods. *Nature (London)* **322,** 261–263.

Miller, A. L., Fluck, R. A., McLaughlin, J. A. and Jaffe, L. F. (1990). Calcium waves spread beneath the furrows of cleaving *Oryzias latipes* and *Xenopus laevis* eggs. *Biol. Bull.* **179,** 224–225.

Miller, D. D., Callaham, D. A., Gross, D. J., and Hepler, P. K. (1992). Free Ca^{2+} gradient in growing pollen tubes of *Lilium. J. Cell Sci.* **101,** 7–12.

Miyazaki, S.-I., Hashimoto, N., Yoshimoto, Y., Kishimoto, T., Igusa, Y., and Hiramoto, Y. (1986). Temporal and spatial dynamics of the periodic increase in intracellular free calcium at fertilization of golden hamster eggs. *Dev. Biol.* **118,** 259–267.

Nakajima-Shimada, J., Iida, H., Tsuji, F. I., and Anraku, Y. (1991). Monitoring of intracellular calcium in *Saccharomyces cerevisiae* with an apoaequorin cDNA expression system. *Proc. Natl. Acad. Sci. U.S.A.* **88,** 6878–6882.

Payne, R., and Fein, A. (1987). Inositol 1,4,5-triphosphate releases calcium from specialized sites within *Limulus* photoreceptors. *J. Cell Biol.* **104,** 933–937.

Poenie, M., Alderton, J., Tsien, R. Y., and Steinhardt, R. A. (1985). Changes of free calcium levels with stages of the cell division cycle. *Nature (London)* **315,** 147–149.

Prasher, D., McCann, R. O., and Cormier, M. J. (1985). Cloning and expression of the cDNA coding for aequorin, a bioluminescent calcium-binding protein. *Biochem. Biophys. Res. Commun.* **126,** 1259–1268.

Read, N. D., Allan, W. T. G., Knight, H., Knight, M. R., Malho, R. Russell, A., Shacklock, P. S., and Trewavas, A. J. (1992). Imaging and measurement of cytosolic free calcium in plant and fungal cells. *J. Microsc.* **166,** 57–86.

Ridgway, E. B., Gilkey, J. C., and Jaffe, L. F. (1977). Free calcium increases explosively in activating medaka eggs. *Proc. Natl. Acad. Sci. U.S.A.* **74,** 623–627.

Rizzuto, R., Simpson, A. W. M., Brini, M., and Pozzan, T. (1992). Rapid changes of mitochondrial Ca^{2+} revealed by specifically targeted recombinant aequorin. *Nature (London)* **358,** 325–327.

Rooney, T. A., Sass, E. J., and Thomas, A. P. (1990). Agonist-induced cytosolic calcium oscillations originate from a specific locus in single hepatocytes. *J. Biol. Chem.* **265,** 10792–10796.

Sanderson, M. J., Charles, A. C., and Dirksen, E. R. (1990). Mechanical stimulation and intercellular communication increases intracellular Ca^{2+} in epithelial cells. *Cell Regul.* **1,** 585–596.

Sheu, Y.-A., Kricka, L. J., and Pritchett, D. B. (1993). Measurement of intracellular calcium using bioluminescent aequorin expressed in human cells. *Analy. Biochem.* **209,** 343–347.

Shimomura, O. (1986). Isolation and properties of various molecular forms of aequorin. *Biochem. J.* **234,** 271–277.

Shimomura, O. (1991). Preparation and handling of aequorin solutions for the measurement of cellular Ca^{2+}. *Cell Calcium* **12,** 635–643.

Shimomura, O., Musicki, B., and Kishi, Y. (1988). Semi-synthetic aequorin: An improved tool for the measurement of calcium ion concentration. *Biochem. J.* **251,** 405–410.

Shimomura, O., Musicki, B., and Kishi, Y. (1989). Semi-synthetic aequorins with improved sensitivity to Ca^{2+} ions. *Biochem. J.* **261,** 913–920.

Shimomura, O., Inouye, S., Musicki, B., and Kishi, Y. (1990). Recombinant aequorin and recombinant semi-synthetic aequorins: Cellular Ca^{2+} ion indicators. *Biochem. J.* **270,** 309–312.

Snowdowne, K. W., and Borle, A. B. (1984). Measurement of cytosolic free calcium in mammalian cells with aequorin. *Am. J. Physiol.* **247,** C396–C408.

Speksnijder, J. E., Miller, A. L., Weisenseel, M. H., Chen, T.-H., and Jaffe, L. F. (1989). Calcium buffer injections block fucoid egg development by facilitating calcium diffusion. *Proc. Natl. Acad. Sci. U.S.A.* **86,** 6607–6611.

Speksnijder, J. E., Sardet, C., and Jaffe, L. F. (1990). Periodic calcium waves cross ascidian eggs after fertilization. *Dev. Biol.* **142,** 246–249.

Steinhardt, R. A., and Alderton, J. (1988). Intracellular free calcium rise triggers nuclear envelope breakdown in the sea urchin embryo. *Nature (London)* **332,** 364–366.

Stricker, S. A., Centonze, V. E., Paddock, S. W., and Schatten, G. (1992). Confocal microscopy of fertilization-induced calcium dynamics in sea urchin eggs. *Dev. Biol.* **149,** 370–380.

Sutherland, P. J., Stephenson, D. G., and Went, I. R. (1980). A novel method of introducing Ca^{2+}-sensitive photoproteins into cardiac cells. *Proc. Aust. Physiol. Pharmacol. Soc.* **11,** 160p.

Takamatsu, T., and Wier, W. G. (1990). Calcium waves in mammalian heart. *FASEB J.* **4,** 1519–1525.

Wick, R. A. (1987). Quantum-limited imaging using microchannel plate technology. *Appl. Optics* **26,** 3210–3218.

Wick, R. A. (1989). Photon counting imaging: Applications in biomedical research. *BioTechniques* **7,** 262–268.

Williams, D. A., and Fay, F. S. (1990a). Imaging of cell calcium. Collected papers and reviews. *Cell Calcium* **11,** 55–249.

Williams, D. A., and Fay, F. S. (1990b). Intracellular calibration of the fluorescent calcium indicator Fura-2. *Cell Calcium* **11,** 75–84.

Wiza, J. L., Henkel, P. R., and Roy, R. L. (1977). Improved microchannel plate performance with a resistive anode encoder. *Rev. Sci. Instrument.* **48,** 1217–1219.

Woodruff, R. I., Miller, A. L., and Jaffe, L. F. (1991). Differences in free calcium concentration between oocytes and nurse cells revealed by corrected aequorin luminescence. *Biol. Bull.* **181,** 349–350.

Woods, N. M., Cuthbertson, K. S. R., and Cobbold, P. H. (1987). Agonist-induced oscillations in cytoplasmic free calcium concentrations in single rat hepatocytes. *Cell Calcium* **8,** 79–100.

Yates, D. W., and Campbell, A. K. (1979). Calcium transport into sarcoplasmic reticulum vesicles measured by obelin luminescence. *In* "Detection and Measurement of Free Calcium in Cells" (C. C. Ashley and A. K. Campbell, eds.), pp. 257–268. Amsterdam: Elsevier/North Holland.

Yoshimoto, Y., and Hiramoto, Y. (1991). Observation of intracellular Ca^{2+} with aequorin luminescence. *Int. Rev. Cytol.* **129,** 45–73.

CHAPTER 14

Targeting Recombinant Aequorin to Specific Intracellular Organelles

Rosario Rizzuto, Marisa Brini, and Tullio Pozzan

Department of Biomedical Sciences and
CNR Center for the Study of Mitochondrial Physiology
University of Padova
35121 Padova, Italy

This chapter describes the use of recombinant aequorins as Ca^{2+} probes targeted to specific subcellular locations. Ca^{2+}-reporting proteins are ideal candidates for this purpose, since polypeptides offer the unique opportunity of being modified to include specific targeting information. On the other hand, the cloning of the cDNA (Inouye *et al.*, 1985) gives the opportunity of recombinant expression, thereby circumventing the need for microinjection with the subsequent limitation to a few cell types. The properties of photoproteins, as well as their use for measuring Ca^{2+} concentrations, are described in detail in other parts of this book. This chapter therefore mainly focuses on other aspects of the methodology, particularly on the procedures adopted for targeting, expressing, and reconstituting the photoprotein in intact mammalian cells.

I. Principles of Subcellular Targeting

In the last decade, our knowledge of how proteins are sorted correctly to the different cellular compartments has improved dramatically. On the one hand, the analysis of a large number of polypeptide sequences has highlighted short amino acids stretches shared by proteins with the same subcellular location. Deletion analysis experiments have, in most cases, shown that these stretches are sufficient and necessary for a correct intracellular sorting of the protein. On the other hand, *in vitro* and *in vivo* studies are starting to clarify the molecular machinery that controls protein traffic and import in the various membrane-bound organelles [mitochondria, endoplasmic reticulum (ER), nucleus].

Exhaustive reviews on protein sorting to different organelles have been published (Hartl *et al.*, 1989; Nothwehr and Gordon, 1990; Garcia-Bustos *et al.*, 1991); nevertheless, we briefly summarize the sequence requirements for protein targeting to three different cell compartments for which aequorin localization has been achieved (mitochondria, nucleus) or is currently under way (ER). In these organelles, targeting signals have been described that account for the proper localization of most, but not all, resident proteins.

Mitochondria. Most mitochondrial proteins (except the 13 polypeptides encoded by the mtDNA) are synthesized on cytosolic free ribosomes and reach the organelle after translation (Hartl *et al.*, 1989). In most cases, mitochondrial import depends on the presence of a cleavable N-terminal extension that is generally rich in basic (mainly arginine) and hydroxylated residues, devoid of acidic ones, and folded into an amphiphilic α-helix (von Heijne, 1986). After import, the leader peptide is removed by specific matrix proteases (Hendrick *et al.*, 1989).

Nucleus. The sequence requirements for nuclear localization are quite limited: a short basic sequence (PKKKRKV) of SV40 large T antigen is sufficient to target a cytoplasmic protein (pyruvate kinase) to the nucleus and commonly is indicated as the prototype of nuclear localization sequences (NLS) (Garcia-Bustos *et al.*, 1991). Similar sequences have been identified in most nuclear proteins but, in contradiction to the "SV40 paradigm," they are not always sufficient for targeting a heterologous polypeptide or the nuclear protein itself to the nucleus. Nuclear localization can be obtained only if the NLS is extended further. A new consensus has been identified that is present in more than 50% of nuclear proteins and is composed of a bipartite motif: two basic residues, a spacer region, and a second basic cluster, which usually bears a high similarity to the SV40 NLS (Dingwall and Laskey, 1991).

Endoplasmic Reticulum. Two signals must be present in a resident ER protein: a leader peptide, which determines the insertion of the nascent polypeptide into the rough ER, and a retention signal, to prevent progression to later compartments (Golgi, vesicles, extracellular medium). For insertion into the ER, both the peptide (a short sequence rich in hydrophobic residues) and the molecular mechanism are fairly well known (Nothwehr and Gordon, 1990). ER localization, however, depends on the additional presence of "positive" signals that promote the retention or the retrieval of the resident proteins (Munro and Pelham, 1987). Although different pathways are certainly present (Sitia and Meldolesi, 1992), the best characterized "ER retention" signal is a noncleavable short C-terminal sequence (KDEL) that (1) is present in most luminal ER proteins, (2) is necessary for ER localization of these proteins, and (3) is sufficient for trapping either heterologous proteins or endogenous proteins destined for secretion in the ER.

II. Construction of Chimeric Aequorin cDNAs

In principle, subcellular targeting of aequorin can be accomplished in two different ways. In the first, a minimal targeting sequence can be added to the photoprotein. In the second, the photoprotein itself can be fused to a resident organellar protein that already includes all the information necessary for its proper localization. As we discuss in the next several paragraphs, we have used both strategies to design chimeric aequorins destined for three different cell compartments: the nucleus, the mitochondria, and the ER. In all cases, a fusion protein is made that should retain the light emission and Ca^{2+} sensitivity of the native polypeptide. Although this evaluation can be done only a posteriori (i.e., when the cDNA is expressed and the chimeric polypeptide produced), useful information on the modifiability of aequorin is already available in the literature. For example, to induce secretion to the periplasmic space of transformed bacteria, Inouye *et al.* (1989) have added an N-terminal signal to aequorin with no effect on its luminescence or Ca^{2+} sensitivity. Similar results were obtained by Casadei *et al.* (1990), who produced a luminescent antibody by expressing a chimeric protein composed of an IgG heavy chain at the N terminus and aequorin at the C terminus. On the contrary, no functional aequorin chimeras have been reported to date in which the photoprotein is located at the N terminus. In particular, Nomura and associates (1991) have shown that minimal modifications of the aequorin C terminus, including deletion or substitution of the final proline or addition of one or more amino acids, almost completely abolish aequorin luminescence. In contrast, Kendall *et al.* (1992) have provided preliminary evidence that a good luminescence signal can be obtained from aequorin carrying KDEL at its C terminus. Currently, we have no explanation for this unexpected result. In general, based on these results, one can reasonably predict that if a localization signal can be placed at the N terminus (thereby not affecting the C terminus of aequorin), the aequorin cDNA (and the encoded polypeptide) can be modified by simply adding the targeting sequence. Conversely, if the signal is localized at the C terminus (and cannot be moved elsewhere in the polypeptide), a simple fusion protein is unlikely to be successful and different targeting strategies must be established.

A. Addition of a Targeting Signal

The simplest example of this approach was the construction of the mitochondrial aequorin probe. In this case, the cDNA for a mitochondrial protein, subunit VIII of human cytochrome *c* oxidase (Rizzuto *et al.*, 1989), was digested with the restriction enzymes *Hin*dIII and *Eco*RI, thus releasing a fragment encoding the cleavable signal peptide and the first 6 amino acids of the mature polypeptide (i.e., presumably preserving the recognition site for the matrix presequence protease). The cDNA for aequorin (Inouye *et al.*, 1985) was digested with the same endonucleases, yielding a fragment extending from the start of the coding sequence (excluding only the first amino acid) to the 3′ noncoding region, downstream of the polyadenylation signal. Since the coding sequences of the two DNA fragments were in frame, they were ligated together in the vector; the chimeric cDNA encoded a mitochondrially targeted aequorin polypeptide (Rizzuto *et al.*, 1992). The accuracy of the construction was controlled by sequencing the cDNA.

A similar approach was used to construct the nuclear probe (Brini *et al.*, 1993). In this case, the targeting signal was obtained from the rat glucocorticoid receptor (GR). In untreated cells, the wild-type GR is a cytosolic protein since the hormone-binding domain, presumably by interacting with a heat-shock protein and masking the NLS, prevents transport to the nucleus. On hormone binding, the inhibition is relieved and translocation occurs. When a minimal portion of the GR that includes the NLS and lacks the hormone-binding domain is fused to a foreign protein, the polypeptide is localized constitutively in the nucleus (Picard and Yamamoto, 1987). We therefore have excised from the GR cDNA (Miesfeld *et al.*, 1986) a fragment that encodes the NLS and have fused it to the previously described *Hin*dIII/*Eco*RI fragment of the aequorin cDNA. Since the fusion occurs in frame, in the encoded polypeptide the GR NLS (sufficient to drive a heterologous protein, such as β-galactosidase, to the nucleus) is placed in front of aequorin.

B. Fusion of Aequorin to a Resident Protein

An alternative approach to aequorin targeting rests on the possibility of fusing the whole photoprotein to a specifically localized polypeptide, that is, adding a Ca^{2+}-reporting flag to a resident protein. We have adopted this strategy to design the ER probe. Although no data are yet available on this fusion protein, we discuss this approach since it may prove useful in targeting aequorin to organelles for which a localization signal cannot be placed at the N terminus or for which the targeting signal is unknown. Moreover, if this strategy is successful, aequorin may be linked to relevant molecules (ion channels, receptors, etc.) and Ca^{2+} concentrations can be monitored in defined cytosolic or organellar environments.

As already discussed, most resident proteins are retained in the ER lumen because of the presence of the C-terminal sequence KDEL. Ideally, an ER aequorin should be constructed by adding a leader peptide at the N terminus (which promotes translocation across the rough ER membrane) and a retention signal at the C terminus. In principle, both modifications appear essential: if the former signal is lacking, the protein will be cytosolic; if the latter is missing, the protein will be inserted in the ER but will progress in the secretory pathway (Golgi, vesicles, etc.) and the Ca^{2+} probe will not be specific to the ER. However, since, based on the results of Nomura et al. (1991), a C-terminal modification of aequorin appears doomed to be unsuccessful (but see Kendall et al., 1992), we have followed a different approach. In particular, we have taken advantage of the detailed knowledge of the sorting process of immunoglobulins to construct a chimera between aequorin and part of the heavy chain of IgM. Immunoglobulins are composed of heavy (HC) and light (LC) chains that are translated and translocated in the ER independently. Before assembly with the LC, the HC is known to be retained in the lumen of the ER because it is bound to the abundant ER protein heavy chain binding protein (BiP). After assembly, BiP is displaced and the whole Ig progresses toward secretion. In the absence of the LC (i.e., when the HC is expressed in a cell type lacking the LC), the HC is bound to BiP and is not translocated further. We have fused the HindIII/EcoRI fragment of the aequorin cDNA to a portion of the HC cDNA that encodes the leader peptide and part of the HC, including the domain (CH1) that is bound by BiP (Neuberger, 1983). The chimeric polypeptide should behave like an Ig HC carrying an aequorin domain; therefore, it should be secreted when transfected into an LC-expressing cell line (e.g., myeloma cells), but be retained in the ER when transfected into a cell line that does not express the LC (mutant myeloma cells or cells of different lineage).

The strategy employed for constructing mitochondrial aequorin, which is discussed in detail in this chapter, is summarized in Fig. 1.

III. Expression of Recombinant Aequorin

Although data are now available on nuclear aequorin also (Brini et al., 1993), in the rest of this chapter we discuss other aspects of this technique by referring to the first chimeric probe, that is, mitochondrial aequorin (mtAEQ), which was analyzed more extensively and thus may represent the prototype of this family of Ca^{2+} probes.

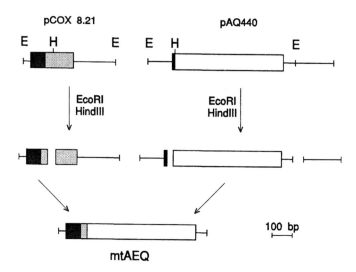

Fig. 1 Construction strategy of mitochondrial aequorin. H and E indicate the positions of *Hind*III and *Eco*RI sites, respectively. pCOX 8.21, cDNA encoding subunit VIII of human cytochrome *c* oxidase; pAQ440, cDNA encoding native aequorin; mtAEQ, chimeric cDNA encoding "mitochondrial" aequorin. (Reprinted, with permission, from Rizzuto *et al.*, 1992.)

The cDNA encoding the mitochondrially targeted aequorin was subcloned in the plasmid pMT2 (Kaufman *et al.*, 1989). pMT2 is a widely used expression vector that contains two restriction sites downstream of the adenovirus late promoter and tripartite leader and of an artificial intron (derived from mouse immunoglobulin). No selectable marker is present in the vector, which allows high levels of transient expression of a cDNA from its own start codon AUG. For stable expression, a co-transfection must be done with a selectable plasmid (see Section III,B). The mtAEQ cDNA was excised from the cloning vector pBS$^+$ (Stratagene, La Jolla, California) by digestion with *Eco*RI and was subcloned in pMT2. After transformation of competent bacteria, positive clones were identified by hybridizing nitrocellulose replicas of the plates with the aequorin cDNA. The correct orientation was verified by restriction mapping of the recombinants. Figure 2 shows the ethidium bromide stain of a diagnostic digestion (*Hind*III) of five different recombinants, which include two "sense," two "antisense," and a double insertion of the mtAEQ cDNA in the vector. One of the "sense" orientations (indicated by the arrow) was termed mtAEQ/pMT2 and was used for all the transfections.

A. Transfection and Transient Expression

We currently use, and here describe, the calcium phosphate procedure for transfecting mammalian cell lines, which has given excellent results with all the tested cells (bovine aortic endothelial, HeLa, primary culture of cardiac

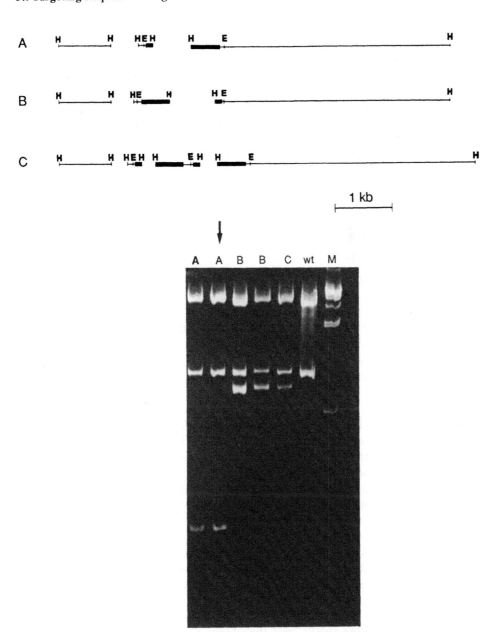

Fig. 2 Restriction mapping of five independent recombinants recovered from the transformation of mtAEQ in pMT2. A diagnostic digestion was made with *Hin*dIII. (*Top*) Scheme showing the restriction fragments obtained in the different cases. (*Bottom*) Ethidium bromide stain of the acrylamide gel. H and E indicate the positions of *Hin*dIII and *Eco*RI sites, respectively. A, B, and C represent single ''sense,'' single ''antisense,'' and double ''sense'' insertions, respectively. The arrow indicates the chosen recombinant.

myocytes, and neurons). However, we have successfully tested different procedures (namely electroporation and charged liposomes), which can be used as an alternative to transfection.

For transient expression, the cells are seeded onto round glass cover slips (13-mm diameter, to fit into the luminometer chamber) laid on the bottom of a 90-mm tissue culture Petri dish. Alternatively, the cells can also be seeded directly onto the Petri dish and replated onto the cover slips after transfection, or detached and used in suspension. Transfections are done when the cells reach 50–70% confluence, using the protocol described at the end of the chapter.

B. Generation of Aequorin-Producing Clones

1. Selection of Stable Transfectants

The transfections are done as for the transient expression, with the following variations:

1. The cells are plated directly onto the tissue culture dish
2. 40 μg DNA/dish is utilized, in the following ratio—36 μg mtAEQ/pMT2, 4 μg selectable marker pSV2neo (Southern and Berg, 1982)—to favor the integration of the nonselectable gene of interest (mtAEQ)

On day 3 of the transfection protocol (described at the end of the chapter), selection with G418 is begun. Since large variations are observed among cell lines, we always test the sensitivity of the line to be transfected by challenging control dishes with a large spectrum of drug doses. Once a calibration curve is obtained, we usually choose for selection a dose 25–50% higher than the minimal effective dose (i.e., the dose for which no spontaneously resistant clone is present in the control dishes). To date, we have used G418 concentrations ranging from 0.3 mg/ml (for PC-12 cells) to 0.8 mg/ml (for endothelial and HeLa cells).

When massive cell death occurs, the medium is changed every 1–2 days; otherwise the medium is changed every 3–4 days. After a few weeks (the lag time is variable, depending on the duplication time of the cells), isolated G418-resistant cellular clones become visible. When these clonal populations reach a diameter of 1–2 mm, the medium is withdrawn, the dish washed with phosphate buffered saline (PBS), and the clones picked with a sterile toothpick (prepared with cotton at the extremity) and transferred onto 24-well culture plates. As an alternative, subcloning can be carried out by limiting dilution. When each well is full, the cells are trypsinized and transferred to a 75-mm culture flask. After splitting, a confluent flask is frozen and the other is utilized for RNA extraction.

2. Screening Clones for High Production

Total RNA is extracted from each flask using the RNAzol B method (Biotecx Laboratories, Houston, Texas), according to the protocol of the producing company, and is electrophoresed onto denaturing 1% agarose gels. Figure 3 shows the results of the Northern analysis of a pool of G418-resistant HeLa clones stably transfected with mtAEQ/pMT2. Among these clones, 59 produced the highest amount of mtAEQ mRNA; a frozen stock was thawed and the cells were analyzed further for aequorin luminescence (see subsequent discussion).

C. Injection of *in Vitro* Synthesized mRNA

When large cells or primary cultures are analyzed, an alternative to transfection is the microinjection of *in vitro* synthesized mRNA. This approach, although technically more cumbersome, may allow a higher production and therefore be the method of choice for single cell analysis. We have not done experiments of this type. However, we have tested the feasibility of the approach, both for the mitochondrial and for the nuclear (nuAEQ) constructs, by producing mtAEQ mRNA *in vitro* and microinjecting *Xenopus laevis* oocytes. Figure 4 shows the ethidium stain of the electrophoretic run of ⅕ of each reac-

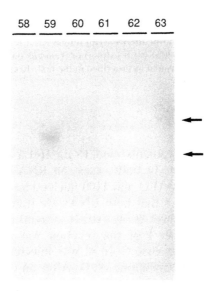

Fig. 3 Northern analysis of a pool of stable transfectants. Total RNA was electrophoresed, blotted, and probed with random-prime radiolabeled (Feinberg and Vogelstein, 1983) aequorin cDNA. Clone 59 represents the highest producer among all clones (as highlighted by the short exposure). Arrows indicate the positions of ribosomal RNAs.

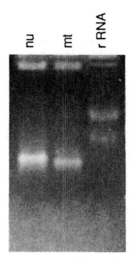

	Wheat germ extract (counts/μg RNA)	X. laevis oocytes (counts/oocyte)
mtAEQ	818	2300
nuAEQ	1109	2700

Fig. 4 *In vitro* transcription and translation of targeted aequorins. (*Top*) Ethidium bromide stain of the electrophoresis gel of *in vitro* synthesized "nuclear" (nuAEQ) and "mitochondrial" (mtAEQ) aequorin mRNAs. HeLa total RNA is shown also for size comparison. (*Bottom*) Result of cell-free (wheat germ extract) or *Xenopus* oocyte mRNA translation. Aequorin was reconstituted and measured as described in the text. 10 cps, approx. 1 pg aequorin.

tion (see protocols) and 15 μg HeLa total RNA, used for size and quantity comparison. In both cases, an RNA of the appropriate size (~800 nucleotides for mtAEQ and 1100 nucleotides for nuAEQ) was produced. The lower panel shows that both RNAs are translatable into functional aequorin, both *in vitro* (wheat germ extract system) and *in vivo* (*X. laevis* oocytes). In the former case, ⅕ of the reaction was translated and reconstituted *in vitro*; in the latter case, ~20 nl was injected into each oocyte, as described elsewhere (Wormington, 1991). After an overnight incubation, the oocytes were disrupted mechanically; after freeze–thawing and centrifugation, the supernatant was analyzed. Aequorin was reconstituted *in vitro* as described in the following section. The data, expressed as light emitted on Ca^{2+} addition, show that, in all cases, the synthetic RNA was translated into functional aequorin.

═══ IV. Reconstitution of Functional Aequorin

The whole active photoprotein is composed of a 22-kDa apoprotein (which, as we have discussed previously, can be modified to allow for specific intracellular targeting) and a hydrophobic prosthetic group, coelenterazine, which plays a direct role in the reaction of light emission. Whereas native aequorin as extracted from the jellyfish is already functional, the protein as produced with molecular biology techniques lacks the prosthetic group and is, therefore, totally inactive alone. Coelenterazine can be synthesized *in vitro* (Inoue *et al.*, 1975) and spent aequorin can be reconstituted into the active form simply by adding coelenterazine in the presence of β-mercaptoethanol (β-ME) (Shimomura, 1991). Similarly, recombinant aequorin can be reconstituted in a variety of cell types (prokaryotes, yeasts, plants, and mammalian cells) by adding coelenterazine to the incubation medium (Knight *et al.*, 1991a,b; Nakajima-Shimada *et al.*, 1991; Rizzuto *et al.*, 1992). Coelenterazine is, in fact, highly hydrophobic and readily permeates cell membranes.

A. *In Vitro*

For each clone, a 75-mm flask of cells is trypsinized, washed once with medium, once with PBS, and resuspended in 0.5 ml 150 mM Tris, pH 7.5,4 mM ethylene glycol bis(β-aminoethylether)-$N,N,N,'N'$-tetraacetic acid (EGTA), 1 mM PMSF. After three cycles of freeze–thawing, the debris is centrifuged and discarded. To a 200-μl aliquot of the soluble supernatant, 2 μl β-ME and 2 μl coelenterazine (250 μM in methanol) are added. After an overnight incubation at 4°C in the dark, the Ca^{2+}-dependent luminescence is measured in a Packard Picolite (Meriden, CT) luminometer by adding 10 mM $CaCl_2$ to a 50-μl aliquot of the sample. The total quantity of reconstituted aequorin is deduced from the comparison with the Ca^{2+}-dependent luminescence of known amounts of purified aequorin.

B. *In Vivo*

The highest producers among the clones are used for the *in vivo* studies. Routinely, the active protein is reconstituted by adding 2.5 μM coelenterazine to the external medium. Optimal reconstitution can be obtained with a 6-hr incubation (longer incubations are usually not needed); in most cases, after 1 hr, sufficient active aequorin is present for an *in vivo* measurement. Various conditions have been tested: the cells have been reconstituted in the presence and absence of serum, in saline solution and medium, at 4°C and 37°C. The best condition, both for reconstitution and for cell viability, for the cells we have studied is the simple addition of coelenterazine to the complete growth medium (plus fetal bovine serum) at 37°C and is, therefore, the one used in the experiments described in the following sections.

V. Biochemical Analysis of Aequorin Compartmentalization

The correct targeting of the photoprotein can be verified with two approaches. The first is the biochemical analysis of the segregation of aequorin and marker enzymes in cell fractionation studies. The second, which is discussed in Section VI, is the challenging of living cells with drugs that selectively interfere with mitochondrial Ca^{2+} accumulation. At first, cytosolic localization can be excluded by the following experiment. We trypsinize 1×10^6 cells, wash them once with medium and once with PBS, and resuspend them in 200 μl Krebs–Ringer solution supplemented with 4 mM EGTA. Digitonin (20 μM) is added and, when the plasma membrane is permeabilized (as revealed by staining with a vital dye such as eosin or Trypan blue), the cells are centrifuged; the supernatant and pellet (resuspended in the same buffer) are analyzed separately. In both cases, in conjunction with aequorin, which is reconstituted and measured as previously described, the activity of lactate dehydrogenase (LDH), a cytosolic marker enzyme, is monitored. As shown previously in endothelial cells (Rizzuto *et al.*, 1992), >95% of aequorin is associated with the pellet whereas >90% of LDH is released in the HeLa clones as well, indicating that the photoprotein is retained within membrane-bound organelles (although this result does not prove that aequorin is in the mitochondria). At this point, a more accurate cell fractionation can be carried out, using the protocol described at the end of the chapter. LDH and aequorin are measured as previously described for the digitonin experiments; for aequorin, the medium is supplemented with 2 mM EGTA to prevent aequorin consumption during reconstitution. In parallel, the activity of a mitochondrial marker enzyme, citrate synthase, is measured also. Although, probably because of some mitochondrial damage, a little aequorin (>10%) is found in the soluble cytosolic fraction also, clearly the segregation of the photoprotein closely parallels that of the mitochondrial enzyme, with a more than 40-fold enrichment in the mitochondrial fraction.

VI. *In Vivo* Measurements

Our detection system does not differ from that constructed and described in detail by Cobbold and Lee (1991). Therefore, we only summarize here the principles, referring to that article for all technical details. The cells are plated onto a 13-mm cover slip that is located in a perfusion chamber (15-mm diameter, 2-mm depth) placed immediately in front of a low noise photomultiplier. The photomultiplier (EMI 9789, with built-in amplifier–discriminator; Thorn EMI, Ruislip, Middlesex, U.K.) is cooled to 4°C by a constant temperature incubator, while the chamber is thermostatted at 37°C via a waterbath and perfused with medium via a Gilson Minipuls3 (Villiers-Le-Bel, France) peristaltic pump. Convection of heat from the warm chamber to the cold photo-tube is prevented by a flow of cold air provided by an aquarium pump placed in the 4°C cabinet. The output of the amplifier–discriminator is captured by an EMI C660 photon-counting board (Thorn EMI) in a 386 IBM-compatible microcomputer and stored for further analysis. On-line monitoring of the signal is provided by connecting the board to a 05.80L Linseis (Selb, Germany) chart recorder.

Before measuring, the cells are loaded with coelenterazine, as previously described. The excess of coelenterazine is washed away during perfusion with a modified Krebs–Ringer buffer containing 125 mM NaCl, 5 mM KCl, 1 mM MgSO$_4$, 1 mM CaCl$_2$, 1 mM Na$_3$PO$_4$, 5.5 mM glucose, 20 mM HEPES (pH 7.4 at 37°C). Challenging with agonists or other substances is done in the same buffer. Three experiments are shown in Figs. 5–7 that refer to measurements of a stably transfected clone, of a transiently transfected cell population, and of the clone after permeabilization of the cell membrane with digitonin, respectively.

Figure 5 shows, in the upper panel, the on-line monitoring of light emission by clone 59 (described in Section III,B). At rest, the signal is close to background; when the cells are challenged with two different agonists (carbachol and histamine) that, via the generation of inositol 1,4,5-trisphosphate (IP$_3$), provoke a rise in [Ca^{2+}]$_i$, a large and rapid increase of aequorin luminescence is observed. Within 1–2 min, the signal returns almost to background. The experiment is completed by the discharge of all unconsumed aequorin, to calculate the total amount of photoprotein present in the cells. Cell disruption and aequorin discharge is accomplished by perfusing the chamber with a hypoosmotic solution (10 mM CaCl$_2$ in H$_2$O). Total aequorin content must be known to convert luminescence data into absolute [Ca^{2+}] values. As described in detail elsewhere (Moisescu and Ashley, 1977), for known conditions of temperature, ionic strength, pH, Mg^{2+}, and so on, a calibration curve can be drawn in which, in the physiological range (0.1–10 μM Ca^{2+}), the Ca^{2+} concentration correlates with the fraction of aequorin consumption (and, thus, with light emission). Therefore, at every moment, provided the total light still to be consumed (L$_{max}$) is known, the Ca^{2+} concentration can be calculated. Cuthbertson (University of Liverpool, UK) has allowed us to use his computer software that calculates, from the output data of the detection system, the fractional rate of aequorin consumption and converts it into free calcium values. In the lower panel of Fig. 5, the luminescence data of the same experiment are converted into Ca^{2+} values, thereby allowing the monitoring of mitochondrial free Ca^{2+} ([Ca^{2+}]$_m$). All our data, including the next experiments discussed in this chapter, routinely are converted to and expressed as [Ca^{2+}]$_m$ values.

Figure 6 shows the [Ca^{2+}]$_m$ variations on maximal agonist stimulation in a population of HeLa cells transiently transfected with mtAEQ/pMT2. The [Ca^{2+}]$_m$ response does not differ, both in amplitude and kinetics, from that of

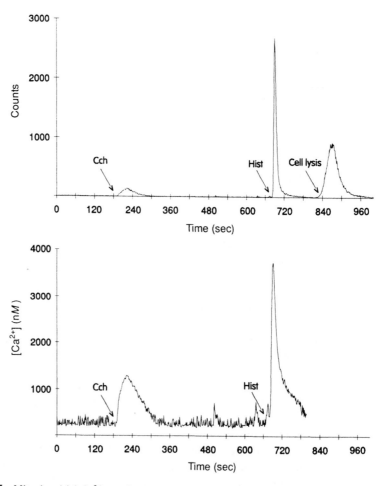

Fig. 5 Mitochondrial Ca^{2+} monitoring in clone 59. (*Top*) Luminescence output data. (*Bottom*) Data converted into free calcium values by the program of R. Cuthbertson (University of Liverpool). Where indicated, carbachol (CCh; 100 μM) and histamine (Hist; 100 μM) were added. Cells were lysed by osmotic rupture in high calcium.

the analyzed clone, which thus can be considered quite "typical" of the population. Moreover, these data show that transiently transfected cells express sufficient recombinant aequorin to allow an accurate [Ca^{2+}]$_m$ measurement. This conclusion has been confirmed by transfecting primary cultures of neurons and cardiac myocytes, with similar results.

Finally, as shown in Fig. 7, permeabilized transfected cells can allow classical experiments of mitochondrial physiology to be performed *in situ*. Cells of clone 59 have been permeabilized with digitonin, thereby equilibrating the cytosol with the external medium. Mitochondrial Ca^{2+} uptake through the

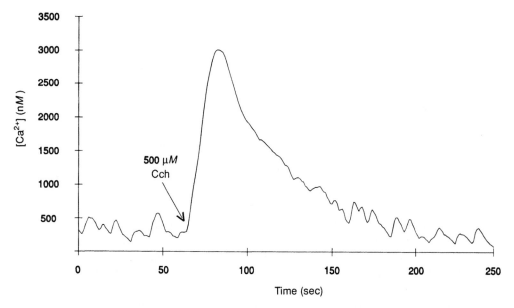

Fig. 6 Mitochondrial Ca^{2+} monitoring in transiently transfected HeLa cells. Ca^{2+} values were obtained as described in Fig. 5. Where indicated, the cells were challenged with 500 μM carbachol (CCh).

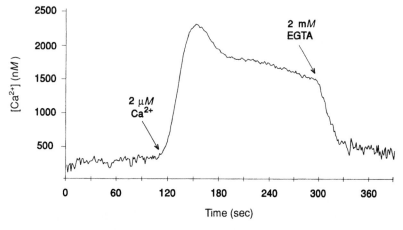

Fig. 7 Monitoring of mitochondrial Ca^{2+} in permeabilized cells of clone 59. The cells, perfused with KRB medium supplemented with 50 μM EGTA, were permeabilized prior to recording by a 90-sec incubation with 100 μM digitonin. Mitochondrial Ca^{2+} uptake was initiated by perfusing the cells with KRB without EGTA (containing approximately 2 μM Ca^{2+}; first addition). Where indicated, perfusion was continued with KRB containing 2 mM EGTA.

uniport occurs when the medium Ca^{2+} concentration is raised from virtually zero (in EGTA buffer) to the micromolar range. As expected, the uptake is inhibited by Ruthenium Red, a blocker of the mitochondrial Ca^{2+} uniporter, and by the collapse of mitochondrial membrane potential (not shown).

VII. Conclusions

We describe in this chapter a new procedure for measuring Ca^{2+} concentrations in defined cell compartments that is based on the specific targeting and recombinant expression of the coelenterate protein aequorin in mammalian cells. We have successfully followed this approach in several cell types (HeLa, endothelial cells, neurons, myocytes) and for two subcellular locations (mitochondria and nucleus), thus proving that it may represent a general method for designing subcellular Ca^{2+} probes. Several issues now must be faced to evaluate the possibility of extending its applications. As discussed in the chapter, the microinjection of the *in vitro* synthesized mRNA must be tested directly, since this technique appears important for extending the methodology to all cells (including primary cultures). Also, new chimeras must be constructed with proteins of interest (receptors, channels), thereby opening the possibility of a more precise "molecular" targeting that may allow the monitoring of Ca^{2+} concentrations in defined cellular microenvironments, for example, that just beneath the plasma membrane. Finally, imaging of aequorin-expressing cells should be attempted for a single cell analysis of subcellular Ca^{2+} homeostasis.

Appendix I: Protocols

Our molecular biology and transfection protocols are quite standard, and do not differ substantially from those described in most gene cloning manuals (e.g., Sambrook *et al.*, 1989).

RNA Extraction

A 75-mm flask, grown to confluence, is washed twice with 5 ml cold PBS (140 mM NaCl, 2 mM KCl, 1.5 mM KH_2PO_4, 8 mM $Na_2HPO_4\cdot2H_2O$, pH 7.4); then 1 ml RNAzol™B (Biotecx Laboratories) is added. The cells are lysed immediately by the solution. The thick homogenate is transferred to a 1.5-ml tube, to which 100 μl chloroform is added and mixed. After centrifugation, the aqueous phase is transferred to a new tube and precipitated (4°C for 15 min) with an equal volume of isopropanol. The RNA, resuspended in 1 mM ethylene diamine tetraacetic acid (EDTA) (pH 8.0), is quantified by OD reading.

Northern Blotting

Stock solutions:

100× MOPS: 2 M MOPS, pH 7.3, with NaOH

100× NaAcetate/EDTA: 0.5 M NaAcetate, 0.1 M EDTA, pH 7.2 with NaOH

Running buffer: 1× MOPS, 1× NaAcetate/EDTA

Gel: 1% agarose, 1× MOPS, 1× NaAcetate/EDTA, 6.6% formaldehyde

Samples: 30 μg RNA, 1× MOPS, 1× NaAcetate/EDTA, 50% formamide, 6.6% formaldehyde

After electrophoresis (100 V for 2–3 hr), the gel is washed for 1 hr in 20× SSC (3M NaCl, 0.3 M NaCitrate, pH 7.6) and transferred overnight onto Nylon membrane by capillary Northern blotting (transfer solution: 10× SSC). The following day, the RNA is fixed to the membrane by UV irradiation (short wavelength, 6 min at 35-cm distance). The membrane is prehybridized (PH buffer: 1 mM EDTA, 20 mM phosphate buffer, pH 7.2, 100 mM NaCl, 50 mM PIPES, pH 6.8, 5% SDS) at 65°C for 15 min, then hybridized (1×10^8 cpm/ 10 ml PH buffer) overnight. The filter is washed 4 times in Wash A (1× SSC, 5% SDS, 10 mM phosphate buffer, pH 7.2), 30 min each, and 3 times in Wash B (0.5% SSC, 0.1% SDS), 15 min each; then the filter is exposed for autoradiography with an intensifying screen.

In Vitro Transcription

The mRNA is synthesized using the Stratagene RNA Transcription kit (Cat. # 200340) as recommended by the producer, except for the addition of 0.4 mM m^{7}G(5')ppp(5')G (cap analog).

Templates: mtAEQ in pBS+, linearized with *Xba*I, and nuAEQ in pBS+, linearized with *Sal*I, treated with Proteinase K, 0.05 mg/ml for 30 min at 37°C, double extracted with phenol/chloroform, precipitated in ethanol and resuspended in DEP-H$_2$O (1 ml DEP/liter H$_2$O)

Reaction: 5 μg template DNA, 0.5 mM each rNTP, 0.4 mM m^7 G(5')ppp(5')G, 30 mM DTT, 40 U RNAsin, 50 U RNA polymerase, 40 mM Tris, pH 8.0, 8 mM MgCl$_2$, 2 mM spermidine, 50 mM NaCl, 1 hr at 37°C

In Vitro Translation

Translate $\frac{1}{5}$ of the transcription reaction *in vitro* with the Boehringer Mannheim Wheat Germ Extract kit (Cat. # 1103 067), as recommended by the producer (Boehringer Mannheim Biochemicals, Indianapolis, Indiana).

Reaction: $\frac{1}{5}$ transcription reaction, 1 mM magnesium acetate, 100 mM potassium acetate, 5 μM each amino acid, 1 mM ATP, 0.02 mM GTP, 8 mM creatine phosphate, 50 μg/ml spermine, 14 mM HEPES, pH 7.6, 40 μg/ml creatine kinase, final volume 75 μl, 1 hr at 30°C

Transfection of Cultured Cells (Calcium Phosphate Procedure)

Day 1

Morning Resuspend the DNA (30 μg/90-mm dish) in 300 μl TE (10 mM Tris, pH 8.0, 1 mM EDTA), add 30 μl NaAcetate (3M pH 5.2), mix, add 690 μl ethanol. Precipitate the DNA (15 min at −70°C, centrifuge 15 min, lyophilize) and resuspend in 450 μl sterile TE. Change the culture medium (fresh, no selective drug) on the Petri dishes.

Evening The protocol described here for one Petri dish can be scaled up appropriately for larger transfections. Transfer the DNA to a 13-ml tissue culture tube and add 50 μl 2.5 M CaCl$_2$. Mix and add the solution dropwise under vortexing to a 13-ml tube containing 0.5 ml 2× HBS (280 mM NaCl, 50 mM HEPES, 1 mM Na$_2$HPO$_4$, pH 7.12 at 25°C). Incubate for 30 min at room temperature; then add the DNA precipitate to the dish, dropwise.

Day 2

Change the medium (fresh, no drug).

Day 3

1. Transient expression—The coverslips are ready to be used for the experiments.

2. Stable expression—Selection can be started with G418.

Cell Subfractionation

The cells plated onto 15-cm Petri dish are allowed to grow to confluence. After one wash with cold PBS, the cells are scraped at 4°C with a rubber policeman (cut at a low angle) into 5 ml PBS, centrifuged at 1000 rpm (Heraeus (Osterode, Germany) Minifuge GL) and resuspended in 2 ml 0.25 M sucrose, 10 mM Tris, pH 7.4, and 0.1 mM EDTA with the aid of a 1-ml pipette tip (10 strokes). Cells are broken by 10 passages through a 22G 1¼-inch needle fitted on a 5-ml plastic syringe. The efficiency of cell breakage is controlled at the microscope ("naked" nuclei are visible). After centrifugation at 2500 rpm (4°C), the pellet (which contains unbroken cells as well as nuclei) is discarded and the supernatant recentrifuged at 10,000 rpm (4°C, Sorvall (DuPont Co., Wilmington, DE) SS34 rotor). The mitochondrial pellet is washed once in 10 ml buffer and resuspended in 1 ml buffer; the supernatant is centrifuged at 40000 rpm for 1 hr (4°C, Beckmann (Palo Alto, CA) 50Ti). The microsomal pellet, resuspended in 1 ml buffer, and the soluble supernatant are used, with the mitochondrial pellet, for the following analysis.

Measurement of Enzymatic Activities

The activities of lactate dehydrogenase (LDH) and citrate synthase (CS) are followed spectrophotometrically.

LDH: wavelength, 340 nm; assay, 50 mM K$_2$HPO$_4$/KH$_2$PO$_4$, pH 7.5, 0.63 mM pyruvate, 200 μM NADH; initiate the reaction with the sample addition

CS: wavelength, 412 nm; assay, 80 mM Tris, pH 8.0, 0.1 mM acetyl CoA, 0.1 mM 5,5'-dithiobis(2-nitrobenzoic acid) (DTNB); initiate the reaction with 0.5 mM oxaloacetate

Acknowledgments

We thank Y. Kishi and O. Shimomura for the gift of purified coelenterazine, Y. Sakaki for the aequorin cDNA, C. Ashley for a sample of purified aequorin, P. Cobbold and R. Cuthbertson for suggestions and help in constructing the detection system and for allowing us to use the data analysis software, and G. Ronconi and M. Santato for technical assistance. This work was supported in part by grants from the Italian Research Council (CNR) Strategic Projects "Biotechnology" and "Oncology", the project "AIDS" of the Italian Health Ministry," "Telethon," and the Italian Association for Cancer Research (AIRC) to T. Pozzan.

References

Brini, M., Murgia, M., Pasti, L., Picard, D., Pozzan, T., and Rizzuto, R. (1993). Nuclear Ca^{2+} concentration measured with specifically targeted recombinant aequorin. *EMBO J.*, in press.

Casadei, J., Powell, M. J., and Kenten, J. H. (1990). Expression and secretion of aequorin as a chimeric antibody by means of a mammalian expression vector. *Proc. Natl. Acad. Sci. U.S.A.* **87**, 2047–2051.

Cobbold, P. H., and Lee, J. A. C. (1991). Aequorin measurement of cytoplasmic free calcium. *In* "Cellular Calcium. A Practical Approach" (J. G. McCormack and P. H. Cobbold. eds.), pp. 55–81. Oxford: Oxford University Press.

Dingwall, C., and Laskey, R. A. (1991). Nuclear targeting sequences—A consensus? *Trends Biochem. Sci.* **16**, 478–481.

Feinberg, A. P., and Vogelstein, B. (1983). A technique for radiolabelling DNA restriction endonuclease fragments to high specific activity. *Anal. Biochem.* **132**, 6–13.

Garcia-Bustos, J., Heitman, J., and Hall, M. N. (1991). Nuclear protein localization. *Biochim. Biophys. Acta* **1071**, 83–101.

Hartl, F.-U., Pfanner, N., Nicholson, D. W., and Neupert, W. (1989). Mitochondrial protein import. *Biochim. Biophys. Acta* **988**, 1–45.

Hendrick, J. P., Hodges, P. E., and Rosenberg, L. E. (1989). Survey of amino-terminal proteolytic cleavage site in mitochondrial precursor proteins: Leader peptide cleaved by two matrix proteases share a three-amino-acid motif. *Proc. Natl. Acad. Sci. U.S.A.* **86**, 4056–4060.

Inoue, S., Sugiura, S., Kakoi, H., and Hasizume, K. (1975). Squid bioluminescence. II. Isolation from *Watasenia scintillans* and synthesis of 2-(p-hydroxybenzyl)-6-(p-hydroxyphenyl)-3,7-dihydroimidazolo[1,2-a]pyrazin-3-one. *Chem. Lett.* 141–144.

Inouye, S., Masato, N., Sakaki, Y., Takagi, Y., Miyata, T., Iwanaga, S., Miyata, T., and Tsuji, F. I. (1985). Cloning and sequence analysis of cDNA for the luminescent protein aequorin. *Proc. Natl. Acad. Sci. U.S.A.* **82**, 3154–3158.

Inouye, S., Aoyama, S., Miyata, T., Tsuji, F. I., and Sakaki, Y. (1989). Overexpression and purification of the recombinant Ca^{2+}-binding protein, aequorin. *J. Biochem.* **105**, 473–477.

Kaufman, R. J., Davies, M. V., Pathak, V. K., and Hershey, J. W. B. (1989). The phosphorylation state of eukaryotic initiation factor 2 alters translational efficiency of specific mRNAs. *Mol. Cell. Biol.* **9**, 946–958.

Kendall, J. M., Dormer, R. L., and Campbell, A. K. (1992). Targeting aequorin to the endoplasmic reticulum of living cells. *Biochem. Biophys. Res. Commun.* **189**, 1008–1016.

Knight, M. R., Campbell, A. K., Smith, S. M., and Trewawas, A. J. (1991a). Recombinant aequorin as a probe for cytosolic free Ca^{2+} in *Escherichia coli. FEBS Lett.* **282**, 405–408.

Knight, M. R., Campbell, A. K., Smith, S. M., and Trewawas, A. J. (1991b). Transgenic plant aequorin reports the effect of touch and cold-shock and elicitors on cytoplasmic calcium. *Nature (London)* **352**, 524–526.

Miesfeld, R., Rusconi, S., Godowski, P. J., Maler, B. A., Okret, S., Wikstrom, A.-C., Gustafsson, J.-A., and Yamamoto, K. R. (1986). Genetic complementation of a glucocorticoid receptor deficiency by expression of cloned receptor cDNA. *Cell* **46**, 389–399.

Moisescu, D. G., and Ashley, C. C. (1977). The effect of physiologically occurring cations upon aequorin light emission. *Biochim. Biophys. Acta* **460**, 189–205.

Munro, S., and Pelham, H. R. B. (1987). A C-terminal signal prevents secretion of luminal ER proteins. *Cell* **48**, 899–907.

Nakajima-Shimada, J., Iida, H., Tsuji, F. I., and Anraku, Y. (1991). Monitoring of intracellular calcium in Saccaromyces cerevisiae with an apoaequorin cDNA expression system. *Proc. Natl. Acad. Sci. U.S.A.* **88**, 6878–6882.

Neuberger, M. S. (1983). Expression and regulation of immunoglobulin heavy chain gene transfected into lymphoid cells. *EMBO J.* **2**, 1373–1378.

Nomura, M., Inouye, S., Ohmiya, Y., and Tsuji, F. I. (1991). A C-terminal proline is required for bioluminescence of the Ca^{2+}-binding protein, aequorin. *FEBS Lett.* **295**, 63–66.

Nothwehr, S. F., and Gordon, J. I. (1990). Targeting of proteins into the eukaryotic secretory pathway: Signal peptide structure/function relationship. *BioEssays* **12**, 479–484.

Picard, D., and Yamamoto, K. R. (1987). Two signals mediate hormone-dependent nuclear localization of the glucocorticoid receptor. *EMBO J.* **6**, 3333–3340.

Rizzuto, R., Nakase, H., Darras, B., Fabrizi, G. M., Mengel, T., Walsh, F., Kadenbach, B., Di Mauro, S., Francke, U., and Schon, E. A. (1989). A gene specifying subunit VIII of human cytochrome *c* oxidase is localized to chromosome 11 and is expressed in both muscle and non-muscle tissues. *J. Biol. Chem.* **264**, 10595–10600.

Rizzuto, R., Simpson, A. W. M., Brini, M., and Pozzan, T. (1992). Rapid changes of mitochondrial Ca^{2+} revealed by specifically targeted recombinant aequorin. *Nature (London)* **358**, 325–327.

Sambrook, J., Fritsch, E. F., and Maniatis, T. (1989). "Molecular Cloning. A Laboratory Manual," 2d Ed. Cold Spring Harbor, New York: Cold Spring Harbor Laboratory Press.

Shimomura, O. (1991). Preparation and handling of aequorin solutions for the measurement of cellular Ca^{2+}. *Cell Calcium* **12**, 635–643.

Sitia, R., and Meldolesi, J. (1992). The endoplasmic reticulum: A dynamic patchwork of specialized subregions. *Mol. Biol. Cell* **3**, 1067–1072.

Southern, P. J., and Berg, P. (1982). Transformation of mammalian cells to antibiotic resistance with a bacterial gene under control of the SV40 early region promoter. *J. Mol. Appl. Genet.* **1**, 327–341.

von Heijne, G. (1986). Mitochondrial targeting sequences may form amphiphilic helices. *EMBO J.* **5**, 1335–1342.

Wormington, M. (1991). Preparation of synthetic mRNAs and analyses of translational efficiency in microinjected *Xenopus* oocytes. *Meth. Cell Biol.* **36**, 167–183.

INDEX

VOLUMES IN SERIES

Founding Series Editor
DAVID M. PRESCOTT

Volume 1 (1964)
Methods in Cell Physiology
Edited by David M. Prescott

Volume 2 (1966)
Methods in Cell Physiology
Edited by David M. Prescott

Volume 3 (1968)
Methods in Cell Physiology
Edited by David M. Prescott

Volume 4 (1970)
Methods in Cell Physiology
Edited by David M. Prescott

Volume 5 (1972)
Methods in Cell Physiology
Edited by David M. Prescott

Volume 6 (1973)
Methods in Cell Physiology
Edited by David M. Prescott

Volume 7 (1973)
Methods in Cell Biology
Edited by David M. Prescott

Volume 8 (1974)
Methods in Cell Biology
Edited by David M. Prescott

Volume 9 (1975)
Methods in Cell Biology
Edited by David M. Prescott

Volume 10 (1975)
Methods in Cell Biology
Edited by David M. Prescott

Advisory Board Chairman
KEITH R. PORTER

Volume 21B (1980)
Normal Human Tissue and Cell Culture, Part B: Endocrine, Urogenital, and Gastrointestinal Systems
Edited by Curtis C. Harris, Benjamin F. Trump, and Gary D. Stoner

Volume 22 (1981)
Three-Dimensional Ultrastructure in Biology
Edited by James N. Turner

Volume 23 (1981)
Basic Mechanisms of Cellular Secretion
Edited by Arthur R. Hand and Constance Oliver

Volume 24 (1982)
The Cytoskeleton, Part A: Cytoskeletal Proteins, Isolation and Characterization
Edited by Leslie Wilson

Volume 25 (1982)
The Cytoskeleton, Part B: Biological Systems and *in Vitro* Models
Edited by Leslie Wilson

Volume 26 (1982)
Prenatal Diagnosis: Cell Biological Approaches
Edited by Samual A. Latt and Gretchen J. Darlington

Series Editor
LESLIE WILSON

Volume 27 (1986)
Echinoderm Gametes and Embryos
Edited by Thomas E. Schroeder

Volume 28 (1987)
***Dictyostelium discoideum:* Molecular Approaches to Cell Biology**
Edited by James A. Spudich

Volume 29 (1989)
Fluorescence Microscopy of Living Cells in Culture, Part A: Fluorescent Analogs, Labeling Cells, and Basic Microscopy
Edited by Yu-Li Wang and D. Lansing Taylor

Volume 30 (1989)
Fluorescence Microscopy of Living Cells in Culture, Part B: Quantitative Fluorescence Microscopy—Imaging and Spectroscopy
Edited by D. Lansing Taylor and Yu-Li Wang

Series Editors
LESLIE WILSON AND PAUL MATSUDAIRA

Printed and bound by CPI Group (UK) Ltd, Croydon, CR0 4YY

03/10/2024

01040318-0004